Protein Structure Prediction

METHODS IN MOLECULAR BIOLOGY™

John M. Walker, SERIES EDITOR

419. Post-Transcriptional Gene Regulation, edited by *Jeffrey Wilusz, 2008*
418. Avidin–Biotin Interactions: *Methods and Applications,* edited by *Robert J. McMahon, 2008*
417. Tissue Engineering, Second Edition, edited by *Hannsjörg Hauser and Martin Fussenegger, 2007*
416. Gene Essentiality: *Protocols and Bioinformatics,* edited by *Andrei L. Osterman, 2008*
415. Innate Immunity, edited by *Jonathan Ewbank and Eric Vivier, 2007*
414. Apoptosis in Cancer: *Methods and Protocols*, edited by *Gil Mor and Ayesha Alvero, 2008*
413. Protein Structure Prediction, Second Edition, edited by *Mohammed Zaki and Chris Bystroff, 2008*
412. Neutrophil Methods and Protocols, edited by *Mark T. Quinn, Frank R. DeLeo, and Gary M. Bokoch, 2007*
411. Reporter Genes for Mammalian Systems, edited by *Don Anson, 2007*
410. Environmental Genomics, edited by *Cristofre C. Martin, 2007*
409. Immunoinformatics: *Predicting Immunogenicity* In Silico, edited by *Darren R. Flower, 2007*
408. Gene Function Analysis, edited by *Michael Ochs, 2007*
407. Stem Cell Assays, edited by *Mohan C. Vemuri, 2007*
406. Plant Bioinformatics: *Methods and Protocols,* edited by *David Edwards, 2007*
405. Telomerase Inhibition: *Strategies and Protocols,* edited by *Lucy Andrews and Trygve O. Tollefsbol, 2007*
404. Topics in Biostatistics, edited by *Walter T. Ambrosius, 2007*
403. Patch-Clamp Methods and Protocols, edited by *Peter Molnar and James J. Hickman, 2007*
402. PCR Primer Design, edited by *Anton Yuryev, 2007*
401. Neuroinformatics, edited by *Chiquito J. Crasto, 2007*
400. Methods in Lipid Membranes, edited by *Alex Dopico, 2007*
399. Neuroprotection Methods and Protocols, edited by *Tiziana Borsello, 2007*
398. Lipid Rafts, edited by *Thomas J. McIntosh, 2007*
397. Hedgehog Signaling Protocols, edited by *Jamila I. Horabin, 2007*
396. Comparative Genomics, *Volume 2,* **edited by** *Nicholas H. Bergman, 2007*
395. Comparative Genomics, *Volume 1,* **edited by** *Nicholas H. Bergman, 2007*
394. Salmonella: *Methods and Protocols,* edited by *Heide Schatten and Abe Eisenstark, 2007*
393. Plant Secondary Metabolites, edited by *Harinder P. S. Makkar, P. Siddhuraju, and Klaus Becker, 2007*
392. Molecular Motors: *Methods and Protocols*, edited by *Ann O. Sperry, 2007*
391. MRSA Protocols, edited by *Yinduo Ji, 2007*
390. Protein Targeting Protocols, Second Edition, edited by *Mark van der Giezen, 2007*
389. *Pichia* Protocols, Second Edition, edited by *James M. Cregg, 2007*
388. Baculovirus and Insect Cell Expression Protocols, Second Edition, edited by *David W. Murhammer, 2007*
387. Serial Analysis of Gene Expression (SAGE): *Digital Gene Expression Profiling,* edited by *Kare Lehmann Nielsen, 2007*
386. Peptide Characterization and Application Protocols, edited by *Gregg B. Fields, 2007*
385. Microchip-Based Assay Systems: *Methods and Applications,* edited by *Pierre N. Floriano, 2007*
384. Capillary Electrophoresis: *Methods and Protocols,* edited by *Philippe Schmitt-Kopplin, 2007*
383. Cancer Genomics and Proteomics: *Methods and Protocols,* edited by *Paul B. Fisher, 2007*
382. Microarrays, Second Edition: *Volume 2, Applications and Data Analysis,* edited by *Jang B. Rampal, 2007*
381. Microarrays, Second Edition: *Volume 1, Synthesis Methods,* edited by *Jang B. Rampal, 2007*
380. Immunological Tolerance: *Methods and Protocols,* edited by *Paul J. Fairchild, 2007*
379. Glycovirology Protocols, edited by *Richard J. Sugrue, 2007*
378. Monoclonal Antibodies: *Methods and Protocols,* edited by *Maher Albitar, 2007*
377. Microarray Data Analysis: *Methods and Applications,* edited by *Michael J. Korenberg, 2007*
376. Linkage Disequilibrium and Association Mapping: *Analysis and Application,* edited by *Andrew R. Collins, 2007*
375. In Vitro Transcription and Translation Protocols: *Second Edition,* edited by *Guido Grandi, 2007*
374. Quantum Dots: edited by *Marcel Bruchez and Charles Z. Hotz, 2007*
373. Pyrosequencing® Protocols, edited by *Sharon Marsh, 2007*
372. Mitochondria: Practical Protocols, edited by *Dario Leister and Johannes Herrmann, 2007*
371. Biological Aging: *Methods and Protocols,* edited by *Trygve O. Tollefsbol, 2007*
370. Adhesion Protein Protocols, *Second Edition,* edited by *Amanda S. Coutts, 2007*
369. Electron Microscopy: *Methods and Protocols, Second Edition,* edited by *John Kuo, 2007*
368. Cryopreservation and Freeze-Drying Protocols, *Second Edition,* edited by *John G. Day and Glyn Stacey, 2007*
367. Mass Spectrometry Data Analysis in Proteomics, edited by *Rune Matthiesen, 2007*
366. Cardiac Gene Expression: *Methods and Protocols,* edited by *Jun Zhang and Gregg Rokosh, 2007*
365. Protein Phosphatase Protocols: edited by *Greg Moorhead, 2007*

METHODS IN MOLECULAR BIOLOGY™

Protein Structure Prediction

Second Edition

Edited by

Mohammed J. Zaki
and
Christopher Bystroff

Rensselaer Polytechnic Institute, Troy, New York, USA

HUMANA PRESS ✳ TOTOWA, NEW JERSEY

999 Riverview Drive, Suite 208
Totowa, New Jersey 07512

www.humanapress.com

This publication is printed on acid-free paper. ∞
ANSI Z39.48-1984 (American Standards Institute) Permanence of Paper for Printed Library Materials

Production Editor: Rhukea Hussain
Cover design by Karen Schulz

For additional copies, pricing for bulk purchases, and/or information about other Humana titles, contact Humana at the above address or at any of the following numbers: Tel.: 973-256-1699; Fax: 973-256-8341; E-mail: humana@humanapr.com; or visit our Website: www.humanapress.com

Printed in the United States of America. 10 9 8 7 6 5 4 3 2 1
eISBN 978-1-59745-574-9

Library of Congress Control Number: 2007933144

Preface

For 40 years we have known the essential ingredients for protein folding – an amino acid sequence and water. But the problem of predicting the three-dimensional structure from its sequence has eluded computational biologists even in the age of supercomputers and high throughput structural genomics. Will we ever solve the "protein folding problem", or will we simply settle for a solution to the "protein prediction problem"? This book covers elements of both the data-driven comparative modeling approach to structure prediction and also recent attempts to simulate folding using explicit or simplified models. Despite the unsolved mystery of how a protein folds, advances are being made in predicting the interactions of proteins with other molecules, such as small ligands, nucleic acids, or other proteins. Also, rapidly advancing are the methods for solving the inverse folding problem, the problem of finding a sequence to fit a structure. This book focuses on the various computational methods for prediction, their successes, and their limitations, from the perspective of their most well-known practitioners. An overview of the chapters in this volume is given below.

Overview of Protein Structure Prediction

In the first chapter, entitled "A Historical Perspective of Template-Based Protein Structure Prediction," Jun-tao Guo, Kyle Ellrott, and Ying Xu give a comprehensive, as well as historical, account of protein structure prediction. They touch upon methods spanning threading, fold recognition, homology modeling, ab initio methods, and their hybrids. They also discuss recent progress in the worldwide blind structure prediction evaluation experiments like CASP and its cousin for automated servers, CAFASP.

In the second chapter "The Assessment of Methods for Protein Structure Prediction," Anna Tramontano, Domenico Cozzetto, Alejandro Giorgetti, and Domenico Raimondo take a critical look at extant methods for protein structure prediction and assess how well they perform. They focus on automatic assessment methods as well as the CASP challenges and discuss their limitations and trade-offs.

Template-Based Methods

In the third chapter "Aligning Sequences to Structures," Liam J. McGuffin discusses the current approaches to template-based fold prediction. The goal here is to align new protein sequences to library of known/template folds. Liam also shows a step-by-step guide to template alignment.

In the fourth chapter "Protein Structure Prediction Using Threading," Jinbo Xu, Feng Jiao, and Libo Yu discuss approaches for protein threading. After setting up the general requirements for protein structure prediction by threading, they specifically focus on their successful new method called RAPTOR, which combines linear programming with machine learning approaches.

Structure Alignment and Indexing

In the fifth chapter "Algorithms for Multiple Protein Structure Alignment and Structure-Derived Multiple Sequence Alignment," Maxim Shatsky, Ruth Nussinov, and Haim J. Wolfson, present methods to recognize the structural core common to a set of proteins through multiple structure alignment. They also discuss how to align multiple sequences with knowledge derived from structural alignment.

In the sixth chapter "Indexing Protein Structures using Suffix Trees," Feng Gao and Mohammed J. Zaki describe a new approach to 3D database searching for protein sub-structures. Given a large set of proteins, they extract local structural features, which are converted into a set of symbols, which can be indexed using a traditional suffix tree. They show how one can rapidly retrieve approximately similar protein substructure matching a query protein.

Protein Features Prediction

In the seventh chapter "Hidden Markov Models for Prediction of Protein Features," Christopher Bystroff and Anders Krogh present a comprehensive overview of Hidden Markov Models (HMMs), which are used extensively in protein structure/sequence algorithms. They specifically focus on the applications of HMMs to predict signal peptides, secondary and local structure, and transmembrane helices.

In the eighth chapter "The Pros and Cons of Predicting Protein Contact Maps," Lisa Bartoli, Emidio Capriotti, Piero Fariselli, Pier Luigi Martelli, and Rita Casadio discuss methods to predict protein contact maps. Contact maps are "simplified" 2D representations of the 3D proteins structure yet, they retain most of the important features for protein folding. They discuss the strengths and weaknesses of the contact map representation and highlight ways to improve contact map predictions.

In the ninth chapter "Road Map Methods for Protein Folding," Mark Moll, David Schwarz, and Lydia E. Kavraki give a comprehensive survey of "roadmap" approaches to protein folding. Roadmap methods, inspired by motion planning techniques in robotics research, provide a model for understanding and predicting the folding mechanism or pathway.

Methods for De Novo Structure Prediction

In the tenth Chapter "Scoring Functions for De Novo Protein Structure Prediction Revisited," Shing-Chung Ngan, Ling-Hong Hung, Tianyun Liu, and Ram Samudrala, provide a thorough review of both physics-based and knowledge-based scoring functions for conformational samples in de novo protein structure prediction.

In the eleventh chapter "Protein–Protein Docking: Overview and Performance Analysis," Kevin Wiehe, Matthew W. Peterson, Brian Pierce, Julian Mintseris, and Zhiping Weng focus on Fast Fourier Transform-based methods for protein docking. They specifically focus on the ZDOCK algorithm and study its performance on benchmark datasets and study its strengths and weaknesses through regression analysis.

In the final chapter "Molecular Dynamics Simulations of Protein Folding," Angel E. Garcia describes the Replica Exchange Molecular Dynamics (REMD) method for molecular dynamics simulation. He illustrates the effectiveness of the REMD method on the folding of a small protein.

Mohammed J. Zaki
Chris Bystroff

Contents

Preface v
Contributors xi

PART I: OVERVIEW OF PROTEIN STRUCTURE PREDICTION

1 A Historical Perspective of Template-Based Protein Structure Prediction
Jun-tao Guo, Kyle Ellrott, and Ying Xu *3*

2 The Assessment of Methods for Protein Structure Prediction
Anna Tramontano, Domenico Cozzetto, Alejandro Giorgetti, and Domenico Raimondo *43*

PART II: TEMPLATE-BASED METHODS

3 Aligning Sequences to Structures
Liam James McGuffin *61*

4 Protein Structure Prediction Using Threading
Jinbo Xu, Feng Jiao, and Libo Yu *91*

PART III: STRUCTURE ALIGNMENT AND INDEXING

5 Algorithms for Multiple Protein Structure Alignment and Structure-Derived Multiple Sequence Alignment
Maxim Shatsky, Ruth Nussinov, and Haim J. Wolfson *125*

6 Indexing Protein Structures Using Suffix Trees
Feng Gao and Mohammed J. Zaki *147*

PART IV: PROTEIN FEATURES PREDICTION

7 Hidden Markov Models for Prediction of Protein Features
Christopher Bystroff and Anders Krogh *173*

8 The Pros and Cons of Predicting Protein Contact Maps
Lisa Bartoli, Emidio Capriotti, Piero Fariselli, Pier Luigi Martelli, and Rita Casadio *199*

9 Roadmap Methods for Protein Folding
Mark Moll, David Schwarz, and Lydia E. Kavraki *219*

PART V: METHODS FOR DE NOVO STRUCTURE PREDICTION

10 Scoring Functions for De Novo Protein Structure Prediction Revisited
Shing-Chung Ngan, Ling-Hong Hung, Tianyun Liu, and Ram Samudrala *243*

11 Protein–Protein Docking: Overview and Performance Analysis
Kevin Wiehe, Matthew W. Peterson, Brian Pierce, Julian Mintseris, and Zhiping Weng *283*

12 Molecular Dynamics Simulations of Protein Folding
Angel E. Garcia *315*

Index *331*

Contributors

LISA BARTOLI • *CIRB/Department of Biology, University of Bologna, Bologna, Italy*
CHRISTOPHER BYSTROFF • *Department of Biology, Rensselaer Polytechnic Institute, Troy NY, USA*
EMIDIO CAPRIOTTI • *CIRB/Department of Biology, University of Bologna, Bologna, Italy*
RITA CASADIO • *CIRB/Department of Biology, University of Bologna, Bologna, Italy*
DOMENICO COZZETTO • *Department of Biochemical Sciences, University of Rome "La Sapienza", Rome, Italy*
KYLE ELLROTT • *Department of Biochemistry and Molecular Biology, University of Georgia, Athens, GA, USA*
PIERO FARISELLI • *CIRB/Department of Biology, University of Bologna, Bologna, Italy*
FENG GAO • *Department of Computer Science, Rensselaer Polytechnic Institute, Troy, NY, USA*
ANGEL E. GARCIA • *Department of Physics, Applied Physics and Astronomy, Rensselaer Polytechnic Institute, Troy, NY 12180*
ALEJANDRO GIORGETTI • *Department of Biochemical Sciences, University of Rome "La Sapienza", Rome, Italy*
JUN-TAO GUO • *Department of Biochemistry and Molecular Biology, University of Georgia, Athens, GA, USA*
LING-HONG HUNG • *Department of Microbiology, University of Washington School of Medicine, Seattle, WA, USA*
FENG JIAO • *School of Computer Science, University of Waterloo, Waterloo, Canada*
LYDIA E. KAVRAKI • *Computer Science Department, Rice University, Houston, TX, USA*
ANDERS KROGH • *The Bioinformatics Centre, Inst. Mol. Biol. and Physiology, University of Copenhagen, Denmark*
TIANYUN LIU • *Department of Microbiology, University of Washington School of Medicine, Seattle, WA, USA*

PIER LUIGI MARTELLI • *CIRB/Department of Biology, University of Bologna, Bologna, Italy*
LIAM JONES MCGUFFIN • *The University of Reading, Reading, UK*
JULIAN MINTSERIS • *Bioinformatics Program, Boston University, Boston, MA, USA*
MARK MOLL • *Information Sciences Institute, University of Southern California, Marina del Rey, CA, USA*
SHING-CHUNG NGAN, • *Department of Microbiology, University of Washington School of Medicine, Seattle, WA, USA*
RUTH NUSSINOV • *Sackler Inst. of Molecular Medicine, Tel Aviv University, Tel Aviv, Israel and Basic Research Program, SAIC-Frederick, Inc., Frederick, MD, USA*
MATTHEW W. PETERSON • *Department of Biomedical Engineering, Boston University, Boston, MA, USA*
BRIAN PIERCE • *Bioinformatics Program, Boston University, Boston, MA, USA*
DOMENICO RAIMONDO • *Department of Biochemical Sciences, University of Rome "La Sapienza", Rome, Italy*
RAM SAMUDRALA • *Department of Microbiology, University of Washington School of Medicine, Seattle, WA, USA*
DAVID SCHWARZ • *Computer Science Department, Rice University, Houston, TX, USA*
MAXIM SHATSKY • *School of Computer Science, Tel Aviv University, Tel Aviv, Israel*
ANNA TRAMONTANO • *Department of Biochemical Sciences, University of Rome "La Sapienza", Rome, Italy*
ZHIPING WENG • *Department of Biomedical Engineering, Boston University, Boston, MA, USA*
KEVIN WIEHE • *Bioinformatics Program, Boston University, Boston, MA, USA*
HAIM J. WOLFSON • *School of Computer Science, Tel Aviv University, Tel Aviv, Israel*
JINBO XU • *Toyota Technological Institute at Chicago, Chicago, IL, USA*
YING XU • *Department of Biochemistry and Molecular Biology, University of Georgia, Athens, GA, USA*
LIBO YU • *Bioinformatics Solutions Inc., Waterloo, Canada*
MOHAMMED J. ZAKI • *Department of Computer Science, Rensselaer Polytechnic Institute, Troy, NY, USA*

I

Overview of Protein Structure Prediction

1

A Historical Perspective of Template-Based Protein Structure Prediction

Jun-tao Guo, Kyle Ellrott, and Ying Xu

Summary

This chapter presents a broad and a historical overview of the problem of protein structure prediction. Different structure prediction methods, including homology modeling, fold recognition (FR)/protein threading, ab initio/de novo approaches, and hybrid techniques involving multiple types of approaches, are introduced in a historical context. The progress of the field as a whole, especially in the threading/FR area, as reflected by the CASP/CAFASP contests, is reviewed. At the end of the chapter, we discuss the challenging issues ahead in the field of protein structure prediction.

Key Words: Structure prediction; fold recognition; protein threading; CASP/CAFASP; meta-server; fragment assembly; energy function; comparative/homology modeling.

1. Introduction

The classic experiment by Anfinsen in the early 1970s demonstrated that all the information a protein needs to fold properly is encoded in its amino acid sequence *(1)*, suggesting that one should be able to predict, at least theoretically, the three-dimensional (3D) conformation of a protein from its sequence alone. Since then, many efforts have been devoted to this fascinating and challenging problem, attempting to tackle this problem from different angles including the ones from biophysics, chemistry, and biological evolution. The problem of predicting a protein's 3D structure from its amino acid sequence has been called the "holy grail of molecular biology" and considered as equivalent to deciphering "the second half of the genetic code" *(2)*. Over the past 30

From: *Methods in Molecular Biology, vol. 413: Protein Structure Prediction, Second Edition*
Edited by: M. Zaki and C. Bystroff © Humana Press Inc., Totowa, NJ

years, particularly since the start of the Human Genome Project (HGP), the problem of protein structure prediction has generated enormous interests among protein structural biologists and computational scientists partly because of the potential impacts to many areas of biology because the knowledge of the tertiary structure is essential to the understanding of the biological function and functional mechanism of a protein. The importance of computational solution to protein structures is increasing owing to the rapid growth in the number of sequenced genomes and the relatively slow growth rate in the number of experimentally determined protein structures. Not surprisingly, protein structure prediction has become a vital part of the world-wide Structural Genomics projects, which are designed to develop capabilities for potentially solving most of the protein structures in nature through effectively integrating experimental techniques and computational prediction and modeling ***(3)***.

Earlier works on the protein structure prediction have been primarily focused on physics-based methods with an attempt to understand the folding process ***(4,5)***. The basic idea of computational protein folding is to find the lowest free-energy structure for an amino acid sequence, based on the thermodynamic hypothesis formulated by Anfinsen ***(1)***, through searching the exceedingly large conformational space of the protein. While this still represents the ultimate goal of protein structure modeling, we are clearly far from achieving this goal because of the enormity and the complexity of the conformational space of a protein compared to the computing resources that are currently available and the inadequacy of the existing energy force fields. An alternative and yet very attractive approach has been to only predict the final structure of a protein-folding process. Such an approach is attractive because it makes the protein structure solution problem more practically solvable. More importantly, focusing only on the "final" and static structure of a complex folding process allows researchers to take full advantage of the wealthy information of previously solved protein structures and make protein structure prediction using a so-called template-based protein structure paradigm ***(6)***. Template-based structure prediction methods, ranging from the de novo methods that use relatively short structural templates to homology modeling methods that use the entire protein structure as template, have made great strides in protein structure prediction in the past 15 years and have been used to make many structure predictions before their experimental structures are available, which have later proved to be highly useful in guiding experimental designs. For example, Bajorath et al. ***(7–9)*** predicted two structural models for human CD40 ligand gp39 by comparative modeling (CM) techniques and used the models to guide a series of mutagenesis experiments to identify the residues of gp39 that are important for the interaction with CD40 and to analyze the locations of naturally occurring gp39 mutations. After the structure of gp39 was solved later by X-ray crystallography ***(10)***, Bajorath did a detailed assessment on the modeling accuracy and the validity of model-based mutagenesis and mapping studies ***(11)***. With the exception of a few prediction errors in the loop regions, the gp39 models were well predicted, including residues important for CD40 binding, and have significantly aided in the design of mutagenesis experiments. These studies highlight the usefulness of structural models in guiding and rationalizing the mutagenesis experiments or experiments in general.

Generally speaking, structure prediction techniques fall into three categories: ab initio prediction, protein threading [or sometimes referred as *fold recognition* (FR)], and homology modeling ***(12)***. Ab initio methods make structure predictions without using any structural information of previously solved protein structures; instead, they are entirely based on the first principles of physics. Structure prediction by homology modeling is based on accurate sequence alignments between a query protein and a template protein with solved structures; hence, the prediction accuracy of this class of methods heavily depends on the sequence similarity between two proteins. Protein threading represents a more general class of prediction techniques than homology modeling as it uses both sequence similarity information when exists and structural fitness information between the query protein and the template structure. Although homology modeling has been mainly used for detailed (e.g., all heavy-atom) structure prediction when a query protein has a close homolog in the Protein Data Bank (PDB) ***(13)***, protein threading is often used for FR and backbone structure prediction when a query protein might have only remote structural homologs or analogs in PDB. We have noticed that the boundaries among these three classes of prediction techniques have started to become blurred because scientists have started to integrate the strengths of different methods to make their prediction methods more effective and more generally applicable ***(14)***.

A unique event in the field of computational structural biology is the biennial contest for protein structure prediction, called Critical Assessment of Structure Prediction (CASP), which was initiated by John Moult and others in 1994 ***(15, 16)***. CASP has been effectively used to assess the overall prediction capability for protein structures by the existing prediction techniques in an objective way and to measure the progress of the field as a whole and to identify the major technical breakthroughs between two consecutive CASP contests. In each CASP contest, the protein sequences of soon to be released structures, solved by either X-ray crystallography or nuclear magnetic resonance (NMR) spectroscopy methods, are made available to all registered predictors through the Internet.

Structural predictions by different predictors on each prediction target are submitted through the Internet and then assessed by a team of independent assessors through comparing the predicted structures to the solved but yet released experimental structures. A companion contest of CASP is Critical Assessment of Fully Automated Structure Prediction (CAFASP), which was initiated in 1998 after CASP3 *(17)* and has been used to assess the performance of fully automated prediction servers on the Internet without any human input.

In this chapter, we present an introduction to template-based prediction methods and discuss the progress of the prediction techniques, mainly based on the data from the CASP/CAFASP contests. We focus our discussion particularly on the advances in the FR category while touching on key developments in homology modeling. At the end of this chapter, we discuss the challenging issues ahead in the field of protein structure prediction, mainly to keep up with the identification rate of genes and their proteins by the world-wide genome sequencing and bioinformatics efforts.

2. A Brief Overview of Template-Based Protein Structure Prediction Methods

The basic premise for template-based protein structure prediction is three-fold: (1) similar sequences adopt similar protein structures *(18,19)*; (2) many unrelated sequences fold into similar structures *(20,21)*; and (3) there are only a relatively small number of unique structural folds, when compared with the number of proteins in nature *(22–27)*. The first observation forms the foundation of homology modeling, whereas the second and the third observations/assumptions are the foundations of FR/protein threading. Before Bowie et al.'s seminal paper *(28)* on protein threading, which started a new wave of "fold recognition" rush since 1991, template-based protein structure prediction was mainly a playground for homology modeling. The first structural model, derived using a template-based approach, was built in 1969 by Browne and colleagues *(19)*. In their work, a wire skeletal model (a real physical model) of α-lactalbumin was constructed based on the X-ray structure of lysozyme. Subsequent developments in computer graphics and distance geometry have provided important tools for comparative model building of protein structures *(29)*. Since the report of the first protein structure model, the structure of many important proteins have been modeled through homology modeling, including relaxins *(30)*, insulin-like growth factors *(31)*, serine proteases *(32)*, renin *(33)*, inflammatory protein C5a *(34)*, angiogenin *(35)*, and immunoglobulins *(36)*. The development of the threading approach [the term "threading" was first introduced by Jones et al. in their *Nature* paper in 1992 *(37)*] is based on the premise that the number of unique structural folds in nature is probably a few orders of magnitude smaller than the number of proteins in nature, possibly ranging from a few hundred to a few thousand *(22–27)*. These theoretical estimates on the number of unique folds in nature have been (partly) supported by the fact that in the past 5 years, less than 10% of the protein structures newly deposited in PDB represent new structural folds *(38)*. People have found that many unrelated protein sequences fold into similar structures and certain structural folds seem remarkably popular among proteins without any apparent sequence similarity, such as triose-phosphate isomerase (TIM) barrels *(39–42)*. On the basis of such observations, the protein-threading technique has been used to address two key issues: "Which structural fold does a given protein adopt among the experimentally solved protein structures if any, and where should each of the residues of the given protein be placed in the identified structural fold if any?" Compared with homology modeling, protein threading represents a more general class of prediction methods and substantially extends the scope of structure modeling using homology modeling-based approaches.

Template-based structure prediction methods generally consist of the following five key steps: (1) identification of structural templates through either sequence-based or structure-based methods; (2) alignment of the target sequence to the identified template structure; (3) model building, including loop and side-chain modeling, based on predicted sequence-structure alignment(s); (4) model evaluation; and (5) model refinement (*see* **Fig. 1**). There are several excellent review papers with details on model building, evaluation, and refinement

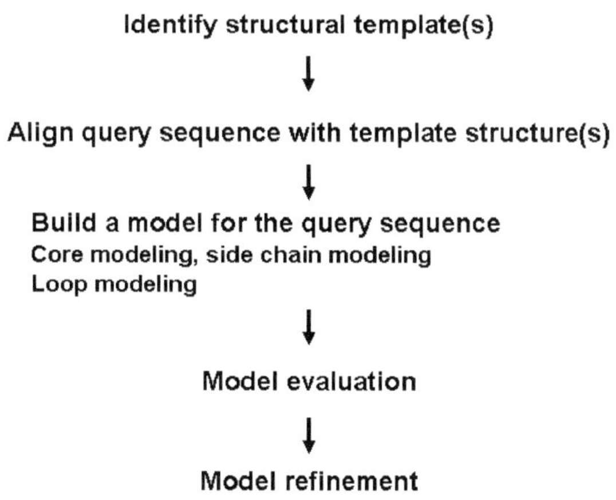

Fig. 1. Five key steps for template-based protein structure prediction.

(43–45). In this chapter, we focus mainly on the first two steps and briefly introduce the model building methods.

There are a number of different methods for the identification of the structural templates or to assign structural folds for a given target sequence, from a collection of experimentally solved protein structures. Existing FR methods generally fall into two classes. Methods of the first class use solely sequence information, whereas the second-class methods, or threading methods, use either structural information alone or combination of structural and sequence information. We introduce the threading approach first. Protein threading is essentially a sequence–sequence comparison method when structural information is not considered. **Figure 2** shows the major milestones over the course of methodology development for template-based structure prediction.

2.1. Protein Threading

Protein threading, introduced in the early 1990s *(28,37)*, has played a key role in protein structure prediction. By using simple measures for fitness of different amino acid types to local structural environments defined in terms of

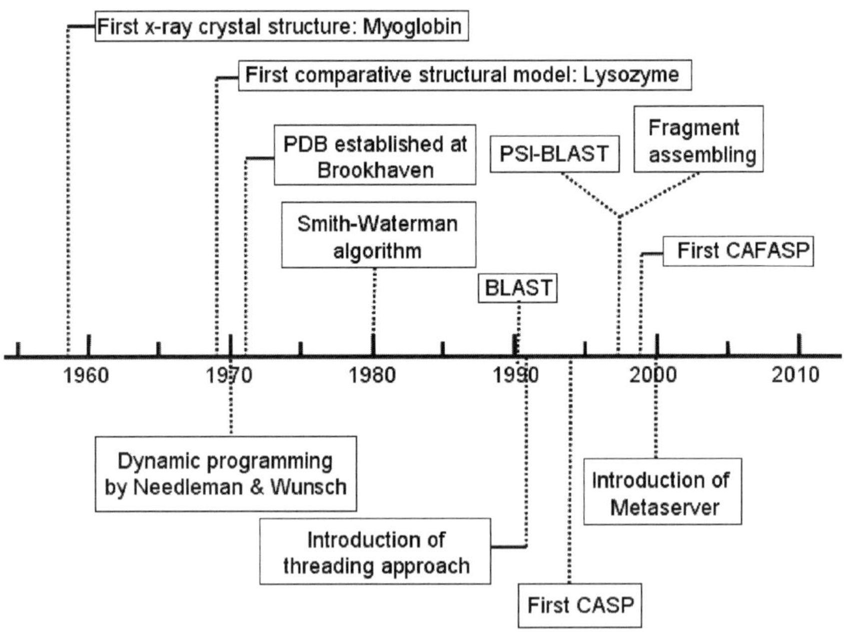

Fig. 2. Major milestones in development of template-based protein structure prediction.

solvent accessibility and protein secondary structure, Bowie et al. developed a novel approach to assessing if a protein sequence fits well with a given protein structural fold, essentially an inverse protein-folding problem *(28)*. As the environment of a particular residue, in which a 3D structure is encoded as a 1D profile, tends to be more highly conserved than the identity of the residue itself, the method is able to detect more distant relationships than the purely sequence-based methods. Realizing that environment-based methods are incapable of detecting structural similarities among proteins as a result of convergent evolution, Jones et al. developed a novel dynamic programing approach to protein-FR by considering specific pair interactions explicitly and introduced the concept of "threading" *(37)*. Therefore, unlike the sequence-based methods, protein threading takes advantage of the structural information of a template. These studies had laid the foundation and led to the development of a large class of threading techniques.

The basic idea of protein threading is to "thread," literally, the amino acids of a query protein, following their sequential order and allowing for insertions and gaps, into structural positions of a template structure in an optimal way measured by a scoring function. This procedure is repeated for each template structure in a database of protein structures. The quality of a sequence-structure alignment is typically assessed using statistical-based energy terms or physical-based energies. The "best" sequence-structure alignment provides a prediction of the backbone atoms of the query protein. The development of a threading-based structure prediction technique generally involves four key issues: (1) development of energy functions for assessing the quality of a sequence-structure alignment or placement; (2) threading algorithms for finding a sequence-structure alignment that optimizes a given energy function; (3) statistical assessment and FR; and (4) development of a structural template library. We discuss the details of each of the areas in the next few subheadings. **Figure 3** shows a timeline of the major milestones in protein threading.

2.1.1. Threading Energy Functions

Unlike the classic physical energy used in protein-folding studies, energy functions employed to score a particular sequence-structure alignment in protein threading are mainly statistics-based, also called knowledge-based. The idea of using knowledge-based potential energies in protein threading is that experimentally determined structures contain great amount of information on the stabilizing forces within proteins. Statistical analyses of protein structures can possibly capture the underlying rules governing the structural stabilities of proteins, which can be realized by relying explicitly or implicitly on

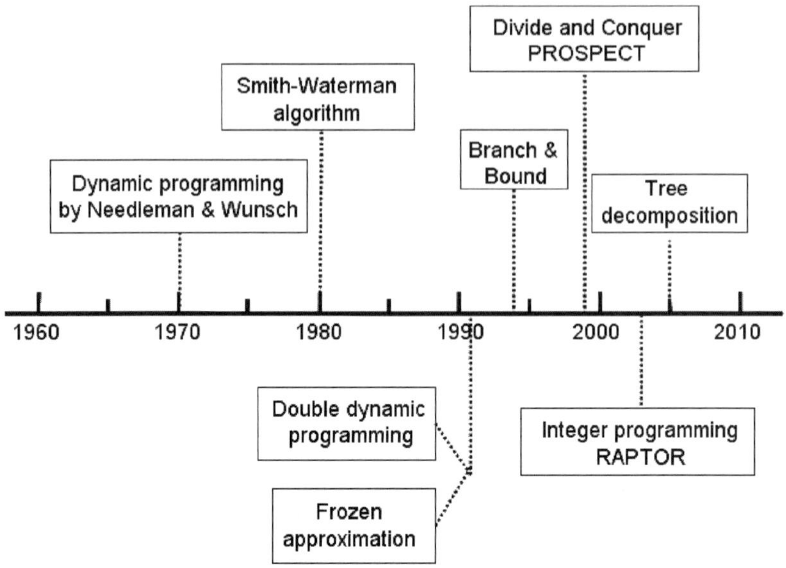

Fig. 3. Key developments in threading algorithms.

Boltzmann's principal: frequently observed states correspond to low energy states of a system ***(46)***. The idea of deriving knowledge-based potentials from known protein structures has a long history even before their first applications in protein threading by Bowie et al. ***(28)*** and Jones et al. ***(37)***. For example, Tanaka and Scheraga ***(47)*** first reported their study on medium and long-range interaction potentials and applications to predicting protein 3D structures 30 years ago, which was followed by numerous related studies by several other groups ***(5,48,49,50)***.

Earlier threading approaches generally ignore sequence similarities between the query sequence and the template protein ***(51)***. Instead, they considered only the preferences of each amino acid of the query protein to the physical–chemical environments of the template structure. These environments generally fall into two classes. One is defined in terms of static measures, such as solvent accessibility and secondary structures as described in the original work by Bowie et al. ***(28)***. This type of potential is also called singleton energy. For example, the solvent accessibility can be described in three states, exposed, buried, and medium exposed. As for the secondary structure environment, one can use three major types of secondary structures, helix (H), strand (E), and loop or coil (C). The combination of these two terms results in, in this case, nine different structural environments. The singleton energy term can be calculated from a (non-redundant) protein structure database using Boltzmann statistics. The basic idea can be explained as follows. If an amino acid type is frequently observed in the interiors across many protein structures, it suggests that the interior of a protein is a favorable place for this amino acid type to be placed. The singleton energy can be possibly written as follows:

$$e_{\text{single}(i,j)} = -\ln(O_{i,j}/E_{i,j})$$

where $O_{i,j}$ is the observed frequency of amino acid type i in structural environment j. $E_{i,j}$ is less obvious, which represents the expected frequency of amino acid type i in structural environment j. If nine different structural environments are used as discussed earlier, a 20×9 matrix will be generated, with each of the 20 rows representing an amino acid type and each of the nine columns representing a structural environment ***(52,53)***. In their original study on the threading methodology, Bowie et al. used 18 different structural environments, made up of six different accessibility states, each of which can be in one of three types of secondary structures, i.e., helix, strand, or coil conformation.

Another class of the energy function is called pair-wise energy, which describes the interactions between two residues. It measures the preference of having two particular types of amino acids spatially close to each other. Jones et al. initially proposed and applied such an energy function developed by Sippl ***(37,50)*** in their FR study. The basic idea of such an energy function, again, comes from statistical mechanics. This knowledge-based potential can be written as

$$g_{ij} = -kT \ln\left(\frac{P_{ij}}{\bar{P}}\right)$$

where k and T are the Boltzmann constant and temperature, respectively. P_{ij} is the observed frequency of residue pairs i and j at a certain distance, where the distance is measured between the C_β atoms of the two residues; and $\bar{P}$ is the reference state.

There are two types of pair-wise energies, distance-dependent and distance-independent. It has been observed that distance-dependent pair-wise interaction energy could provide more accurate threading results than that of a distance-independent energy as outlined above. A distance-dependent energy could be estimated as follows:

$$\bar{u}(i,j,r) = -\ln\left(\frac{N_o(i,j,r)}{N_B(i,j,r)}\right)$$

where r is the distance between residues i and j; $N_o(i, j, r)$ is the observed number of pairs of residues (i, j) within a distance range from $r - \Delta r/2$ to $r + \Delta r/2$ in a database of protein structures for some width Δr, and $N_B(i, j, r)$ is the expected number of pairs (i, j) within the same distance range. The challenging issue in accurately estimating the interacting energy $\bar{u}(i, j, r)$ is how to estimate $N_B(i, j, r)$. Under the assumption that we are dealing with an ideal infinite liquid-state system within a volume V and residues are distributed uniformly, called a *uniform distribution model* ***(50,54–56)***, $N_B(i, j, r)$ can be estimated using

$$N_B(i, j, r) = N_i N_j \left(\frac{4\pi\Delta r}{V} \right)$$

where N_i and N_j are the numbers of amino acid types i and j in the protein structure database, respectively. Realizing that this model is not accurate when dealing with finite systems such as a protein structure, Zhou et al. ***(57)*** developed a new energy model called DFIRE for distance-scaled finite ideal gas reference state, which uses the following formula:

$$\bar{u}(i, j, r) = \begin{cases} -\eta \ln \dfrac{N_o(i, j, r)}{\left(\dfrac{r}{r_{\text{cut}}} \right)^{\alpha} \dfrac{\Delta r}{\Delta r_{\text{cut}}} N_o(i, j, r_{\text{cut}})} & , r \le r_{\text{cut}} \\ 0 & , r > r_{\text{cut}} \end{cases}$$

where constant η is related to the system temperature and can be determined empirically. In this model, the authors made two corrections to the uniform distribution model. First, DFIRE used r^{α} instead of r^2, considering that the number of interaction pairs in a finite system could not actually reach the level of r^2 as in an infinite system, where α is determined through minimizing the distribution fluctuation of interaction distances on a training set. Second, DFIRE only considers short-range pair-wise interactions, that is, interaction energy becomes zero when the distance between the interacting pairs is beyond a cutoff distance r_{cut}. Another unique feature of DFIRE is that it is less sensitive to database composition than other distance-dependent potentials ***(58)***. The DFIRE energy and its derivatives have made great strides in protein threading prediction as evidenced in the CASP6 experiment ***(59)***.

In addition to these simple forms of potentials, there are other sophisticated potential functions developed with an attempt to improve the performance of FR. However, these potentials are too complicated to be used in real threading programs. Generally, these potentials are tested using a self-recognition method ***(60)***. For example, Kocher et al. ***(61)*** introduced a torsion angle term in addition to a residue–residue interaction term and an accessibility term. Nishikawa and Matsuo ***(62)*** developed a more sophisticated FR potential with four terms, a side-chain–side-chain interaction potential, a hydration potential, a hydrogen-bonding potential, and a local conformation (ψ, ϕ) potential. Again, the potential was tested by the standard self-recognition test and showed improvement over conventional sequence alignment-based methods.

Probably, the most sophisticated potential for protein threading is the side-chain packing and orientation-dependent statistical potentials ***(63,64)***. Miyazawa and Jernigan applied a uniform distribution and ignored high-frequency contributions of the observed distribution of contacts in deriving their orientational dependence of side-chain packing. Their potential has been shown to significantly improve the ability to recognize the native structural folds from the decoys ***(64)***. Despite the ability of these potentials in self-recognition studies, their performance in real threading has yet to be tested.

Threading has been a coarse-grained structure-prediction method, in which the residues of the query sequence are placed on the backbone of a structural template. As we do not know the exact coordinates of non-backbone atoms and the physical energies are very sensitive to small variations, it would be difficult to apply the physics-based energy function directly to the threading framework unless we can solve the backbone threading and the side-chain packing at the same time, which represents the tremendous computational challenge and requires novel algorithms to solve the problem. On the basis of performance from the recent CASP contests ***(65,66)***, we begin to suspect that residue-based potential energy functions outlined above are probably reaching their limits. Hence, we expect that more physics-based energy functions will emerge in near future.

One of the major breakthroughs in residue-based energy function development in the past ten years came as a result of integrating evolutionary information into the energy functions. It has been observed that it improves the threading accuracy by using all homologous sequences of a query protein instead of using the query sequence alone during threading ***(53)***. One simple way to incorporate the multiple sequence information is through using the sequence profile, derived from the aligned homologous sequences, hence generalizing the sequence-structure alignment problem to a sequence profile-structure alignment problem. An even more general way for measuring the fitness score of a threading alignment is through comparing the sequence profiles of the query protein and the template protein, also known as profile-profile alignment, which is very successful in detecting distantly related homologs (see more detailed discussion on sequence profiles in **Subheading 2.2.**) ***(67,68)***.

Other types of energy functions have also been used in the existing threading programs. These include match scores between the predicted secondary structures of the query sequence and the secondary structures in the template, and gap penalties.

The total energy of an alignment can be calculated using a weighted sum as follows:

$$E_{\text{total}} = \omega_{\text{m}} E_{\text{mutation}} + \omega_{\text{s}} E_{\text{singleton}} + \omega_{\text{p}} E_{\text{pairwise}} + \omega_{\text{g}} E_{\text{gap}} + \omega_{\text{ss}} E_{\text{ss}}$$

where E_{mutation}, $E_{\text{singleton}}$, E_{pairwise}, E_{gap}, and E_{ss} represent mutation energy, singleton energy, pair-wise energy, gap penalty, and secondary structure match energy, respectively. The weight (ω) of each energy term can be practically derived through optimizing the threading performance on a set of query–template pairs, both of which have their structures solved.

2.1.2. Sequence-Structure Alignment Algorithms

The basic goal of protein threading is to find an alignment (or placement) between a query protein sequence and a template structure that optimizes the aforementioned threading energy function. The importance of finding the best alignment between a sequence and a template structure cannot be overstated as currently it dictates the quality of the model. Generally, threading algorithms can be grouped into two major classes: heuristic algorithms and rigorous algorithm. One of the most popular heuristic threading algorithms is dynamic programing. Owing to its simplicity and efficiency, many of the earlier threading programs used dynamic programing or its variations. If we do not consider pair-wise interactions in our threading energy function, the sequence-structure alignment problem is essentially a sequence-sequence alignment problem, which can be solved rigorously using a dynamic programing approach. For example, in their original threading paper, Bowie et al. defined 18 different structural environments and represented a template structure as a sequence of structural environments. Hence, a threading problem is essential to find an optimal alignment between a query sequence and a sequence of structural environments, which can be solved using a dynamic programming approach. Despite the obvious computational advantages by representing a template 3D structure as a sequence of structural environments, it has been well documented that the threading performance based on such a formulation of a threading problem could be substantially improved by including residue–residue pair-wise interaction energies ***(52,69)***.

When a pair-wise potential is considered in a threading energy function, the simple dynamic programing strategies used previously will no longer guarantee to find the sequence-structure alignment that achieves the minimum energy simply because pair-wise interactions substantially complicate the computation for an optimal threading alignment. It has been proved that the general threading problem is NP-hard under some generalized (possibly unrealistic) assumption *(70)*, suggesting the intrinsic difficulty in computationally solving the problem. Researchers have tried various heuristic techniques to "fix" the inadequacy problem of the dynamic programing approach when considering pair-wise interactions by adding different types of "patches" to the overall framework of dynamic programing. Jones et al. employed a double-dynamic programing approach derived from structural superposition methods *(37)*. Their approach applies dynamic programing at two levels, a high-level scoring matrix and a low-level matrix for each element of the high-level matrix. For each F_{ij} of the high-level matrix, the likelihood of i being aligned to j is calculated by a low-level optimal alignment with the constraint that F_{ij} is part of the alignment. "Frozen approximation" is another popular method, which assumes that the interaction between two residues of the query protein placed in two nearby (template) structural positions can be approximated by an interaction between one query residue placed in one structural position and the original residue in the other position of the template structure *(71–74)*. Specifically, when the algorithm assigns an amino acid from the query protein to a structural position of the template from the beginning to the end of the query protein, it calculates the relevant interaction energy using this newly assigned residue and query residues already assigned to its nearby positions plus the template residues in nearby positions yet to be assigned, where a cutoff is used to define "nearby" structural positions. Intuitively, the algorithm should work to some degree in capturing some of the interaction "patterns" encoded in the query protein sequence as some of the position-equivalent residues between the native structure and the native-like template structure should have similar physicochemical properties, suggesting the validity of the frozen approximation scheme.

Several other heuristic threading programs have also been developed. For example, GenTHREADER uses a classical sequence-sequence alignment algorithm to generate query-template alignments and then evaluates the alignments using a threading potential in a post-processing step *(75)*. The program 3D-position-specific scoring matrix (PSSM) also employs a dynamic programing scheme to find the best alignment between the sequence profile of a query protein and the sequence profile of the template protein *(76)*. FUGUE utilizes the environment-specific amino acid substitution tables and structure-dependent gap penalties *(77)*.

Although heuristic algorithms "solve" a threading problem fast, they achieve the computational efficiency at the expense of prediction accuracy. Since the middle of the 1990s, a number of rigorous threading algorithms have been developed that guarantee to find the globally optimal threading alignments, measured in terms of energy functions outlined in the previous section. The first rigorous threading algorithm that considers pair-wise interactions was a branch and bound algorithm developed by Lathrop and Smith ***(78)***, although its actual computing time and the practical usefulness have not been well documented. Xu et al. developed a threading program, PROSPECT, which solves rigorously the globally optimal threading problem, using a divide-and-conquer strategy ***(52)***. Its practical usefulness and the value in rigorously solving the threading problem were demonstrated through the prediction server of the program ***(79)***. The threading problem was later formulated as a linear integer programing (LIP) problem and was implemented as a computer program, RAPTOR ***(80)***. The authors of RAPTOR took advantage of the extensive research results in the area of LIP to make the program run much faster than PROSPECT, although the same set of energy function is used. It was convincingly demonstrated, through applications of programs such as PROSPECT and RAPTOR at the CASP contests, that threading programs with guaranteed global optimality do have an advantage over programs without this property ***(81,82)***. The only disadvantage though is that these rigorous algorithms tend to be slower when compared with heuristic algorithms such as dynamic programing-based threading algorithms.

Tree decomposition-based algorithm seems to represent another powerful technique for solving the threading problem rigorously. This technique is currently being actively investigated, which is based on the idea of tree-decomposition of an interaction graph representing possible alignments between a query sequence and a template structure ***(83,84)***. In a sense, this type of technique represents a generalization of the divide-and-conquer idea, whose framework allows taking full advantage of powerful graph-theoretic results to make the threading algorithm much faster than any of the previous rigorous threading algorithms. In this formulation, both the template structure and the query sequence are represented as graphs; vertices denote core secondary structures, and edges represent interactions between the cores. A sequence-structure alignment problem essentially corresponds to finding an isomorphic mapping from the structure graph to a subgraph of the sequence graph. The efficiency of the alignment hinges on the *tree width* of the structure graph. Intuitively, the *tree width* of a graph measures how much the graph is "tree-like." The "tree-like" representation for graphs is called a *tree decomposition*. Given a tree decomposition of a structure graph with tree width t, a dynamic programing algorithm can be employed to find the globally optimal sequence-structure alignment in time $O(k^t N^2)$, where k is a small integer and N is the number of amino acids in the template structure ***(83,84)***. The efficiency of the algorithm can be illustrated as follows. When using 7.5 Å C_β–C_β distance as a cutoff for defining pair-wise interactions, among 3890 non-redundant protein tertiary structure templates compiled using PISCES ***(85)***, only 0.8% of them have tree width $t > 10$ and 92% have $t < 6$. Computational results have indicated that the tree decomposition-based threading algorithm runs substantially faster than both the divide-and-conquer and the integer programing-based threading programs without compromising alignment accuracy ***(84)***.

2.1.3. FR/Statistical Significance of Threading Alignments

To solve the FR problem, it is not enough to have a powerful threading algorithm, which can only find the best possible alignment between a query sequence and a specific structural template. We still need to find which of the structural templates, from a large collection of solved protein structures, represents the correct, that is, native-like, structural fold. One simple way to determine the correct structural fold is to rank the templates based on the threading scores as reported in Jones et al.'s original threading study ***(37)***. However, doing so is problematic as the alignment scores between a query sequence and different template structures are in general not comparable directly with each other. Some structures might tend to have higher baseline threading scores than the others, no matter what query sequences are used. Some normalization techniques have been applied in early threading program to help identify the true protein fold. Bowie et al. normalized their threading scores using the score distribution obtained by aligning many sequences of similar length to each of the structure templates ***(28)***. Others do it by shuffling the query sequence many times (keeping the same length and composition as the query sequence) and align each of the sequence to each of the template structures ***(86)***.

In a sense, the FR problem is similar to the homology search problem through sequence comparison against a database of sequences. In sequence-sequence alignments, there have been a number of models developed for assessing the statistical significance of the scores. For example, the statistical significance of an alignment score can be estimated using methods such as calculating the p-value (a probability of a score occurring by chance) or the e-value (expected number of times the score will be seen given the size of the database) ***(87–90)***. Developing rigorous and effective statistical models for protein threading has proven to be more challenging than the sequence alignment problem. There have been a number of attempts to use empirical methods to assess the significance of alignment scores. One popular practice is to use a z-score scheme. The z-score is the threading scores in standard deviation unit relative to the average of the threading score distribution of random sequences with the same amino acid composition and length as a query sequence ***(91)***. In practice, the average and the standard deviation are estimated by threading between a template and a large number of randomly shuffled query sequences. The z-score of a particular alignment can be defined as follows:

$$z = \frac{E - \bar{E}}{\sigma}$$

where $\bar{E}$ and σ are the average and the standard deviation of the energy distribution resulted from threading alignments between the template and the sequences from the re-shuffled query sequence. This approach is effective to some degree. But its limitation is also obvious based on practical applications. One of the key reasons for the limited success of the z-score scheme is that the underlying assumption for the z-score scheme to be effective is that the threading scores should follow a normal distribution, which is in general not true for threading scores.

It has been shown that like the optimal sequence-sequence alignment scores, optimal structure-structure alignment scores generally follow an extreme-value distribution ***(92)***. On the basis of this observation, Sommer et al. developed a scheme for p-value estimation for threading scores, in which parameters of the model are estimated by fitting the threading scores against an extreme value distribution ***(93)***. A key characteristic of the work is that it attempts to have a unified model for threading problems involving different lengths and compositions of the proteins involved.

There have been various studies attempting to derive empirical and effective models for assessing the statistical significance of a calculated threading score. The basic idea is to derive a "normalized" threading score, based on a training data set, which consists of query-template pairs with different lengths and different compositions. A neural network and a support vector machine (SVM) could be trained to best mimic the degree of correctness of each threading alignment by a threading program on the training set, using various parameters collected from the threading program such as singleton energy value, pair-wise energy value, the lengths of the query and the template, their compositions. A number of such normalization schemes have been developed and employed in various threading programs ***(52,53,75,80,94)***. For example, by using a SVM, the RAPTOR program, which employs an integer programming approach, has significantly improved its FR performance ***(95)***.

2.1.4. Template Structure Library

The performance of a threading program largely depends on the "completeness" of the template structure library it uses. If the template structure library does not contain a homologous protein to a query protein, no matter how good the threading energy, algorithm and assessment capabilities are, it is not going to predict the structure correctly. Actually, the accuracy of a predicted structure also largely depends on how close the template structure is to the actual structure of the query protein. Hence, generally speaking, the more comprehensive a template structure library is, the more accurate we can expect a threading prediction will be. Therefore, to get the best performance of a threading program, the ideal template library should include all the structures in PDB *(13)*. However, it is presently impractical to include all PDB structures in a template structure library because of the amount of time required to compute one sequence-structure alignment. In addition, many protein structures in PDB are redundant in principle, one can use one representative from each protein family, defined by SCOP *(40)*, as the template library for threading. To be on the safe side, it is advisable to include several members from each protein family, considering the possible structural variations among the "equivalent" proteins from different organisms. Inclusion of multiple members of the same family in the template library could also help to derive more accurate threading alignments, based on the consensus of the multiple sequence-structure alignments between the query and the multiple homologous structures.

To construct a representative template library, one can use single chains or protein domains as members. There are several ways to select the representatives, by either sequence identity-based or structural similarity-based methods. Three popular protein structure classification databases, SCOP *(40)*, CATH *(42)*, and FSSP *(96)*, are domain-based classifications using both the sequence and structure information. The disadvantage of these three databases is that they are not updated very often. PISCES *(85)*, on the contrary, is a program that is suitable for selecting sequence-based representative data set, which can be updated as needed.

2.2. Sequence-Based Alignment Methods

As discussed earlier, protein threading is essentially a sequence-sequence comparison method when structural information is not considered. Sequence-based alignment methods have been the primary tools for earlier homology-based modeling (also called CM) efforts *(43)*. When there are close homologs in PDB for a query sequence, the easiest method for structure prediction is to use a sequence-sequence alignment method, such as Smith–Waterman algorithm ***(97)***, BLAST ***(98)*** or FASTA ***(99)***. Earlier versions of BLAST do not allow gapped local sequence alignment, which have limited its applications. Most recent versions of BLAST do allow gapped alignments, in which a gap penalty function is applied. One of the attractive properties of BLAST is that it provides a reliable way for assessing the statistical significance of an alignment result.

However, when the sequence identity is less than 30%, especially when the evolution relationship is not obvious between a structural template and a query protein, more sophisticated methods might be required. Currently, profile-based methods are the most popular ones for detecting more distant evolutionary relationships. The core of these profile-based methods is the position-specific profiles derived from the alignment of multiple sequences among proteins from the same family or even super-family. PSI-BLAST ***(100)*** and hidden Markov Models (HMMs) ***(101,102)*** represent two popular methods for generating a sequence profile based on the multiple sequence alignments among homologous proteins of a query sequence. In a profile, each position is represented as a vector describing the relative frequency of each of the 20 amino acid types in this aligned position. Profiles have been used to represent both the query proteins and the template proteins, leading to the development of more sensitive homology detection tools through profile-profile comparison ***(103–105)***. In profile-profile alignment approach, the similarity score of two positions can simply be calculated as the dot product of the two vectors.

Although traditional homology modeling methods rely primarily on sequence-only approach and early FR methods rely primarily on the protein-threading approach, the introduction of PSI-BLAST and HMM has somehow blurred the boundary between homology modeling and FR. Currently, sequence-based profile methods are the major approach to derive the sequence-structure alignment in both the CM and FR categories. The popularity of sequence-based profile methods in FR is largely because the performance of this type of methods is almost on par with threading-based approaches for detecting remote homologous proteins, which prompted the question if structural information or the threading approaches in general, is useful ***(16)***. Nevertheless, single-method servers that incorporate structural information, including RAPTOR (rigorously treats pair-wise interactions) ***(80)***, SP3 (uses depth-dependent structural alignments) ***(106)***, and SPARKS (contains a backbone torsion term, a buried surface term, and a contact-energy term) ***(107)*** have performed very well in both CM and FR categories as demonstrated in recent CASP contests ***(59,82)***. On the contrary, there is a clear limitation for sequence-based profile methods for detecting remote homologous relationship. For example, the profile methods alone are not going to do well when sequence profiles are not possible to construct because there are not enough known sequences that are clearly related to the query sequence or potential templates.

2.3. Comparative/Homology Model Building

Once a sequence-structure alignment has been built using either sequence-based approach or threading method as describe above, 3D models can be constructed using various methods. The very first homology model, a wire and plastic model of α-lactalbumin, was constructed in 1969 by Browne et al. using a method called rigid body assembly *(19)*. Since then, many different homology modeling methods and programs have been developed. Generally, these model-building methods can be grouped into four classes: (1) rigid body assembly, (2) segment matching, (3) spatial restraint, and (4) artificial evolution model building. **Figure 4** shows a timeline of the major milestones in the area of comparative/homology modeling.

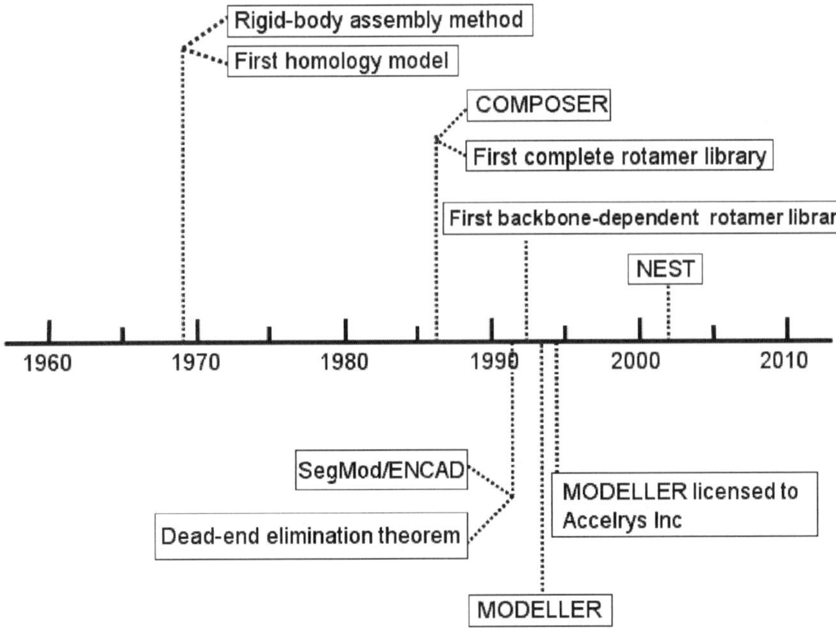

Fig. 4. Timeline of major developments in the area of comparative/homology modeling.

The rigid-body assembly method was first introduced in 1969 and is still widely used *(19,108)*. It starts with the identification of the conserved core and variable regions and then assembles a model from a number of rigid bodies obtained from the template structures *(43,45)*. This method has been implemented in several computer programs, including COMPOSER *(109)*, PriSM *(110)*, 3D-JIGSAW *(111)*, and SWISS-MODEL *(112)*.

Segment-matching method is developed based on the observation that most hexapeptide segments of protein structures can be clustered into about 100 structural classes *(113)*. The model construction is done through using a subset of atomic positions of templates as "guiding" points, which in general are the conserved segments in a sequence-structure alignment and, assembling short, all-atom segments derived either by scanning all the known protein structures *(114)* or by a conformational search guided by an energy function *(115)*, to fit these guiding positions. The unique feature of segment-matching method is that it can be used to model both side-chain atoms and loops. SegMod/ENCAD is the first program implemented using segment-matching approach *(116)*.

The most popular homology modeling program is MODELLER *(117)*, which uses spatial restraints derived from the template structure(s) (the approach is called *satisfaction of spatial restraints*) to guide the model building process. MODELLER was introduced in 1993 and has been part of the Insight Package since 1994. MODELLER models a structure typically by generating many restraints first, assuming that the corresponding distances and angles between aligned residues in the sequence-structure alignment are similar. A model is then derived by minimizing the violations of all restraints including homology-derived restraints and other stereochemical restraints from a molecular mechanics force field, such as bond length, bond angles, dihedral angles, and non-bond atom–atom contacts *(43)*. One of the advantages of using satisfaction of spatial restraints method is that it can incorporate various restraints from experiments, such as NMR experiments, site-directed mutagenesis and cross-linking experiments.

The artificial evolution method attempts to build structural models by simulating the natural process of structural evolution from a template structure to the target model. For example, the sequence-structure alignment can be broken down as a series of evolutional operations, such as mutation, insertion and deletion. The structural model can then be built from the template structure by changing one evolutionary event at a time *(45)*. NEST, the core program within the JACKAL Modeling Package, uses this approach *(118)*.

In any homology modeling program, side-chain prediction and loop modeling are the two key components. These are usually done on fixed backbone coordinates. The accurate prediction of loops and side-chains has great bearing in applications as they are often of functional importance and can play key roles in forming enzyme active sites, antigen-antibody recognition, ligand–receptor binding, and the binding between metal ions and metal-binding proteins ***(119)***. There are two main approaches to loop modeling, database-based methods ***(120–122)*** and ab initio methods ***(123–125)***. The database-based approach to loop prediction is done by finding a segment of main chain that fits the two ends of a loop whereas the ab initio methods involve the generation of a large number of candidate conformations and the evaluation of the conformations using an energy function. In addition to the two basic methods, procedures that combine the above two basic approaches have also been described ***(126)***.

In his pioneering modeling study, Greer used a simple algorithm to insert loops from homologous proteins into the target protein ***(120)***. This database-based approach is accurate and efficient when modeling short loops. However, it is limited by the exponential increase in the number of possible conformations for longer loops. Studies have shown that only segments of seven residues or less have most of their possible conformations found in known protein structures ***(127)***. Van Vlijmen and Karplus partially solved this database completeness problem by combining database search and restrained energy optimization, which extends the loop modeling range up to nine residues with candidate segments from a database ***(126)***. The ab initio loop prediction method is based on a conformational search guided by a scoring function. It does not have the database completeness problem for long loops, but the accuracy for modeling long loops is still low. Various ab initio methods, which exploit different scoring functions, protein representation, and optimization algorithms, have been described ***(119)***. Realizing that most approaches try to find the lowest energy conformation without considering conformational entropy effects, Xiang et al. implemented a procedure called "colony energy" that considers the shape of the energy well, which improved the accuracy ***(125)***. Jacobson et al. used an Optimized Potential for Liquid Simulations (OPLS) all-atom force field, the Surface Generalized Born implicit solvent model, and a hierarchical optimization strategy to achieve a better performance [e.g., 1.0 Å root mean square deviation (rmsd) for eight-residue loops] ***(128)***.

Side-chain prediction represents another challenge in homology modeling. Nearly all of the side-chain prediction methods are based on one of the rotamer libraries with discrete side-chain conformations, either backbone-dependent or backbone-independent. While the concept of rotamer was around as early as 1970 ***(129)***, the first rotamer library with a list of all likely conformations of side-chains and their average dihedral angles, variances, and frequencies was not introduced until 1987 by Ponder and Richards ***(130)***. In 1993,

Dunbrack and Karplus ***(131)*** presented the first backbone-dependent rotamer library. A backbone-independent rotamer library developed by Richardson and colleagues ***(132)*** probably represents one of the most accurate libraries. In deriving the library, they used much stricter criteria including removing side-chains (1) with high B-factors, (2) with clashes with any other atom, and (3) with uncertain amide or histidine ring orientations after an optimization procedure ***(133)***. Xiang and Honig ***(134)*** have done a detailed study on the prediction accuracy with different rotamer libraries. They showed that using an extensive library in which bond lengths and bond angles were taken from the database rather than using idealized values yields RMSD values of only 0.62 Å for core residues ***(134)***.

Given a rotamer library and a defined energy function, the side-chain prediction problem becomes a combinatorial optimization problem. Two different search strategies have been widely applied for side-chain predictions, exact algorithms that guarantee to find the side-chain conformations with global minimum energy and approximation algorithms, including Monte Carlo simulation-based methods ***(135)*** and cyclical search method ***(131)***. The first exact algorithm for side-chain prediction is called dead-end elimination (DEE) algorithm ***(136)***, which provides a powerful deterministic approach to finding the global minimum energy conformation (GMEC) by comparing the energy distributions of different candidate rotamers at a given position and identifying certain rotamers, which cannot exist in the GMEC. Different versions of DEE have been developed later to improve the performance of the original DEE algorithm ***(137–141)***. The side-chain prediction problem has also been formulated as a graph-theoretic problem and solved by combinatorial optimization algorithms, such as the biconnected graph in SCWRL program ***(142)*** and tree-decomposition algorithm in SCATD ***(143)***. These new approaches run much faster without compromising the prediction accuracy and can be used for large-scale predictions. On the contrary, a study by Xiang and Honig ***(134)*** in 2001 demonstrated that the combinatorial problem in side-chain conformation prediction does not appear to be very important, which is supported by other studies. Desmet et al. developed a new method, called Fast and Accurate Side-chain Topology and Energy Refinement (FASTER), for global optimization of protein side-chain conformations ***(144)***. They showed that low-order local minima may be as accurate as the global minima. The FASTER algorithm is 100–1000 times faster than the DEE method, at the same time; it produces nearly identical results ***(144)***.

Recently, two independent comparative studies of homology modeling programs concluded that no single modeling program outperforms the others in all tests. However, some programs perform better than the others, such as MODELLER ***(117)***, NEST ***(118)***, and SegMod/ENCAD ***(116)*** in one study ***(145)***, and Prime (Schrodinger, LLC), DSModeler (commercial version of MODELLER), and Sybyl ***(109,146)*** in another study ***(147)***. It should be noted that most of the programs tested in these two studies are different.

2.4. Fragment Assembly

When a query sequence does not have apparent structural homologs or analogs, the aforementioned FR approaches will not be able to make a good structure prediction as these methods suffer from the fundamental limitation of being only able to recognize known folds. Different techniques will be needed. While theoretically speaking, ab initio approaches could be used for tackling such problems, it is well known that existing ab initio folding techniques, along with the existing force fields, are probably not ready for folding proteins accurately in general. A new class of structure prediction methods, called fragment assembly-based methods, is emerging, which have played key roles in making structure predictions when a protein does not have a native-like structural fold in the PDB database. The basic idea is stimulated by the observation that the backbone structure of a protein can be constructed from a number of fragment taken from other proteins ***(121,148)***. Generally, a fragment assembly method consists of two major steps. It first identifies structural fragments from a fragment structure library, to which segments of a query protein might fold into and then the method assembles the fragments into a whole structure using some energy minimization technique. FRAGFOLD, which exploited this strategy to greatly narrow the search of the conformational space and had some success in the CASP2 experiment in 1996, was the first such effort that has demonstrated the power of fragment assembly-based approach for structure prediction, which could potentially predict structures with a novel fold ***(149,150)***. ROSETTA, developed by David Baker's group, probably represents the most successful program using the fragment assembly strategy ***(151)***, based on its prediction results at the previous two CASP contests ***(65,66)***. ROSETTA method first divides a query sequence into overlapping sequence fragments approximately nine residues in length. Then, the protein structure database is searched for sequence fragments that are similar to each fragment of the query sequence and in their secondary structures. Protein conformations are then built up using these corresponding fragment structures. The best structures are selected and then refined as the final prediction. For smaller structures, ROSETTA can generate one or a few relatively accurate structures ***(152)***.

Upon the successful showing of ROSETTA in the past a couple of CASP contests, more programs have been developed using fragment assembly approach exclusively or as part of the prediction package, which include UNDERTAKER ***(153)***, ABLE ***(154)***, SIMFOLD ***(155,156)***, PROFESY ***(157)***. Although these methods use similar strategy, they differ in several key areas: (1) the length of the fragment, (2) fragment assemble method, and (3) energy function. Typically, the lengths of the structural fragments may range from a short peptide as used in ROSETTA, SIMFOLD, and ABLE to super-secondary structural unit ***(158)***. Structural fragments are often selected based on the compatibility between a fragment and a sub-sequence of the query protein. In practice, many different sub-structures or fragments could be selected for each short sequence as a short sequence may adopt different conformations in different structural environments. Most programs use Monte Carlo-simulated annealing search strategy to assemble the fragments. The best structure is then selected from a large set of possible structures, which is usually done through clustering of candidate structures using knowledge- or physics based potentials ***(159)***.

These fragment-based methods for protein structure prediction clearly represent a new breed of structure prediction technique, which combines strengths of template-based structure prediction methods (in the stage of initial fragment structure identification) and ab initio prediction techniques (in the stage of piecing together fragment structures to form one whole structure). Because of their general applicability, we expect that this class of structure prediction techniques could potentially become the dominating technique for protein structure prediction.

3. CASP/CAFASP

Starting from 1994, protein structure predictors have been having their own "Olympic games," called CASP, every 2 years. A companion contest, CAFASP, was introduced in 1998 ***(17)*** to assess the performance of fully automated prediction servers without any human input. Although each contest is held during the summer of every other year, the contest results will not be released until the CASP meeting in the winter of the same year. The predictors share the similar experience but with different emotions—joy, disappointment, excitement, or surprise when the contest results are announced by the contest assessors at the CASP meetings. Interestingly, Roland Dunbrack, the assessor of the FR category of CASP6, employed a scoring system similar to that used in diving and gymnastics events in the Olympic games, where the lowest and the highest scores from six structure comparison programs are removed before being summed up or being averaged ***(66)***. The goals of the experiments are to evaluate the successes and failures of structure prediction techniques as a whole, to identify the bottlenecks, and to provide the directions for future improvements ***(15,16)***.

For assessment purpose, CASP protein targets (i.e., protein sequences) are divided into three categories with an increasing level of prediction difficulty: CM, FR, and new folds (NF). CM and FR are further divided into a few subcategories. For example, CM is further divided into CM-easy (targets whose structural templates can be identified easily by BLAST) and CM-hard. FR targets are grouped into FR/H for remote homology detection and FR/A for analogous FR ***(160)***.

CASP has attracted many structure predictors to participate in this fun and exciting event, in which participants range from graduate students to world-class structural biologists and modelers. At CASP1, only 35 groups took part in the experiment. This number continues to increase steadily over the course of CASP. In 2004, over 200 prediction teams from 24 countries participated in CASP6. In FR prediction category, only nine groups participated in CASP1, whereas 165 teams made predictions for the targets in this category in CASP6 (*see* **Fig. 5**). CASP has clearly provided a big stage for different prediction groups to showcase their best prediction tools. As the field evolves, CASP contests have also evolved accordingly. For example, in the first three CASP contests, prediction targets were divided into three

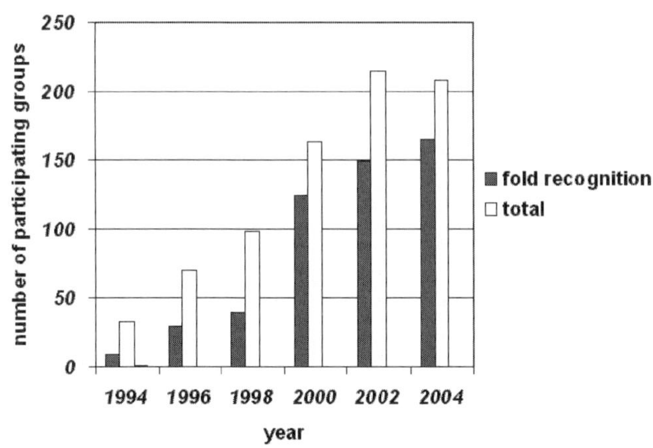

Fig. 5. Number of groups participating CASP and the number of groups participating the fold recognition category from CASP1 (1994) to CASP6 (2004).

categories: CM, FR/threading, and ab initio prediction. At CASP4, ab initio was reclassified as "new fold methods" in recognition of the fragment assembly methods. Now ab initio is reserved for prediction methods that employ the first principles only.

After the first decade of CASP experiments, the obvious question we want to ask is how much progress we have made in terms of our overall capabilities in protein structure prediction in 10 years. Much has been written about the CASP assessments (published as special issues by *Proteins: Structure, Function, and Bioinformatics*). So here, we briefly highlight the progress made and the challenging issues, especially in FR category. For homology modeling and ab initio prediction, the general consensus has been that progress has been quite limited in the past several years. Although there have been some steady improvements in the homology modeling category, especially in alignment accuracy, it is widely believed that the improvements are partly attributed to the increase in size of both the sequence and structure databases. Previously identified problems in this category remain unsolved. For example, model refinement has been identified as the major bottleneck ***(16)***, which includes loop modeling and side-chain prediction that are highly correlated with the prediction accuracy of the backbone conformation. We have yet to see substantial improvements in this area. As for ab initio predictions, while still highly important, they seem to be gradually being replaced by the emerging fragment assembly-based de novo approaches, at least for practical purposes. In that regard, the success of fragment assembly-based approaches might have an undesirable side effect in overshadowing the work on the classical ab initio folding studies ***(161)***, having lured young scientists to the more practically useful and potentially more fundable work on de novo structure prediction. Clearly, studies on ab initio folding should continuously get support as empirical methods could not provide nearly as much insight into the mechanisms of protein folding as the ab initio folding studies will. Even on the practical side, ab initio methods still hold the key to improved capabilities in model refinements. Scheraga group's successful prediction of a 48-residue protein (T0215) using an ab initio method at CASP6 was clearly highly encouraging to the studies of ab initio techniques for folding proteins.

Compared with other areas of protein structure prediction, rapid progress has been made in the area of FR ***(160)***. Fewer than 10 teams and about 30 teams participated in the predictions in the FR area at the CASP1 and CASP2, respectively. Most teams use threading methods, which differ mainly in two areas, the choice of the potentials and the methods for finding the best sequence-structure alignment. The potentials used ranges from simple structural environments to residue pair interactions ***(28,37)***. The alignment methods range from a simple dynamic programing to modified dynamic programming approach, such as double-dynamic programing ***(37)*** and frozen approximation ***(72)***. In addition, several rigorous threading algorithms, such as PROSPECT (CASP3) and RAPTOR (CASP5), were introduced at various stages of CASP competitions. RAPTOR was ranked the number 1 individual prediction server at CAFASP3 ***(80,82,162)***, indicating the power of using rigorous threading algorithms for structure prediction. The development of PSI-BLAST ***(100)*** and HMMs ***(101,102)*** and the application of these powerful sequence-based homology detection techniques in FR since CASP3 represent a major advancement in fold-recognition methodology development. Nowadays, many threading programs combine the sequence-based method and structure-based method for better prediction ***(53,75,80)***.

One interesting observation from the CASP contests is that human experts who utilize diverse sources of information are more successful than automated prediction servers. Human predictors can collect as much information as possible from the literature and use their own expertise to help them select the correct template from a prediction program. The similar idea and the observation that even though no single server is able to recognize all structural folds correctly, collectively different prediction servers can correctly identify the structural folds for most of the prediction targets has led to a new breed of prediction server called meta-server ***(163)***. The first such successful attempt is a semiautomatic meta-server, CAFASP-CONSENSUS, debuted at CASP4 ***(164)***. CAFASP-CONSENSUS filed predictions after collecting models from the CAFASP automatic servers and selecting the high-scoring folds from multiple servers based on a majority-rule voting scheme. The performance of CAFASP-CONSENSUS was ranked above any of the individual servers at CASP4. As a separately registered prediction team at CASP4, CAFASP-CONSENSUS was ranked 7th in the FR category ***(164)*** among all participating teams. This program was later developed into the first fully automatic meta-server, Pcons ***(165)***. Pcons uses a simple approach for making its predictions. First, it compares the predicted models collected from the individual participating severs, and it counts the number of occurrences of each unique structural fold from the involved families, super-families or folds by different prediction servers. A neural network is then trained and used to combine the assigned score of each predicted model plus the number of occurrences of the model's fold. The strength of this meta-server is mainly attributed to the structural clustering of the initial models as Fischer et al. showed in one scenario that one particular SCOP fold was selected more frequently than all others while no server produced a significant hit ***(164)***. The practice of combining the prediction results from different prediction methods has been long used. For example, over 30 years ago, Schulz et al. ***(166)*** and Matthews ***(167)*** tried to combine different prediction results to obtain a joint prediction of secondary structures for adenyl kinase and bacteriophage T4 lysozyme, respectively. The approach was later implemented as a computer program by Argos et al. ***(168)***.

The success of CAFASP-CONSENSUS at CASP4 has started a new trend in the FR prediction. Many meta-servers have been developed after the introduction of Pcons ***(169)***. At CASP6, many human prediction teams used the predicted models by meta-servers as the starting points of their predictions. Although the overall ideas employed are similar, these meta-servers differ in several areas: (1) how the initial models are selected; (2) how the final model is generated; and (3) how the scores from individual servers are used. Although meta-servers have been a bright spot in recent CASPs as most of them perform better than individual servers, it should be noted that meta-servers have not given us much new insights yet into the fundamentals of protein structure and their prediction. Currently, their values remain on the practical side.

We clearly see two new directions taken by the template-based structure modelers: sequence-based FR and model building using fragment assembly-based structure prediction ***(66)***, whereas the classical threading methods seem to lose some of its steam. We expect that as the prediction methods continue to evolve, some hybrid techniques combining protein threading and de novo methods will prove to be desired, as we start to see that at the CASP6 contests, such as TASSER developed by Skolnick's group, which performed very well at the CASP6 ***(150,170–173)***. Some research groups have tried to use more sophisticated energy functions for a better sequence-structure alignment ***(59)*** and more sophisticated predictions algorithms, such as tree decomposition-based threading techniques ***(84)***. We believe that the best is yet to come.

Although the overall capabilities in protein FR have been steadily improving since CASP1 there are several areas that could clearly use enhanced efforts. First, a correct template may not be identified using the existing FR techniques, for proteins that may have only remote homologs or just structural analogs in PDB. For example, the best individual threading program at CAFASP3, RAPTOR, can only identify about 45% of the targets in the FR category ***(82,162)***. Quite often, the existing methods may rank the correct template as one of the top candidates, but they may have difficulty to rank it as the best one ***(65)***, suggesting that better statistical significance measures should be developed to solve this problem. A rigorous model similar to the one used in sequence-sequence comparison is clearly needed urgently for protein fold predictions. Second, even when a structural fold is correctly identified, the sequence-structure alignment could still be poor. It represents a very challenging problem to find a completely correct alignment between a query sequence and the correctly identified structural fold because of various reasons. For example, the imperfect threading energy functions and the lack of effective ways to refine a sequence-structure alignment have limited the alignment performance. Third, for multi-domain proteins, identification of domain boundaries remains a highly challenging problem based on the CASP6 assessment ***(174)***, which directly effect the performance of many prediction programs. At CASP3, two prediction targets, each consisting of two small domains, were mistakenly predicted by all prediction teams as a large single-domain protein ***(175)***. The average size of a protein domain is about 150 amino acids ***(176)***. Hence, any query protein consisting of more than 150 amino acids is likely to be a multi-domain protein. Correct identification of the domain boundaries will clearly play an important role in making the structure prediction more accurate. Finally, for targets that have large parts without structural equivalents in a template structure, it represents a challenging problem to predict the structures of missing parts, using template-based methods. For such a case, fragment assembly-based approaches or ab initio approaches could help to fill the void.

5. Conclusions

Protein structure prediction methods not only play a significant role in structural genomics projects but also have the potential to have significant impact on many areas of biology. We have witnessed the progress in the past few years over the course of CASP. As the prediction methods continue to evolve, we expect that the distinction between different prediction methods such as homology modeling, FR, and NF prediction will continue to become more blurred. Through a systematic evaluations of the existing prediction techniques by CASP, several major technical hurdles have been identified to make the existing prediction technique substantially more accurate, which include model refinement, improving sequence-structure alignments, reliable discrimination of the correct templates from incorrect templates, and reliable discrimination of the correct models from a pool of structures generated by template-free methods. With the rapid accumulation of structural data through the structural genomics efforts and with the advent of new prediction methodologies at an accelerated rate in the past few years, we remain highly optimistic about the prospect of accurate structure prediction for most of the (soluble) proteins within the next decade.

Acknowledgments

The work is, in part, supported by National Institutes of Health (R01 AG18927), National Science Foundation (DBI-0354771/ITR-IIS-0407204), and by the Georgia Cancer Coalition (a "Distinguished Cancer Scholar" grant).

References

1. Anfinsen, C. B. (1973) Principles that govern the folding of protein chains. *Science* **181,** 223–30.
2. Kolata, G. (1986) Trying to crack the second half of the genetic code. *Science* **233,** 1037–9.
3. Burley, S. K., Almo, S. C., Bonanno, J. B., Capel, M., Chance, M. R., Gaasterland, T., Lin, D. W., Sali, A., Studier, F. W., and Swaminathan, S. (1999) Structural genomics: beyond the Human Genome Project. *Nat Genet* **23,** 151–7.
4. Levitt, M., and Warshel, A. (1975) Computer-simulation of protein folding. *Nature* **253,** 694–8.
5. Levitt, M. (1976) Simplified representation of protein conformations for rapid simulation of protein folding. *J Mol Biol* **104,** 59–107.
6. Contreras-Moreira, B., Ezkurdia, I., Tress, M. L., and Valencia, A. (2005) Empirical limits for template-based protein structure prediction: the CASP5 example. *FEBS Lett* **579,** 1203–7.
7. Bajorath, J., Chalupny, N. J., Marken, J. S., Siadak, A. W., Skonier, J., Gordon, M., Hollenbaugh, D., Noelle, R. J., Ochs, H. D., and Aruffo, A. (1995) Identification of residues on CD40 and its ligand which are critical for the receptor-ligand interaction. *Biochemistry* **34,** 1833–44.
8. Bajorath, J., Marken, J. S., Chalupny, N. J., Spoon, T. L., Siadak, A. W., Gordon, M., Noelle, R. J., Hollenbaugh, D., and Aruffo, A. (1995) Analysis of gp39/CD40 interactions using molecular models and site-directed mutagenesis. *Biochemistry* **34,** 9884–92.
9. Bajorath, J., Seyama, K., Nonoyama, S., Ochs, H. D., and Aruffo, A. (1996) Classification of mutations in the human CD40 ligand, gp39, that are associated with X-linked hyper IgM syndrome. *Protein Sci* **5,** 531–4.
10. Karpusas, M., Hsu, Y. M., Wang, J. H., Thompson, J., Lederman, S., Chess, L., and Thomas, D. (1995) 2 A crystal structure of an extracellular fragment of human CD40 ligand. *Structure* **3,** 1031–9.
11. Bajorath, J. (1998) Detailed comparison of two molecular models of the human CD40 ligand with an x-ray structure and critical assessment of model-based mutagenesis and residue mapping studies. *J Biol Chem* **273,** 24603–9.
12. Schonbrun, J., Wedemeyer, W. J., and Baker, D. (2002) Protein structure prediction in 2002. *Curr Opin Struct Biol* **12,** 348–54.
13. Berman, H. M., Westbrook, J., Feng, Z., Gilliland, G., Bhat, T. N., Weissig, H., Shindyalov, I. N., and Bourne, P. E. (2000) The Protein Data Bank. *Nucleic Acids Res* **28,** 235–42.

14. Jones, D. T. (1997) Progress in protein structure prediction. *Curr Opin Struct Biol* **7,** 377–87.
15. Moult, J., Pedersen, J. T., Judson, R., and Fidelis, K. (1995) A large-scale experiment to assess protein structure prediction methods. *Proteins* **23,** ii–v.
16. Moult, J. (2005) A decade of CASP: progress, bottlenecks and prognosis in protein structure prediction. *Curr Opin Struct Biol* **15,** 285–9.
17. Fischer, D., Barret, C., Bryson, K., Elofsson, A., Godzik, A., Jones, D., Karplus, K. J., Kelley, L. A., MacCallum, R. M., Pawowski, K., Rost, B., Rychlewski, L., and Sternberg, M. (1999) CAFASP-1: critical assessment of fully automated structure prediction methods. *Proteins* **Suppl 3,** 209–17.
18. Chothia, C., and Lesk, A. M. (1986) The relation between the divergence of sequence and structure in proteins. *EMBO J* **5,** 823–6.
19. Browne, W. J., North, A. C., Phillips, D. C., Brew, K., Vanaman, T. C., and Hill, R. L. (1969) A possible three-dimensional structure of bovine alpha-lactalbumin based on that of hen's egg-white lysozyme. *J Mol Biol* **42,** 65–86.
20. Sippl, M. J., and Flockner, H. (1996) Threading thrills and threats. *Structure* **4,** 15–9.
21. Fischer, D., Rice, D., Bowie, J. U., and Eisenberg, D. (1996) Assigning amino acid sequences to 3-dimensional protein folds. *FASEB J* **10,** 126–36.
22. Zhang, C., and DeLisi, C. (1998) Estimating the number of protein folds. *J Mol Biol* **284,** 1301–5.
23. Wang, Z. X. (1998) A re-estimation for the total numbers of protein folds and superfamilies. *Protein Eng* **11,** 621–6.
24. Wang, Z. X. (1996) How many fold types of protein are there in nature? *Proteins* **26,** 186–91.
25. Chothia, C. (1992) Proteins. One thousand families for the molecular biologist. *Nature* **357,** 543–4.
26. Govindarajan, S., Recabarren, R., and Goldstein, R. A. (1999) Estimating the total number of protein folds. *Proteins* **35,** 408–14.
27. Zhang, C. T. (1997) Relations of the numbers of protein sequences, families and folds. *Protein Eng* **10,** 757–61.
28. Bowie, J. U., Luthy, R., and Eisenberg, D. (1991) A method to identify protein sequences that fold into a known three-dimensional structure. *Science* **253,** 164–70.
29. Ripka, W. C. (1986) Computer-assisted model building. *Nature* **321,** 93–4.
30. Isaacs, N., James, R., Niall, H., Bryant-Greenwood, G., Dodson, G., Evans, A., and North, A. C. (1978) Relaxin and its structural relationship to insulin. *Nature* **271,** 278–81.
31. Blundell, T. L., Bedarkar, S., Rinderknecht, E., and Humbel, R. E. (1978) Insulin-like growth factor: a model for tertiary structure accounting for immunoreactivity and receptor binding. *Proc Natl Acad Sci USA* **75,** 180–4.
32. Greer, J. (1981) Comparative model-building of the mammalian serine proteases. *J Mol Biol* **153,** 1027–42.

33. Blundell, T., Sibanda, B. L., and Pearl, L. (1983) Three-dimensional structure, specificity and catalytic mechanism of renin. *Nature* **304,** 273–5.
34. Greer, J. (1985) Model structure for the inflammatory protein C5a. *Science* **228,** 1055–60.
35. Palmer, K. A., Scheraga, H. A., Riordan, J. F., and Vallee, B. L. (1986) A preliminary three-dimensional structure of angiogenin. *Proc Natl Acad Sci USA* **83,** 1965–9.
36. Chothia, C., Lesk, A. M., Levitt, M., Amit, A. G., Mariuzza, R. A., Phillips, S. E., and Poljak, R. J. (1986) The predicted structure of immunoglobulin D1.3 and its comparison with the crystal structure. *Science* **233,** 755–8.
37. Jones, D. T., Taylor, W. R., and Thornton, J. M. (1992) A new approach to protein fold recognition. *Nature* **358,** 86–9.
38. Chandonia, J. M., and Brenner, S. E. (2006) The impact of structural genomics: expectations and outcomes. *Science* **311,** 347–51.
39. Orengo, C. A., Jones, D. T., and Thornton, J. M. (1994) Protein superfamilies and domain superfolds. *Nature* **372,** 631–4.
40. Murzin, A. G., Brenner, S. E., Hubbard, T., and Chothia, C. (1995) SCOP: a structural classification of proteins database for the investigation of sequences and structures. *J Mol Biol* **247,** 536–40.
41. Govindarajan, S., and Goldstein, R. A. (1996) Why are some proteins structures so common? *Proc Natl Acad Sci USA* **93,** 3341–5.
42. Orengo, C. A., Michie, A. D., Jones, S., Jones, D. T., Swindells, M. B., and Thornton, J. M. (1997) CATH–a hierarchic classification of protein domain structures. *Structure* **5,** 1093–108.
43. Marti-Renom, M. A., Stuart, A. C., Fiser, A., Sanchez, R., Melo, F., and Sali, A. (2000) Comparative protein structure modeling of genes and genomes. *Annu Rev Biophys Biomol Struct* **29,** 291–325.
44. Al-Lazikani, B., Jung, J., Xiang, Z., and Honig, B. (2001) Protein structure prediction. *Curr Opin Chem Biol* **5,** 51–6.
45. Xiang, Z. (2007) Homology-based modeling of protein structure. In *Computational Methods for Protein Structure Prediction and Modeling* (Xu, Y., Xu, D., and Liang, J., Eds.), Vol. **1:**319–357, Springer.
46. Sippl, M. J. (1995) Knowledge-based potentials for proteins. *Curr Opin Struct Biol* **5,** 229–35.
47. Tanaka, S., and Scheraga, H. A. (1976) Medium- and long-range interaction parameters between amino acids for predicting three-dimensional structures of proteins. *Macromolecules* **9,** 945–50.
48. Miyazawa, S., and Jernigan, R. L. (1985) Estimation of effective interresidue contact energies from protein crystal-structures - quasi-chemical approximation. *Macromolecules* **18,** 534–52.
49. Eisenberg, D., and McLachlan, A. D. (1986) Solvation energy in protein folding and binding. *Nature* **319,** 199–203.

50. Sippl, M. J. (1990) Calculation of conformational ensembles from potentials of mean force. An approach to the knowledge-based prediction of local structures in globular proteins. *J Mol Biol* **213,** 859–83.
51. Miller, R. T., Jones, D. T., and Thornton, J. M. (1996) Protein fold recognition by sequence threading: tools and assessment techniques. *FASEB J* **10,** 171–8.
52. Xu, Y., and Xu, D. (2000) Protein threading using PROSPECT: design and evaluation. *Proteins* **40,** 343–54.
53. Kim, D., Xu, D., Guo, J. T., Ellrott, K., and Xu, Y. (2003) PROSPECT II: protein structure prediction program for genome-scale applications. *Protein Eng* **16,** 641–50.
54. DeWitte, R. S., and Shakhnovich, E. I. (1996) SMoG: de novo design method based on simple, fast, and accurate free energy estimates. 1. Methodology and supporting evidence. *J Am Chem Soc* **118,** 11733–44.
55. Lu, H., and Skolnick, J. (2001) A distance-dependent atomic knowledge-based potential for improved protein structure selection. *Proteins* **44,** 223–32.
56. Samudrala, R., and Moult, J. (1998) An all-atom distance-dependent conditional probability discriminatory function for protein structure prediction. *J Mol Biol* **275,** 895–916.
57. Zhou, H., and Zhou, Y. (2002) Distance-scaled, finite ideal-gas reference state improves structure-derived potentials of mean force for structure selection and stability prediction. *Protein Sci* **11,** 2714–26.
58. Zhang, C., Liu, S., Zhou, H., and Zhou, Y. (2004) The dependence of all-atom statistical potentials on structural training database. *Biophys J* **86,** 3349–58.
59. Zhou, H., and Zhou, Y. (2005) SPARKS 2 and SP3 servers in CASP6. *Proteins* **61 (Suppl 7),** 152–6.
60. Jones, D. T., and Thornton, J. M. (1996) Potential energy functions for threading. *Curr Opin Struct Biol* **6,** 210–6.
61. Kocher, J. P., Rooman, M. J., and Wodak, S. J. (1994) Factors influencing the ability of knowledge-based potentials to identify native sequence-structure matches. *J Mol Biol* **235,** 1598–613.
62. Nishikawa, K., and Matsuo, Y. (1993) Development of pseudoenergy potentials for assessing protein 3-D-1-D compatibility and detecting weak homologies. *Protein Eng* **6,** 811–20.
63. Buchete, N. V., Straub, J. E., and Thirumalai, D. (2004) Orientational potentials extracted from protein structures improve native fold recognition. *Protein Sci* **13,** 862–74.
64. Miyazawa, S., and Jernigan, R. L. (2005) How effective for fold recognition is a potential of mean force that includes relative orientations between contacting residues in proteins? *J Chem Phys* **122,** 024901.
65. Kinch, L. N., Wrabl, J. O., Krishna, S. S., Majumdar, I., Sadreyev, R. I., Qi, Y., Pei, J., Cheng, H., and Grishin, N. V. (2003) CASP5 assessment of fold recognition target predictions. *Proteins* **53 (Suppl 6),** 395–409.

66. Wang, G., Jin, Y., and Dunbrack, R. L., Jr. (2005) Assessment of fold recognition predictions in CASP6. *Proteins* **61 (Suppl 7),** 46–66.
67. Rychlewski, L., Jaroszewski, L., Li, W. Z., and Godzik, A. (2000) Comparison of sequence profiles. Strategies for structural predictions using sequence information. *Protein Sci* **9,** 232–41.
68. Ginalski, K., Grishin, N. V., Godzik, A., and Rychlewski, L. (2005) Practical lessons from protein structure prediction. *Nucleic Acids Res* **33,** 1874–91.
69. Jones, D., and Thornton, J. (1993) Protein fold recognition. *J Comput Aided Mol Des* **7,** 439–56.
70. Lathrop, R. H. (1994) The protein threading problem with sequence amino acid interaction preferences is NP-complete. *Protein Eng* **7,** 1059–68.
71. Godzik, A., Kolinski, A., and Skolnick, J. (1992) Topology fingerprint approach to the inverse protein folding problem. *J Mol Biol* **227,** 227–38.
72. Godzik, A., and Skolnick, J. (1992) Sequence-structure matching in globular proteins: application to supersecondary and tertiary structure determination. *Proc Natl Acad Sci USA* **89,** 12098–102.
73. Westhead, D. R., Collura, V. P., Eldridge, M. D., Firth, M. A., Li, J., and Murray, C. W. (1995) Protein fold recognition by threading: comparison of algorithms and analysis of results. *Protein Eng* **8,** 1197–204.
74. Flockner, H., Braxenthaler, M., Lackner, P., Jaritz, M., Ortner, M., and Sippl, M. J. (1995) Progress in fold recognition. *Proteins* **23,** 376–86.
75. Jones, D. T. (1999) GenTHREADER: an efficient and reliable protein fold recognition method for genomic sequences. *J Mol Biol* **287,** 797–815.
76. Akmaev, V. R., Kelley, S. T., and Stormo, G. D. (2000) Phylogenetically enhanced statistical tools for RNA structure prediction. *Bioinformatics* **16,** 501–12.
77. Shi, J., Blundell, T. L., and Mizuguchi, K. (2001) FUGUE: sequence-structure homology recognition using environment-specific substitution tables and structure-dependent gap penalties. *J Mol Biol* **310,** 243–57.
78. Lathrop, R. H., and Smith, T. F. (1994) A branch and bound algorithm for optimal protein threading with pairwise (contact potential) interaction preferences. In *Proceedings of the 27th Hawaii International Conference on System Sciences* (Hunter, L. and Shriver, B., Eds), pp. 365–74. IEEE Computer Soc. Press, Los Alamitos, CA.
79. Guo, J.-T., Ellrott, K., Chung, W. J., Xu, D., Passovets, S., Xu, Y. (2004) PROSPECT-PSPP: an automatic computational pipeline for protein structure prediction. *Nucleic Acids Res.* **32**, W522–5.
80. Xu, J., Li, M., Kim, D., and Xu, Y. (2003) RAPTOR: optimal protein threading by linear programming. *J Bioinform Comput Biol* **1,** 95–117.
81. Xu, D., Crawford, O. H., LoCascio, P. F., and Xu, Y. (2001) Application of PROSPECT in CASP4: characterizing protein structures with new folds. *Proteins* **Suppl 5,** 140–8.

82. Xu, J., and Li, M. (2003) Assessment of RAPTOR's linear programming approach in CAFASP3. *Proteins* **53 (Suppl 6),** 579–84.
83. Xu, J., Jiao, F., and Berger, B. (2005) A tree-decomposition approach to protein structure prediction. In *2005 IEEE Computational Systems Bioinformatics Conference*, pp. 247–56, Stanford, CA.
84. Song, Y., Guo, J.-T., Ellrott, K., Xu, Y., and Cai, L. (2007) Efficient algorithms for protein threading via tree decomposition (submitted).
85. Wang, G., and Dunbrack, R. L., Jr. (2003) PISCES: a protein sequence culling server. *Bioinformatics* **19,** 1589–91.
86. Bryant, S. H., and Lawrence, C. E. (1993) An empirical energy function for threading protein sequence through the folding motif. *Proteins* **16,** 92–112.
87. Karlin, S., and Altschul, S. F. (1990) Methods for assessing the statistical significance of molecular sequence features by using general scoring schemes. *Proc Natl Acad Sci USA* **87,** 2264–8.
88. Altschul, S. F., Boguski, M. S., Gish, W., and Wootton, J. C. (1994) Issues in searching molecular sequence databases. *Nat Genet* **6,** 119–29.
89. Altschul, S. F., and Gish, W. (1996) Local alignment statistics. *Methods Enzymol* **266,** 460–80.
90. Pearson, W. R. (1998) Empirical statistical estimates for sequence similarity searches. *J Mol Biol* **276,** 71–84.
91. Bryant, S. H., and Altschul, S. F. (1995) Statistics of sequence-structure threading. *Curr Opin Struct Biol* **5,** 236–44.
92. Levitt, M., and Gerstein, M. (1998) A unified statistical framework for sequence comparison and structure comparison. *Proc Natl Acad Sci USA* **95,** 5913–20.
93. Sommer, I., Zien, A., von Ohsen, N., Zimmer, R., and Lengauer, T. (2002) Confidence measures for protein fold recognition. *Bioinformatics* **18,** 802–12.
94. McGuffin, L. J., and Jones, D. T. (2003) Improvement of the GenTHREADER method for genomic fold recognition. *Bioinformatics* **19,** 874–81.
95. Xu, J. (2005) Fold recognition by predicted alignment accuracy. *IEEE/ACM Trans Comput Biol Bioinform* **2,** 157–65.
96. Holm, L., and Sander, C. (1996) Mapping the protein universe. *Science* **273,** 595–603.
97. Smith, T. F., and Waterman, M. S. (1981) Identification of common molecular subsequences. *J Mol Biol* **147,** 195–7.
98. Altschul, S. F., Gish, W., Miller, W., Myers, E. W., and Lipman, D. J. (1990) Basic local alignment search tool. *J Mol Biol* **215,** 403–10.
99. Pearson, W. R., and Lipman, D. J. (1988) Improved tools for biological sequence comparison. *Proc Natl Acad Sci USA* **85,** 2444–8.
100. Altschul, S. F., Madden, T. L., Schaffer, A. A., Zhang, J., Zhang, Z., Miller, W., and Lipman, D. J. (1997) Gapped BLAST and PSI-BLAST: a new generation of protein database search programs. *Nucleic Acids Res* **25,** 3389–402.
101. Eddy, S. R. (1998) Profile hidden Markov models. *Bioinformatics* **14,** 755–63.

102. Karplus, K., Barrett, C., and Hughey, R. (1998) Hidden Markov models for detecting remote protein homologies. *Bioinformatics* **14,** 846–56.
103. Yona, G., and Levitt, M. (2002) Within the twilight zone: a sensitive profile-profile comparison tool based on information theory. *J Mol Biol* **315,** 1257–75.
104. Wang, G., and Dunbrack, R. L., Jr. (2004) Scoring profile-to-profile sequence alignments. *Protein Sci* **13,** 1612–26.
105. Ohlson, T., Wallner, B., and Elofsson, A. (2004) Profile-profile methods provide improved fold-recognition: a study of different profile-profile alignment methods. *Proteins* **57,** 188–97.
106. Zhou, H., and Zhou, Y. (2005) Fold recognition by combining sequence profiles derived from evolution and from depth-dependent structural alignment of fragments. *Proteins* **58,** 321–8.
107. Zhou, H., and Zhou, Y. (2004) Single-body residue-level knowledge-based energy score combined with sequence-profile and secondary structure information for fold recognition. *Proteins* **55,** 1005–13.
108. Blundell, T. L., Sibanda, B. L., Sternberg, M. J., and Thornton, J. M. (1987) Knowledge-based prediction of protein structures and the design of novel molecules. *Nature* **326,** 347–52.
109. Sutcliffe, M. J., Haneef, I., Carney, D., and Blundell, T. L. (1987) Knowledge based modelling of homologous proteins. Part I: Three-dimensional frameworks derived from the simultaneous superposition of multiple structures. *Protein Eng* **1,** 377–84.
110. Yang, A. S., and Honig, B. (1999) Sequence to structure alignment in comparative modeling using PrISM. *Proteins* **Suppl 3,** 66–72.
111. Bates, P. A., Kelley, L. A., MacCallum, R. M., and Sternberg, M. J. (2001) Enhancement of protein modeling by human intervention in applying the automatic programs 3D-JIGSAW and 3D-PSSM. *Proteins* **Suppl 5,** 39–46.
112. Peitsch, M. C., and Jongeneel, C. V. (1993) A 3-D model for the CD40 ligand predicts that it is a compact trimer similar to the tumor necrosis factors. *Int Immunol* **5,** 233–8.
113. Unger, R., Harel, D., Wherland, S., and Sussman, J. L. (1989) A 3D building blocks approach to analyzing and predicting structure of proteins. *Proteins* **5,** 355–73.
114. Claessens, M., Van Cutsem, E., Lasters, I., and Wodak, S. (1989) Modelling the polypeptide backbone with 'spare parts' from known protein structures. *Protein Eng* **2,** 335–45.
115. van Gelder, C. W., Leusen, F. J., Leunissen, J. A., and Noordik, J. H. (1994) A molecular dynamics approach for the generation of complete protein structures from limited coordinate data. *Proteins* **18,** 174–85.
116. Levitt, M. (1992) Accurate modeling of protein conformation by automatic segment matching. *J Mol Biol* **226,** 507–33.
117. Sali, A., and Blundell, T. L. (1993) Comparative protein modelling by satisfaction of spatial restraints. *J Mol Biol* **234,** 779–815.

118. Petrey, D., Xiang, Z., Tang, C. L., Xie, L., Gimpelev, M., Mitros, T., Soto, C. S., Goldsmith-Fischman, S., Kernytsky, A., Schlessinger, A., Koh, I. Y., Alexov, E., and Honig, B. (2003) Using multiple structure alignments, fast model building, and energetic analysis in fold recognition and homology modeling. *Proteins* **53 (Suppl 6),** 430–5.
119. Fiser, A., Do, R. K., and Sali, A. (2000) Modeling of loops in protein structures. *Protein Sci* **9,** 1753–73.
120. Greer, J. (1980) Model for haptoglobin heavy chain based upon structural homology. *Proc Natl Acad Sci USA* **77,** 3393–7.
121. Jones, T. A., and Thirup, S. (1986) Using known substructures in protein model building and crystallography. *EMBO J* **5,** 819–22.
122. Wojcik, J., Mornon, J. P., and Chomilier, J. (1999) New efficient statistical sequence-dependent structure prediction of short to medium-sized protein loops based on an exhaustive loop classification. *J Mol Biol* **289,** 1469–90.
123. Fine, R. M., Wang, H., Shenkin, P. S., Yarmush, D. L., and Levinthal, C. (1986) Predicting antibody hypervariable loop conformations. II: Minimization and molecular dynamics studies of MCPC603 from many randomly generated loop conformations. *Proteins* **1,** 342–62.
124. Moult, J., and James, M. N. (1986) An algorithm for determining the conformation of polypeptide segments in proteins by systematic search. *Proteins* **1,** 146–63.
125. Xiang, Z., Soto, C. S., and Honig, B. (2002) Evaluating conformational free energies: the colony energy and its application to the problem of loop prediction. *Proc Natl Acad Sci USA* **99,** 7432–7.
126. van Vlijmen, H. W., and Karplus, M. (1997) PDB-based protein loop prediction: parameters for selection and methods for optimization. *J Mol Biol* **267,** 975–1001.
127. Fidelis, K., Stern, P. S., Bacon, D., and Moult, J. (1994) Comparison of systematic search and database methods for constructing segments of protein structure. *Protein Eng* **7,** 953–60.
128. Jacobson, M. P., Pincus, D. L., Rapp, C. S., Day, T. J., Honig, B., Shaw, D. E., and Friesner, R. A. (2004) A hierarchical approach to all-atom protein loop prediction. *Proteins* **55,** 351–67.
129. Chandrasekaran, R., and Ramachandran, G. N. (1970) Studies on the conformation of amino acids. XI. Analysis of the observed side group conformation in proteins. *Int J Protein Res* **2,** 223–33.
130. Ponder, J. W., and Richards, F. M. (1987) Tertiary templates for proteins. Use of packing criteria in the enumeration of allowed sequences for different structural classes. *J Mol Biol* **193,** 775–91.
131. Dunbrack, R. L., Jr., and Karplus, M. (1993) Backbone-dependent rotamer library for proteins. Application to side-chain prediction. *J Mol Biol* **230,** 543–74.
132. Lovell, S. C., Word, J. M., Richardson, J. S., and Richardson, D. C. (2000) The penultimate rotamer library. *Proteins* **40,** 389–408.

133. Lovell, S. C., Word, J. M., Richardson, J. S., and Richardson, D. C. (1999) Asparagine and glutamine rotamers: B-factor cutoff and correction of amide flips yield distinct clustering. *Proc Natl Acad Sci USA* **96,** 400–5.
134. Xiang, Z., and Honig, B. (2001) Extending the accuracy limits of prediction for side-chain conformations. *J Mol Biol* **311,** 421–30.
135. Holm, L., and Sander, C. (1991) Database algorithm for generating protein backbone and side-chain co-ordinates from a C alpha trace application to model building and detection of co-ordinate errors. *J Mol Biol* **218,** 183–94.
136. Desmet, J., Demaeyer, M., Hazes, B., and Lasters, I. (1992) The dead-end elimination theorem and its use in protein side-chain positioning. *Nature* **356,** 539–42.
137. Desmet, J., De Maeyer, M., and Lasters, I. (1997) Theoretical and algorithmical optimization of the dead-end elimination theorem. *Pac Symp Biocomput* 122–33.
138. Goldstein, R. F. (1994) Efficient rotamer elimination applied to protein side-chains and related spin glasses. *Biophys J* **66,** 1335–40.
139. Lasters, I., De Maeyer, M., and Desmet, J. (1995) Enhanced dead-end elimination in the search for the global minimum energy conformation of a collection of protein side chains. *Protein Eng* **8,** 815–22.
140. Lasters, I., and Desmet, J. (1993) The fuzzy-end elimination theorem: correctly implementing the side chain placement algorithm based on the dead-end elimination theorem. *Protein Eng* **6,** 717–22.
141. Pierce, N. A., Spriet, J. A., Desmet, J., and Mayo, S. L. (2000) Conformational splitting: a more powerful criterion for dead-end elimination. *J Comput Chem* **21,** 999–1009.
142. Canutescu, A. A., Shelenkov, A. A., and Dunbrack, R. L., Jr. (2003) A graph-theory algorithm for rapid protein side-chain prediction. *Protein Sci* **12,** 2001–14.
143. Xu, J. (2005) Rapid protein side-chain packing via tree decomposition. *RECOMB* 423–39.
144. Desmet, J., Spriet, J., and Lasters, I. (2002) Fast and accurate side-chain topology and energy refinement (FASTER) as a new method for protein structure optimization. *Proteins* **48,** 31–43.
145. Wallner, B., and Elofsson, A. (2005) All are not equal: a benchmark of different homology modeling programs. *Protein Sci* **14,** 1315–27.
146. Sutcliffe, M. J., Hayes, F. R., and Blundell, T. L. (1987) Knowledge based modelling of homologous proteins. Part II: Rules for the conformations of substituted sidechains. *Protein Eng* **1,** 385–92.
147. Nayeem, A., Sitkoff, D., and Krystek, S., Jr. (2006) A comparative study of available software for high-accuracy homology modeling: from sequence alignments to structural models. *Protein Sci* **15,** 808–24.
148. Kolodny, R., Koehl, P., Guibas, L., and Levitt, M. (2002) Small libraries of protein fragments model native protein structures accurately. *J Mol Biol* **323,** 297–307.

149. Jones, D. T. (1997) Successful ab initio prediction of the tertiary structure of NK-lysin using multiple sequences and recognized supersecondary structural motifs. *Proteins* **Suppl 1,** 185–91.
150. Jones, D. T., Bryson, K., Coleman, A., McGuffin, L. J., Sadowski, M. I., Sodhi, J. S., and Ward, J. J. (2005) Prediction of novel and analogous folds using fragment assembly and fold recognition. *Proteins* **61 (Suppl 7),** 143–51.
151. Simons, K. T., Kooperberg, C., Huang, E., and Baker, D. (1997) Assembly of protein tertiary structures from fragments with similar local sequences using simulated annealing and Bayesian scoring functions. *J Mol Biol* **268,** 209–25.
152. Rohl, C. A., Strauss, C. E., Misura, K. M., and Baker, D. (2004) Protein structure prediction using Rosetta. *Methods Enzymol* **383,** 66–93.
153. Karplus, K., Karchin, R., Draper, J., Casper, J., Mandel-Gutfreund, Y., Diekhans, M., and Hughey, R. (2003) Combining local-structure, fold-recognition, and new fold methods for protein structure prediction. *Proteins* **53 (Suppl 6),** 491–6.
154. Ishida, T., Nishimura, T., Nozaki, M., Inoue, T., Terada, T., Nakamura, S., and Shimizu, K. (2003) Development of an ab initio protein structure prediction system ABLE. *Genome Inform* **14,** 228–37.
155. Chikenji, G., Fujitsuka, Y., and Takada, S. (2003) A reversible fragment assembly method for de novo protein structure prediction. *J Chem Phys* **119,** 6895–903.
156. Fujitsuka, Y., Chikenji, G., and Takada, S. (2006) SimFold energy function for de novo protein structure prediction: consensus with Rosetta. *Proteins* **62,** 381–98.
157. Lee, J., Kim, S. Y., Joo, K., Kim, I., and Lee, J. (2004) Prediction of protein tertiary structure using PROFESY, a novel method based on fragment assembly and conformational space annealing. *Proteins* **56,** 704–14.
158. Jones, D. T., and McGuffin, L. J. (2003) Assembling novel protein folds from super-secondary structural fragments. *Proteins* **53 (Suppl 6),** 480–5.
159. Bujnicki, J. M. (2006) Protein-structure prediction by recombination of fragments. *Chembiochem* **7,** 19–27.
160. Kryshtafovych, A., Venclovas, C., Fidelis, K., and Moult, J. (2005) Progress over the first decade of CASP experiments. *Proteins* **61 (Suppl 7),** 225–36.
161. Cozzetto, D., Di Matteo, A., and Tramontano, A. (2005) Ten years of predictions ... and counting. *FEBS J* **272,** 881–2.
162. Fischer, D., Rychlewski, L., Dunbrack, R. L., Jr., Ortiz, A. R., and Elofsson, A. (2003) CAFASP3: the third critical assessment of fully automated structure prediction methods. *Proteins* **53 (Suppl 6),** 503–16.
163. Sippl, M. J., Lackner, P., Domingues, F. S., Prlic, A., Malik, R., Andreeva, A., and Wiederstein, M. (2001) Assessment of the CASP4 fold recognition category. *Proteins* **Suppl 5,** 55–67.
164. Fischer, D., Elofsson, A., Rychlewski, L., Pazos, F., Valencia, A., Rost, B., Ortiz, A. R., and Dunbrack, R. L., Jr. (2001) CAFASP2: the second critical assessment of fully automated structure prediction methods. *Proteins* **Suppl 5,** 171–83.

165. Lundstrom, J., Rychlewski, L., Bujnicki, J., and Elofsson, A. (2001) Pcons: a neural-network-based consensus predictor that improves fold recognition. *Protein Sci* **10,** 2354–62.
166. Schulz, G. E., Barry, C. D., Friedman, J., Chou, P. Y., Fasman, G. D., Finkelstein, A. V., Lim, V. I., Pititsyn, O. B., Kabat, E. A., Wu, T. T., Levitt, M., Robson, B., and Nagano, K. (1974) Comparison of predicted and experimentally determined secondary structure of adenyl kinase. *Nature* **250,** 140–2.
167. Matthews, B. W. (1975) Comparison of the predicted and observed secondary structure of T4 phage lysozyme. *Biochim Biophys Acta* **405,** 442–51.
168. Argos, P., and Schwarz, J. (1976) An assessment of protein secondary structure prediction methods based on amino acid sequence. *Biochim Biophys Acta* **439,** 261–73.
169. Bujnicki, J. M., and Fischer, D. (2004) "Meta" approaches to protein structure prediction. In *Practical Bioinformatics* (Bujnicki, J. M., Ed.), Vol. 15, pp. 23–34, Springer, Berlin.
170. Zhang, Y., Arakaki, A. K., and Skolnick, J. (2005) TASSER: an automated method for the prediction of protein tertiary structures in CASP6. *Proteins* **61 (Suppl 7),** 91–8.
171. Jones, D. T. (2001) Predicting novel protein folds by using FRAGFOLD. *Proteins* **Suppl 5,** 127–32.
172. Skolnick, J., Kolinski, A., Kihara, D., Betancourt, M., Rotkiewicz, P., and Boniecki, M. (2001) Ab initio protein structure prediction via a combination of threading, lattice folding, clustering, and structure refinement. *Proteins* **Suppl 5,** 149–56.
173. Zhang, Y., Kolinski, A., and Skolnick, J. (2003) TOUCHSTONE II: a new approach to ab initio protein structure prediction. *Biophys J* **85,** 1145–64.
174. Tai, C. H., Lee, W. J., Vincent, J. J., and Lee, B. (2005) Evaluation of domain prediction in CASP6. *Proteins* **61 (Suppl 7),** 183–92.
175. Shortle, D. (1999) Structure prediction: the state of the art. *Curr Biol* **9,** R205–9.
176. Guo, J. T., Xu, D., Kim, D., and Xu, Y. (2003) Improving the performance of DomainParser for structural domain partition using neural network. *Nucleic Acids Res* **31,** 944–52.

2

The Assessment of Methods for Protein Structure Prediction

Anna Tramontano, Domenico Cozzetto, Alejandro Giorgetti, and Domenico Raimondo

Summary

Methods for protein structure prediction are flourishing and becoming widely available to both experimentalists and computational biologists. But, how good are they? What is their range of applicability and how can we know which method is better suited for the task at hand? These are the questions that this chapter tries to address, by describing automatic evaluation methods as well as the world-wide Critical Assessment of Techniques for Protein Structure Prediction (CASP) initiative and focusing on the specific problems of assessing the quality of a protein 3D model.

Key Words: Protein structure prediction; accuracy of protein structure models; CASP; structure prediction servers; metapredictors.

1. Introduction

Protein structure prediction is a field that has attracted enormous interest since the very beginning of protein structural biology. The first model of a protein was produced only about 10 years after the first protein structure was solved and at a time when only two protein structures were available *(1)*. The model was a physical one (no molecular graphics available at the time), but it was a rather good one; it was later established that the root mean square deviation (rmsd) between the alpha carbons of the model and those of the subsequently determined experimental structure was around 1 Å, a result that would be considered interesting even today.

From: *Methods in Molecular Biology, vol. 413: Protein Structure Prediction, Second Edition*
Edited by: M. Zaki and C. Bystroff © Humana Press Inc., Totowa, NJ

As of today, hundreds of servers and tools are widely available for producing a structural model of the protein of interest. The model can then be used as a structural framework for designing further experiments, interpreting functional data, assigning its molecular function to the protein, or as a target for drug design or even as a tool for solving the experimental structure of the protein and more. However, the quality of a model dictates its possible applications, and therefore, the admittedly complex problem of assessing beforehand the quality of models produced by different methods is of outstanding interest.

The issue is obvious: if one produces a model of a protein of known structure, the suspicion might arise that, unwittingly, data extracted from that structure are used in some of the steps of the procedure, and therefore, it would not be correct to extrapolate the results obtained on a test set composed of proteins of known structure to proteins of as yet unknown structure. On the contrary, predicting the structure of a protein for which no structural experimental data are available does not allow the effectiveness of the method to be assessed in a reasonable and predictable time frame.

The solution is to predict a protein structure "just in time" that is soon before the experimental structure of the protein is made available or before any method had a chance of taking the structure of the protein into account for optimizing its parameters.

The former strategy is used by the Critical Assessment of Techniques for protein structure prediction (CASP) experiment *(2)* and the latter by automatic evaluation servers such as EValuation of Automatic protein structure prediction (EVA) *(3)* and Livebench *(4)*.

We will describe these experiments, give some advice about how to make the best use of the data they produce, and discuss their problems and limitations.

2. Materials

The models submitted to each of the CASP experiments and data related to their evaluation are available at http://www.predictioncenter.org. A discussion forum about most of the issues discussed in this chapter can be found at http://www.forcasp.org.

The EVA and Livebench automatic evaluation servers make their data available at http://cubic.bioc.columbia.edu/eva/ and http://bioinfo.pl/meta/livebench.pl respectively.

3. Methods

Predicting the structure of a protein is both an intellectual challenge and a practical issue, especially in light of the recent genomics and structural genomics efforts. The problem is far from being solved in general terms, but it can be addressed using several heuristic strategies. During evolution, proteins tend to preserve their structure. It is therefore possible to derive information about a protein structure on the basis of the structure of an evolutionarily related protein, which, in turn, can be identified by sequence analysis [comparative modeling (CM)] ***(5)***. Even when no sequence similarity between two proteins can be detected, they might share structural similarity. In this case, the problem is to correctly recognize the compatibility of the sequence of the target protein with a known fold [fold recognition (FR)] ***(6,7)***. Finally, a protein might share neither sequence nor structural similarity with any known protein [new fold (NF)], and the prediction of its structure has to rely on different approaches. In many cases, when an NF is discovered, it is observed that it is composed of common structural motifs at the fragment or super-secondary structural level. This prompted the development of methods, known under the name of "fragment-based" ***(8,9)***, which try and assemble fragments of proteins of known structure to reconstruct the complete structure of a target protein.

4. The Difficulty of Evaluating a Prediction

At first sight, it might seem that the evaluation of the correctness of a model is a straightforward task once the experimental structure is available, but matters are not so easy.

First of all, the problem of finding the optimal superposition between two structures, that is, the superposition that minimizes some "distance" measure, does not have a unique solution. The difference between two superimposed structures depends on the fraction of the structures that is superimposed ***(10)***. It is entirely possible that one region of a model is very similar to the corresponding region of the target protein but that the similarity is masked if the whole structure is taken into account in the structural superposition. In other words, there is a relationship between the quality of a structural superposition and the fraction of superimposed structure. The identification of well-predicted regions not only is an issue related to the evaluation of the model but also might have important biological implications if they correspond to, say, the active site of the protein.

Furthermore, the measure traditionally used to evaluate structural similarity, the rmsd, is a quadratic measure. It is defined as the square root of the squared differences between the coordinates of corresponding atoms, and therefore, it will weight more regions that are not well superimposed with respect to the rest. From a biological perspective, if a region of a protein is incorrectly predicted, do we really care by how much or would we rather just like to say that the predicted and experimental regions are more far apart than it is acceptable to derive meaningful insights from the model? This implies that the number of atom pairs of the model and the structure that are within an acceptable distance threshold is probably a better measure for the task of protein structure prediction evaluation.

Proteins are not static objects, they have a dynamic behavior and some regions are more flexible than others. We need to make sure that our quality measure takes this into account and does not penalize a model if it does not reproduce correctly regions of the experimental structure that have significant experimental uncertainty.

Furthermore, proteins are often composed of domains, and an evolutionary relationship between two proteins can be limited to one of the domains and not to the overall protein sequence.

5. The CASP Experiment

In 1994, John Moult proposed a world-wide experiment named CASP *(2)* aimed at establishing the current state of the art in protein structure prediction, identifying what progress has been made, and highlighting where future effort may be most productively focused.

Experimental structural biologists who are about to solve a protein structure are asked to make the sequence of the protein available, together with a tentative date for the release of the final coordinates. In the past 13 years, structural genomics consortia have significantly contributed to the set of CASP targets.

Predictors produce and deposit models for these proteins (the CASP targets) before the structures are made available. Another experiment, synchronized with CASP and called CAFASP *(4)*, has been testing publicly available servers on the same set of targets, providing a unique opportunity for evaluating how much human expert knowledge is important to obtain better models. Recently, this task has been taken over by CASP itself *(11)*. For testing server predictions, sequences are automatically sent to participating servers, and the models received within a short time frame, 48 h, are collected and stored. These models are also made available to human predictors, who have more time at their disposal, to avoid duplication of efforts, because many human predictors make use of automatic server results in their model-building procedure.

Finally, a panel of three assessors compares the models with the structures as soon as they are available and tries to evaluate the quality of the models and to draw some conclusions about the state of the art of the different methods. The experiment is run blindly, that is, the assessors do not know who the predictors are until the very end of the experiment.

Each of the routes to the prediction of a protein structure described before has traditionally been mirrored by a CASP "category," evaluated by one of the three assessors. The categories have some degree of overlap: CM targets for which evolutionary relationships are very hard to identify before knowing their structure can also be considered in the FR category; NFs can share some similarity with existing folds and be considered in both FR and NF categories. Recently, some modifications have been proposed, and the target categories will be reduced to two: template based and non-template based; but a special analysis will be performed on the best models to evaluate the accuracy of details of protein structure predictions, such as positioning of side chains and correct prediction of loop structures. The reasons for this rearrangement will become clear later.

The results of the comparison between the models and the target structures are discussed in a meeting where assessors and predictors convene; the conclusions are made available to the whole scientific community through the World Wide Web and through the publication of a special issue of the journal "Proteins: Structure, Function, and Bioinformatics."

There are several other categories that have been introduced in CASP throughout the years, such as prediction of function, of domain boundaries and of disordered regions, but we will not discuss them here.

The CASP experiment has been extremely successful. It has been repeated every 2 years since its first edition, and there is no sign that it is going to be discontinued in the near future *(12)*. It is a very important experiment, which has the merit of having raised the issue of objective evaluation of structure prediction methods, of prompting the development of the automatic assessment methods that will be described later and of fostering the development of similar initiatives in other fields such as the prediction of protein—protein interaction, gene finding, and scientific literature mining.

6. CASP Measures

As we mentioned, there are two problems with the measure of the similarity between a model and a protein structure: the dependence of the solution on the fraction of superimposed structure and the quadratic form of the rmsd. One solution to the first problem is to use a graph such as the one shown in **Fig. 1**, where the x-axis indicates the fraction of the model that has been superimposed to the target structure and the y-axis reports the corresponding rmsd value (or any other similarity measure) *(13)*.

In the last edition of CASP, there were almost 30,000 submitted 3D models *(14)*, and it is not possible for any assessor or user to visually inspect all the

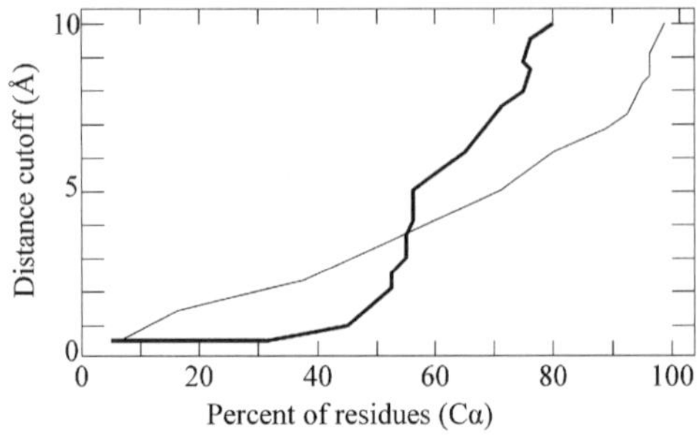

Fig. 1. A plot describing the quality of two predictions in CASP6 for target T0196 (an hypothetical protein from *Pyrococcus furiosus*, PDB code 1XE1). The x-axis indicates the percentage of aligned residues of the target and experimental structure that are closer than the threshold reported on the y-axis. As it can be seen from the plot, one of the models (indicated by the thick line) is closer to the experimental structure for about 60% of the structure, whereas the other turns out to be closer when larger fractions of the modeled and experimental structures are superimposed.

generated plots; so, it is necessary to convert the information into a numerical value, for example, a rough estimate of the area under the curve. The Global Distance Test (GDT-TS), used in CASP, is such a measure. It is defined as the average percentage of Cα atom pairs under a distance cutoff of 1, 2, 4, and 8 Å.

This measure is reasonably satisfactory for highlighting the overall quality of the prediction of the backbone of the protein, but it does not capture the details of the structure, for example, the correct prediction of the conformation of side chains. The latter is evaluated using the number of chi angle values within a threshold (usually set to 30°).

The next problem is related to the experimental uncertainty of the protein structure. CASP provides data for the complete model structure but also for subsets including, for example, all atoms that have a B-factor lower than a threshold (usually 20 Å), residues whose chi angles can be assigned reliably by X-ray crystallography, residues buried in the core, and so on.

Last but not least, CASP also analyzes the predictions of each domain of the target proteins separately.

7. The Problem of Evaluating the Overall Performance of a Method

The final aim of CASP is to highlight which methods work better, and therefore, it is essential to devise a comprehensive measure of the performance of a method on the basis of the results that the method achieved on several targets. And here, things get tricky.

First of all, not all methods are applied or applicable to all CASP targets, and therefore, a comparison between two methods needs to take into account how many models have been submitted using that method, but most importantly, which models. In fact, not all protein structures are equally difficult to predict, so that the relative difficulty of a target should be taken into account. The same problem arises when one wants to ask the obvious question of whether there has been any improvement of the methods in different editions of the experiment: each experiment has its own set of targets; therefore, the performance in one edition should be compared to the performance in another one taking into account the relative difficulties of the targets. The problem, as we will discuss in the **Subheading 8**, is a very complex one, but also extremely important for protein structure prediction evaluation.

8. Evaluating the Difficulty of a Prediction Target

The difficulty of predicting the structure of a given protein can be evaluated a posteriori, analyzing how well it has been predicted on average. In some cases, it is also possible to estimate the difficulty a priori. For example, in CM, one can see how difficult it is to identify the evolutionary relationship between the target protein and the protein of known structure that can be used as template for building the model and how easy it is to obtain a reasonable sequence alignment using standard methods. In FR predictions, one can measure how strong is the sequence-structure fitness signal. In both cases, one can also take into account, in evaluating the difficulty of modeling a protein, how well automatic methods perform the task.

It should be mentioned upfront that none of these strategies is faultless. For example, a posteriori evaluation cannot be used to compare two different CASP experiments, because, hopefully, methods have improved during the two intervening years, and the same is likely to be true for sequence alignment methods. Another effect, even more difficult to take into account, is the increased size of databases.

Traditionally, the difficulty of producing a comparative model for a protein has been measured on the basis of the percent of sequence identity or similarity between the target protein and the protein of known experimental structure used as template for modeling. However, although this measure takes into account the structural effect of the accumulation of mutations in the protein, it is not equally effective for estimating the difficulty of detecting the relationship and of obtaining a correct sequence alignment, that is, of detecting the right correspondence between the amino acids of the target and template proteins. In fact, most methods for the detection of sequence similarities rely on multiple sequence alignment, that is, on information provided by many sequences of the proteins of the same evolutionary family. The increased size of the database can therefore be directly responsible for the improvement in the detection of evolutionary relationships and in the sequence alignment step, which are the essence of the quality of a model.

In CASP, the difficulty of a prediction is estimated on the basis of both its sequence and structural similarity with the potential templates. The former is defined as the fraction of structurally aligned residues (within 5 Å) that are identical between the target and the template, the second as the fraction of pairs of target–template Cα atoms within 5 Å after optimal superposition ***(15)***. When an 1D scale for target difficulty is needed, the average of the two values described above are used.

Another possibility is illustrated in **Fig. 2**. The multiple sequence alignment for each target available at the time of each experiment can be used to calculate the pair-wise sequence identity between each pair of sequences and to construct a graph similar to that shown in the figure. Each node represents one of the sequences in the multiple sequence alignment, and the lengths of the edges are proportional to the distance (inversely proportional to the percent of identity) between the connected nodes. The multiple sequence alignment is a path in the graph that includes all the sequences. In first approximation, the difficulty of aligning the target and template sequences depends on the availability of intermediate sequences, and this is determined by the most difficult pair-wise alignment that we need to perform to go from the target to the template. In other words, we might end up aligning a target and a template sequence only sharing

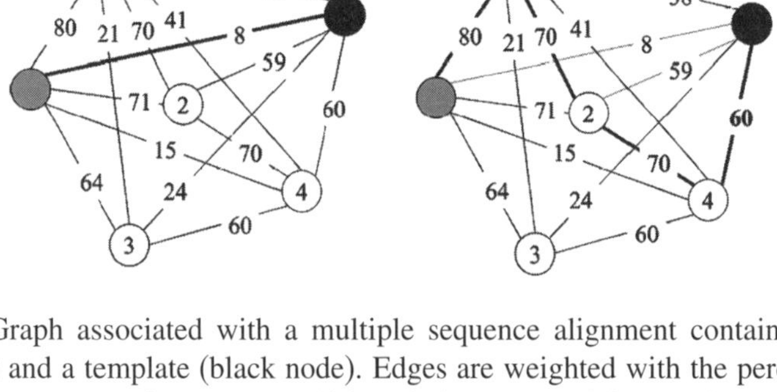

Fig. 2. Graph associated with a multiple sequence alignment containing a target (gray node) and a template (black node). Edges are weighted with the percent identity between the sequences they connect. Although the target and the template only share 8% of identical residues, the recruitment of homologous sequences allows to progressively align pairs of sequences sharing at least 60% sequence identity.

a very low sequence identity, but we might achieve this by aligning pairs of very similar intermediate sequences, starting from the target and "jumping" from one sequence to another until we reach the template much in the same way as we might cross a large river jumping from one emerging stone to the next. The difficulty of crossing the river is not proportional to its width but to the longest jump that we need to make.

Therefore, given all possible paths including target and template, we are interested in the one(s) where the maximum distance between each pairs of traversed nodes is minimal. Once such a path is found, the longest edge in the path, that is, the sequence similarity between the two most diverse sequences in the path is an estimate of the difficulty of aligning target and template, given the distribution of sequences in the multiple sequence alignment ***(16)***.

This approach gives, in first approximation, a measure of the difficulty of aligning the target and template sequence for each target in different experiments, given the database available at the time of the prediction, and can be used to ask whether the alignment of targets and templates of equivalent difficulty has become more accurate with time. **Figure 3** shows a plot of the percent of correctly aligned residues (a residue is considered correctly aligned if, after superposition of the experimental and modeled structure, its Cα atom falls within 3.8 Å of the corresponding experimental atom, and there is no other Cα atom of the experimental structure that is nearer) achieved in the last three

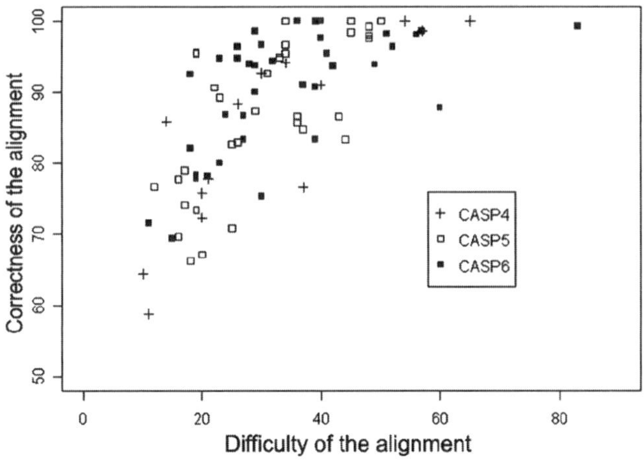

Fig. 3. Scatter plot of the alignment quality obtained in the last three editions of the CASP experiment as a function of the difficulty of the alignment, computed through the method depicted in **Fig. 2**.

CASP experiments for CM targets as a function of the difficulty parameter defined above.

As it can be seen, there has been no major improvement in methods for aligning sequences in the most recent CASP editions, and targets of similar difficulties are aligned with the same level of accuracy. This is somewhat disappointing and urges for novel ideas in the area.

Traditional methods for CM are based on the assumption that each of the modeling steps, including template selection, and alignment, can be optimized separately. It is easy to argue that a better approach would be to optimize all the parameters simultaneously. Clearly, this is beyond our present computational capabilities. However, it is worth noting that the most successful groups in recent CASP experiments used the strategy of constructing several models for each target protein and selecting the most likely one only at the end of the complete model-building procedure. In other words, rather than optimizing each of the steps of the comparative modeling procedure independently, they chose to also funnel sub- optimal intermediate results into each subsequent step. This represents a first degree approximation to a full multi-parameter optimization procedure, and we argue that this type of strategy should be pursued even more aggressively in the future.

It should also be mentioned, however, that predictors in CASP are not necessarily in an ideal position to produce the best models because of the time limitation imposed by the experiment. Also, the fact that the results are public and very visible might stop predictors from trying "risky" innovations.

9. New Challenges

There is no doubt that modeling methods are extremely powerful. At present, experimental structures are known for less than 1% of identified proteins, whereas relatively reliable models can be produced for up to 20% of proteins. In addition, models play an important part in a number of methods for obtaining structural data.

On the contrary, genomic efforts are producing the sequences of an impressive number of proteins, and there is no hope that all of them can be studied experimentally in the foreseeable future. Scientists do need to rely more and more on protein models to understand the function of this plethora of proteins, and, consequently, the required level of accuracy of a model, especially in the details of the structure, is increasing. CASP has highlighted a number of substantial improvements in modeling techniques, such as the development of FR and fragment-based methods, but, unfortunately, improvements in accurately predicting the details of a protein structure (such as positioning of side chain and of structurally divergent regions, i.e., regions of the target protein that deviate substantially from the template) have not been equally satisfactory *(12)*.

The overall conclusion that can be drawn from the analysis of the thousands of model submitted by hundreds of groups is that, rarely, a comparative model is closer to the experimental structure than the template used to build it or to reliably predict structural divergent regions. Furthermore, there seem to be no method able to consistently improve the accuracy of an initial model. An important goal is therefore to foster the development of modeling methods aimed at reaching an accuracy approaching the experimental error *(17)*.

This is the rationale behind the emergence of a new category in CASP, aimed, as we mentioned, at evaluating the quality of the details of the models rather than their overall accuracy. It will be included in the next round of the experiment, and, hopefully, it will be as effective in pushing the field farther as the other CASP categories have been in the past.

10. Automatic Evaluation Servers

CASP is aimed at evaluating the state of the art in prediction methods; however, not all experimentalists interested in obtaining a model of their protein of interest have access to collaborations with outstanding modeling groups. The most common route to prediction for the majority of scientists relies on publicly available automatic servers. It is clearly important to evaluate the accuracy of these servers on a large set of data and in a continuous fashion.

This need has prompted the development of automatic systems that continuously evaluate automatic prediction methods. They collect the predictions returned by different servers for new protein structures before any method had a chance to use them in the training set.

EVA *(3)* is one of the servers that performs this useful service to the community. Every day, EVA downloads the newest protein structures from the Protein Data Bank (PDB) archive *(18)*, extracts the sequences for every protein chain, and sends them to each prediction server registered for the experiment. The collected results are then evaluated and made public.

EVA covers several methods that predict solvent accessibility, secondary structure, and complete 3D modeling. The proteins used in the experiment are such that no pair of them has more than 33% identical residues over more than 100 residues aligned.

Another continuous benchmarking server is Livebench *(19)* that limits itself to the evaluation of 3D models of proteins not sharing a significant sequence similarity (and therefore deemed to be non-homologous) to any protein of known structure. Every week, new entries in the PDB database with a length comprised between 100 and 500 residues are submitted to participating servers and their returned predictions collected and analyzed.

The results of both servers, together with some statistical evaluation of their significance, are publicly available through Internet, and they represent extremely useful tools that should be consulted before using any prediction server.

The possibility of automatically collecting the results of several prediction servers also prompted the development of the so-called metapredictors *(20)*. These are gateways to various methods for protein structure prediction, which "outsource" the prediction task to publicly available servers, collect the results, and evaluate them. Some metapedictors just score the predictions and provide the user with a ranked list, whereas some others combine the predictions returning a single model. They usually perform better than single servers and probably represent the best solution to automatic prediction of protein structure as of today.

11. State of the Art of Structure Prediction Methods: The Usefulness of Protein Models

We said in the introduction that the quality of a model dictates its usefulness for several applications. As we discussed, estimating the quality of a model is not an easy task. However, some rules of thumb can still be provided, with the caveat that they are just indications and that each protein modeling experiment has a story of its own.

Comparative models built on the basis of a significant sequence identity between target and template, above 50–60% are certainly accurate in their overall structure and can be reliably used to analyze the conserved regions of the protein, such as its active site. As we mentioned, apart from special cases *(21)*, the predictions of structurally divergent regions is likely of being much less accurate than the rest of the protein, and it is rather risky to derive biological conclusions from their conformation *(22)*. For very high sequence identity, above 90%, there are usually very few structurally divergent regions, but here, the devil is in the positioning of the side chains. It has been shown that even models of high accuracy would fail if used as targets for drug design because the positioning of the side chain would not be sufficiently accurate *(23)*.

For comparative models, a user should always take into account that the accuracy of the model is not uniform throughout the structure and that functionally important regions are likely to be better conserved, at least for orthologous proteins, than the rest of the structure. Comparative models based on distant evolutionary relationships have been often instrumental in deriving functional properties of the protein, because these are usually brought about by the most conserved parts of the structure, which, in turn, are those predicted more accurately ***(24)***.

Models based on low sequence identity (below 30%), FR methods, and fragment-based methods should only be used as structural frameworks to think about the protein and certainly not for deriving detailed measures of distances or energies. Remember that, if the model is built by comparative modeling, we can at least be sure that the overall topology of the protein is correct, whereas this might or might not be true for fold recognition and fragment-based models. In these cases, only experimental verifications of the features predicted by the model can increase the confidence in a model.

Models can also be used for speeding up the experimental determination of a protein structure. For example, models with a GDT-TS value above 84 are consistently able to solve the phase problem in crystallography, that is, to be used as a tool to estimate the phases of the X-ray diffracted waves, a major problem in X-ray crystallography ***(25)***. Models can also be useful in speeding up the solution of the structure of proteins by nuclear magnetic resonance spectroscopy.

The impressive thrust of biological and computational methods makes it very difficult to predict what we can expect even in the near future. Nevertheless, more and more protein sequences and structures will become available, and there is no doubt that the sheer power of the data will help building more accurate protein structure models. On the contrary, if we look at the history of the past few years, we cannot but expect that new prediction methods will appear. It follows that the possibility of exploring the complete space of protein structure is, finally, within our reach.

References

1. Browne, W. J., North, A. C., Phillips, D. C., Brew, K., Vanaman, T. C. and Hill, R. L. (1969) A possible three-dimensional structure of bovine alpha-lactalbumin based on that of hen's egg-white lysozyme. *J Mol Biol*, **42,** 65–86.
2. Moult, J., Pedersen, J., Judson, R. and Fidelis, K. (1995) A large-scale experiment to assess protein structure prediction methods. *Proteins*, **23**(3), ii–v.
3. Koh, I. Y., Eyrich, V. A., Marti-Renom, M. A., Przybylski, D., Madhusudhan, M. S., Eswar, N., Grana, O., Pazos, F., Valencia, A., Sali, A., et al. (2003) EVA: evaluation of protein structure prediction servers. *Nucleic Acids Res*, **31**(13), 3311–3315.

4. Fischer, D., Elofsson, A. and Rychlewski, L. (2000) The 2000 Olympic Games of protein structure prediction; fully automated programs are being evaluated vis-a-vis human teams in the protein structure prediction experiment CAFASP2. *Protein Eng*, **13**(10), 667– 670.
5. Chothia, C. and Lesk, A. (1986) The relation between the divergence of sequence and structure in proteins. *EMBO J*, **5**(4), 823–826.
6. Sippl, M. J. and Weitckus, S. (1992) Detection of native-like models for amino acid sequences of unknown three-dimensional structure in a data base of known protein conformations. *Proteins*, **13**(3), 258–271.
7. Gribskov, M., McLachlan, A. D. and Eisenberg, D. (1987) Profile analysis: detection of distantly related proteins. *Proc Natl Acad Sci USA*, **84**(13), 4355–4358.
8. Simons, K. T., Bonneau, R., Ruczinski, I. and Baker, D. (1999) Ab initio protein structure prediction of CASP III targets using ROSETTA. *Proteins*, **S3,** 171–176.
9. Bayley, M. J., Jones, G., Willett, P. and Williamson, M. P. (1998) GENFOLD: a genetic algorithm for folding protein structures using NMR restraints. *Protein Sci*, **7**(2), 491–499.
10. Eidhammer, I., Jonassen, I. and Taylor, W. R. (2005) *Protein Bioinformatics: An Algorithmic Approach to Sequence and Structure Analysis*. Wiley & Sons, Chichester.
11. Kryshtafovych, A., Milostan, M., Szajkowski, L., Daniluk, D. and Fidelis, K. (2005) CASP6 data processing and automatic evaluation at the protein structure prediction center. *Proteins*, **S7,** 19–23.
12. Cozzetto, D. and Tramontano, A. (2005) Ten years of predictions . . . and counting. *FEBS J*, **272,** 881–882.
13. Hubbard, T. J. (1999) RMS/coverage graphs: a qualitative method for comparing three-dimensional protein structure predictions. *Proteins*, **S3,** 15–21.
14. Moult, J., Fidelis, K., Rost, B., Hubbard, T. and Tramontano, A. (2005) Critical assessment of methods of protein structure prediction (CASP) - round 6. *Proteins*, **S7,** 3–7.
15. Zemla, A. (2003) LGA-a method for finding 3D similarities in protein structures. *Nucleic Acids Res*, **31,** 3370– 3374.
16. Cozzetto, D. and Tramontano, A. (2005) Relationship between multiple sequence alignments and quality of protein comparative models. *Proteins*, **58**(1), 151–157.
17. Valencia, A. (2005) Protein refinement: a new challenge for CASP in its 10th anniversary. *Bioinformatics*, **21**(3), 277–277.
18. Berman, H. M., Westbrook, J., Feng, Z., Gilliland, G., Bhat, T. N., Weissig, H., Shindyalov, I. N. and Bourne, P. E. (2000) The Protein Data Bank. *Nucleic Acids Res*, **28**(1), 235–242.
19. Bujnicki, J. M., Elofsson, A., Fischer, D. and Rychlewski, L. (2001) LiveBench-1: continuous benchmarking of protein structure prediction servers. *Protein Sci*, **10**(2), 352–361.

20. Fischer, D., Elofsson, A., Rychlewski, L., Pazos, F., Valencia, A., Rost, B., Ortiz, A. R. and Dunbrack R. L. Jr. (2001) CAFASP2: the second critical assessment of fully automated structure prediction methods. *Proteins*, **S5,** 171–183.
21. Chothia, C., Lesk, A. M., Tramontano, A., Levitt, M., Smith Gill, S. J., Air, G., Sheriff, S., Padlan, E. A., Davies, D. and Tulip, W. R. (1989) Conformations of immunoglobulin hypervariable regions. *Nature*, **342**(6252), 877–883.
22. Tramontano, A., Leplae, R. and Morea, V. (2001) Analysis and assessment of comparative modeling predictions in CASP4. *Proteins*, **S5,** 22–38.
23. DeWeese-Scott, C. and Moult, J. (2004) Molecular modeling of protein function regions. *Proteins*, **55**(4), 942–961.
24. Pizzi, E., Tramontano, A., Tomei, L., La Monica, N., Failla, C., Sardana, M., Wood, T. and De Francesco, R. (1994) Molecular model of the specificity pocket of the hepatitis C virus protease: implications for substrate recognition. *Proc Natl Acad Sci USA*, **91**(3), 888–892.
25. Giorgetti, A., Raimondo, D., Miele, A. E. and Tramontano, A. (2005) Evaluating the usefulness of protein structure models for molecular replacement. *Bioinformatics*, **21**(2), ii72–ii76.

II

Template-based Methods

3

Aligning Sequences to Structures

Liam James McGuffin

Summary

Most newly sequenced proteins are likely to adopt a similar structure to one which has already been experimentally determined. For this reason, the most successful approaches to protein structure prediction have been template-based methods. Such prediction methods attempt to identify and model the folds of unknown structures by aligning the target sequences to a set of representative template structures within a fold library. In this chapter, I discuss the development of template-based approaches to fold prediction, from the traditional techniques to the recent state-of-the-art methods. I also discuss the recent development of structural annotation databases, which contain models built by aligning the sequences from entire proteomes against known structures. Finally, I run through a practical step-by-step guide for aligning target sequences to known structures and contemplate the future direction of template-based structure prediction.

Key Words: Structural genomics; comparative modeling; sequence homology; fold recognition; alignment quality; structural annotation; fold templates.

1. Introduction

Perhaps the most important aim of molecular biology is to determine how proteins, encoded by genes within the genome of a given organism, are involved in biochemical processes. By sequencing genes, we can determine sequences of the proteins they encode, which can, in turn, help us determine protein structures. The premise being that the solution of protein structures will then help us to interpret their possible function and how they interact in cellular processes. For instance, by understanding how a chain of amino acids is folded in three dimensions, we can infer which residues may be involved in binding

From: *Methods in Molecular Biology, vol. 413: Protein Structure Prediction, Second Edition*
Edited by: M. Zaki and C. Bystroff © Humana Press Inc., Totowa, NJ

to other molecules. The solution of structures of normally functioning proteins will also help us to improve our understanding of how faulty protein structures may cause disease. The structures of disease-related proteins may then be used to help develop and discover new and more effective drugs and diagnostic methods. Structural data also allow us to infer distant evolutionary relationships between proteins, which may not be obvious from the sequence data. Therefore, we can use protein structure to assign functions to proteins, which have no known sequence homologs. The role of protein structure determination is central to the comprehensive functional annotation of genomes.

Continuing efforts have been made to refine computational methods for protein structure prediction mainly because of the difficulty, expense, and time taken to resolve structures experimentally. The ultimate goal of protein structure prediction has been to accurately model the 3D structure or fold of a protein given its amino acid sequence. The most successful methods have been those that build models by aligning the target sequence to a template fold with an experimentally determined structure. The success of template-based approaches is based on the fact that most sequences are known to adopt a similar structure to one that has already been determined. Techniques for aligning target sequences to known structures have become widely used in structural genomics projects because of their central role in target selection. A great deal of time, expense, and effort that may have been spent solving structures experimentally have been saved through the development of accurate modeling using sequence to structure alignments.

2. The Traditional Techniques for Predicting Structure from Sequence

It is, of course, unnecessary to determine the structure of every protein within the genome of an organism using purely experimental methods. Computational methods for predicting protein structures have been developed, which allow us to interpret 3D structure directly from the proteomic sequence data that have already been obtained. Several different methods are used to predict protein structure from protein sequence data. Blind community-wide experiments, namely Critical Assessment of techniques for protein Structure Prediction (CASP), are carried out biennially to assess the progress of different methods of protein structure prediction (see Chapter 2 and supplements of *PROTEINS: Structure, Function and Genetics* for details; S1, 1997; S3, 1999; S5, 2001; S6, 2003; S7, 2005). Traditionally, each prediction method taking part in the trials has been placed into one of three categories reflecting the extent to which the method relies on knowledge of known structures. These categories are homology or comparative modeling, fold recognition, and *ab initio* or "new fold" prediction. **Table 1** summarizes a simple overview of each prediction category.

Table 1
The Different Approaches to Modeling Protein Folds Traditionally Fall into Three Main Categories Depending on the Level of Information Known About the Target Sequence

Method category	Requirements	Relative computational difficulty	Relative speed	Theoretical sequence coverage
Homology/comparative modeling	Clear homology (> 30% sequence ID) to a template fold of known structure within the PDB	Easy	Fast	Minimum
Fold recognition/threading	A template fold of known structure within the PDB	Medium	Medium	Medium
Ab initio/new fold	The target sequence and/or fragment library	Hard	Slow	Maximum

PDB, Protein Data Bank.

Each method is used in succession depending on the level of information discovered, for instance, if no homologous structure exists, then a fold recognition technique would be required to align sequence to structure, if no fold template can be found then an *ab initio* or "new fold" prediction algorithm would be the only option.

2.1. Comparative Modeling

Comparative modeling (also referred to as homology modeling) involves methods that predict the structure of a sequence by comparing it to known structures with similar sequences. These methods rely on knowledge of known structures and on the premise that similar sequences will have similar structures *(1)*. Target protein sequences are aligned against a library of template protein sequences for which the structures have been determined. The target sequence is then assigned the structure of the template or templates to which it optimally aligns.

Needleman and Wunsch *(2)* originally described an algorithm for optimal pairwise alignment of biological sequences, using a dynamic programing approach. This idea was later extended by Smith and Waterman *(3)* who modified the approach to calculate optimal local alignments. The Smith–Waterman local alignment algorithm was able to match isolated regions of local similarity and was therefore able to align proteins with multiple domains, repeats, or hypervariable regions more accurately.

Although dynamic programing approaches allow us to find the optimal alignment between two sequences many times faster than by exhaustive searching through every possible alignment, such methods are relatively slow in comparison with more modern sequence comparison methods. Methods such as FASTA *(4)* and later Basic Local Alignment Search Tool (BLAST) *(5)* were developed to perform rapid searches for sequence homologs in large sequence databases. These methods produce relatively accurate approximate sequence alignments by quickly finding sub-sequences or "tuples" shared between the target and the template proteins. Although they are comparatively fast, these methods are not as rigorous or as sensitive as dynamic programing methods, and they rely on finding high numbers of matching sub sequences. The effectiveness of sequence searching can be improved by the use of an amino acid substitution matrix. Matrices such as PAM *(6)*, GCB *(7)*, JTT *(8)*, BLOSUM62 *(9)*, STR *(10)* and more recently OPTIMA *(11)* are used to score the alignment of different pairs of amino acids with different weightings. These weightings account for the different physical, chemical, and structural properties shared by each pair of amino acids, for example, a leucine–isoleucine match is scored higher than a leucine–tryptophan.

In the late 1990s, the ability of sequence searching methods to detect more distant evolutionary relationships was improved through the use of sequence profiles from comparisons of multiple aligned sequences and iterative searching. The availability, speed, and sensitivity of the sequence profile-based method Position-Specific Iterative (PSI)-BLAST *(12)* have allowed it to become universally adopted as the standard benchmark, against which newly developed sequence-based searching methods are compared. Indeed, using suitable parameters, PSI-BLAST searching can greatly outperform Smith–Waterman searching in the detection of remote sequence homologs *(13)*. Arguably, the most recent major innovation in sequence searching has been the introduction of profile–profile alignments *(14)*. Improvements in distant homology searching are discussed further in **Subheading 3.1**.

Once a sequence homolog of known structure can be found for a given target, the next step is to use the 3D structure as a template to generate a model of the target. Various homology modeling methods have been developed that build models of a target using one or sometimes many related structures (for a review of comparative modeling methods, see **ref.** ***15***). Generally, the template structures with the highest sequence identity to the targets indicate the most related proteins and are therefore chosen to build models from. However, other considerations such as the resolution of the template and the protein "environments" in which the template and the target are thought to be found should also be taken into account. As fast sequence searching methods such as those discussed above produce "approximate" alignments, the chosen target and templates are often realigned using dynamic programing to find the optimal alignments. Where template and target have low sequence identity, the structure of the template is often used to improve alignments, for instance to avoid the insertion of gaps into regions containing helices or strands *(16,17)*.

Homology models are mostly built from the templates using rigid body assembly *(18)*, although other methods include modeling by segment matching *(19)* and modeling by satisfaction of spatial constraints *(20)*. The accuracy of the built models is evaluated by checking stereochemistry and the compatibility of the target sequence and the modeled structure *(21,22)*. Automatic comparative/homology modeling Web servers such as SWISS-MODEL *(23)* and 3D-JIGSAW *(24)* have become popular means to allow non-experts to produce accurate models of newly determined sequences. Although such automation of comparative modeling has obvious benefits for automatic genome annotation, these methods are only effective when target and template are accurately aligned and when the sequence identity is above the so-called "twilight zone" at approximately 25–30% *(25)*.

2.2. Fold Recognition

It is estimated that, for up to 70% of new protein sequences, there will be a structure with a similar fold in the Protein Data Bank (PDB), from which a suitable model could be constructed *(17)*. Indeed, it has been found that just nine different folds (termed superfolds) may account for up to 30% of the known structures *(26)*. However, for many of these protein targets, no templates

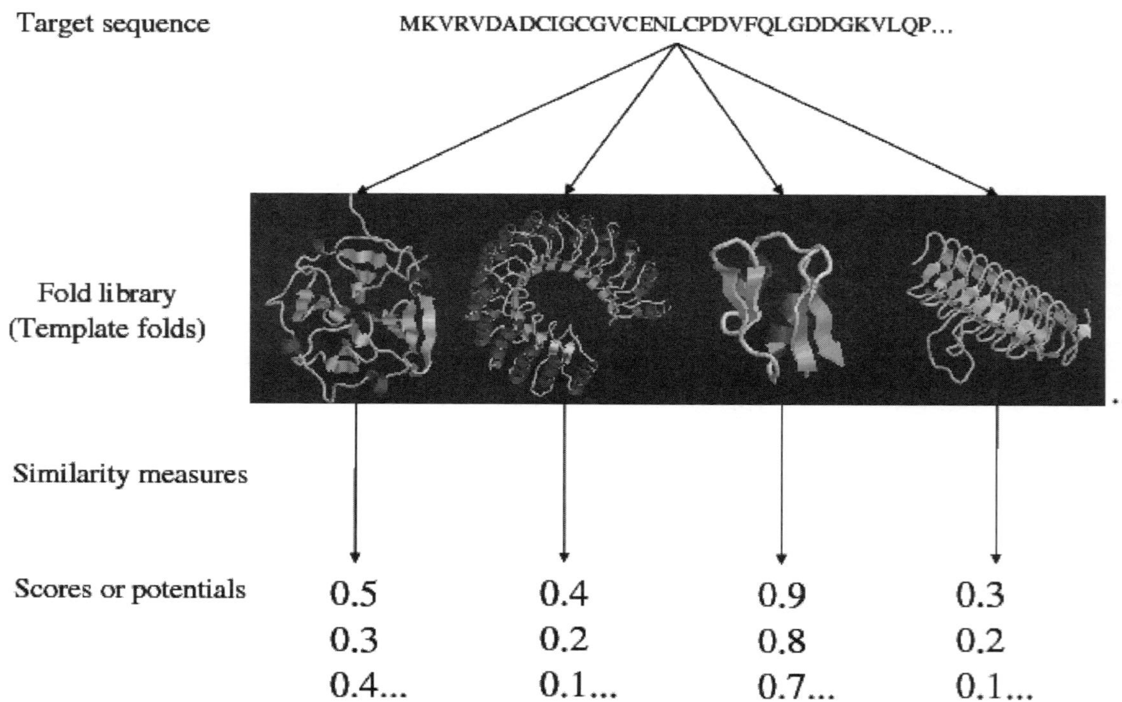

Fig. 1. A simple outline of the concept of protein fold recognition.

will be found by sequence searching methods because of low sequence identity to any known structure. Fold recognition—which was often referred to as threading—attempts to assign folds to sequences that show very low sequence identity to a known structure. **Figure 1** shows a simple outline of how fold recognition methods generally work. The target sequence is compared against a library of fold templates using a scoring scheme, and the template with the highest similarity score (or lowest energy potential) is assumed to be the fold of the target protein. In reality, fold recognition methods often produce more than one measure of similarity, which need to be interpreted by a human expert. This reliance on human interpretation made the full automation of some fold recognition methods more complicated. The automation of fold recognition is discussed further in **Subheading 3**.

Although previous attempts to relate sequence to fold in the absence of sequence homology had been made by Ponder and Richards in 1987 *(27)* and Bowie et al. in 1990 *(28)*, perhaps the first real method for fold recognition was that developed by Bowie et al. in 1991 *(29)*. The method described by Bowie et al. attempted to measure the compatibility of sequence with fold in terms of structural environments, which involved calculating the amino acid preferences for solvent accessibility, contact with polar atoms, and secondary structure type. The premise was that the structural environment of an amino acid would be more conserved than the actual amino acid type itself. The 3D structures were converted to 1D strings—relating to the structural environments of the amino acids—which could then be aligned using a conventional dynamic programing algorithm.

In 1992, Jones et al. developed a more successful method called THREADER *(30)*, which was built upon the fold recognition concept of Bowie et al. *(29)*. Jones' pioneering THREADER method differed from the method of Bowie et al. in that it considered the detailed network of pairwise interactions between individual residues rather than just assigning them to a basic environmental class. Typically, threading methods such as THREADER work by attempting to fit (or "thread") a target sequence directly on the backbone coordinates of known protein structures by using a double dynamic programing algorithm similar to that of Taylor and Orengo *(31)*. The best-fitting model can then be determined from energy potentials—derived from statistical analysis of proteins of known structure, similar to that carried out by Sippl *(32)*—where the best structural match to the target is the template with the lowest energy.

In general, the most successful traditional fold recognition methods were those that were similar to Jones' threading approach *(33)*. A number of related methods for fold recognition were developed throughout the 1990s *(34–36)*. Most of these methods employed some variation of iterative dynamic programing algorithm to build models combined with an analysis of pairwise interactions between structurally adjacent residues.

Fold recognition methods such as threading can be powerful but often owe much of their success to expert human interpretation of results *(37)*. They are often computationally intensive and may be limited in use to the recognition of single-domain folds. Perhaps the single most important limitation of fold recognition, however, is the fact that methods rely on the discovery of known folds as templates. It is therefore impossible to build accurate models for targets, which have novel structures using fold recognition. The "holy grail" of protein structure prediction would be a method for predicting structure directly from the amino acid sequence without any reliance on fold templates.

2.3. **Ab Initio** *and "New Fold" Methods*

Ab initio prediction involves methods that attempt to assemble 3D structures "from first principles" and that do not rely directly on knowledge of known structures. Most *ab initio* prediction methods have traditionally relied on the generation of different chain conformations and evaluation of each conformation using an energy function. Such approaches are extremely computationally intensive because of the large search space that is required. The huge conformational search space required is exemplified by the Levinthal paradox *(38)*, which demonstrates that the time taken for a relatively small protein of 100 residues to exhaustively search through every chain conformation would exceed the estimated age of the Universe. As most proteins in nature fold on a timescale on the order of milliseconds, it is quite clear that they must be using some sort of folding pathway *(38)*. Owing to their reliance on searching conformational space, *ab initio* methods have been limited to the prediction of very short amino acid sequences. Recent projects have been initiated in attempt to reduce the computational time taken, through the utilization of distributed molecular dynamics by Zagrovic et al. *(39)* (http://folding.stanford.edu/) and through the construction of a petaflop supercomputer that can carry out one quadrillion (10^{15}) operations per second, by the Blue Gene project team at IBM (http://www.research.ibm.com/bluegene/) *(40)*. Other computational approaches toward a solution involve theory concerning the possible folding pathways used by proteins in nature. The main problem with *ab initio* methods is that we do not understand how to model the folding of a protein and generally more theory than useful methods has arisen from the *ab initio* prediction field *(41)*.

Recently, methods which explore the possibilities of using fragment libraries of "supersecondary" structural motifs have been developed. These methods are based on the assumption that novel folds will be made up of common structural motifs, representing local minima in the polypeptide chain. By assembling combinations of these template fragments of structure, we can greatly narrow the search space required to fold the protein. The concept of fragment assembly was originally introduced by Jones in the CASP2 experiment in 1996 ***(42)***. Jones' FRAGFOLD method has since undergone a number of improvements and has proved competitive in subsequent CASP experiments ***(42–44)***. Perhaps the most successful fragment assembly method in recent years has been the ROSETTA method, developed by Baker and colleagues ***(45–47)***. Despite the apparent progress in this area, given the CPU time presently required to fold even a short sequence, it still remains impractical to attempt fragment assembly—or indeed any other New Fold or ab intio method—on a genomic scale. However, some groups have begun to make use of peer-to-peer networks to carry out new fold prediction across multiple clusters of computers throughout the world. Both the Predictor@home (http://predictor.scripps.edu) and now also the Rosetta@home (http://boinc.bakerlab.org/rosetta) projects make use of the BOINC software (http://boinc.berkeley.edu) to distribute the load of new fold predictions. As these projects become more popular and the number of users increase, the speed at which new fold predictions can be made should also increase.

3. Contemporary Methods for Sequence to Structure Alignment

The development of rapid and reliable, fully automatic methods for aligning sequences to structures is necessary for comprehensive annotation of proteome sequences to be practical. Optimal sequence alignment methods that use dynamic programing techniques, such as the Smith–Waterman algorithm, are computationally intensive and rely on close homology to a known structure. Similarly, fully automating some traditional fold recognition methods such as optimal sequence threading is also problematic. As mentioned in the previous section, optimal sequence threading is relatively computationally intensive, mostly limited in use to single domains, and expert human interpretation of results is heavily relied on. There has been some effort at making the interpretation of threading results from the THREADER method more intuitive through the addition of a graphical user interface ***(48)*** and by reducing the rather "user unfriendly" table of outputted similarity scores to a single score. Nevertheless, reducing results into a single score automatically has proved much less effective than using human interpretation.

In early CASP experiments, it was sufficient to predict the correct fold and threading methods became popular because of their success with this task. However, as model quality became increasingly important in later experiments, threading methods were criticized for not producing good sequence to structure alignments. On thecontrary, although sequence-based search methods were good at aligning sequences to structures, they were poor at finding distant homologs to be used as templates. Therefore, over the past few years, predictors have focused their efforts on the weaknesses of both homology modeling and fold recognition methods, resulting in the incremental improvement of automatic methods.

Some of the most successful fully automated methods for aligning sequences to structures have combined aspects of both comparative modeling and fold recognition, blurring the traditional boundaries between techniques. Lately, we have seen the development of fully automatic servers which are able to carry out sequence-based searches to detect very distant homologs. We have also seen the development of so-called "hybrid" methods that have employed aspects of traditional fold recognition combined with evolutionary information from distant homology searches.

3.1. Improvements in Sequence Searching

Traditional pairwise sequence alignment methods can be used to assign folds to sequences with obvious evolutionary relationships to a known structure. Brenner et al. ***(49)*** have assessed the reliability of popular pairwise sequence comparison methods such as FASTA ***(4)*** and BLAST ***(5)***, by benchmarking their ability to recognize distant evolutionary relationships. For sequences with identities approximately >30%, fast sequence searching methods such as FASTA and WU-BLAST ***(50)*** compare in accuracy to the slower, Smith–Waterman ***(3)***-based method SSEARCH ***(4)***. However, when sequence identities fall to <30%, conventional pairwise sequence comparison methods fail to detect relationships ***(49)***; therefore, accurately annotating genes that produce proteins with sequences of no discernible sequence identity to any known protein structure is problematic.

Sequence searching was improved beyond pairwise comparisons with the introduction of methods such as PSI-BLAST ***(12)***, Intermediate Sequence Searches (ISS) ***(51)***, SAM-T98 ***(52)***, and Fold and Function Assignment System (FFAS) ***(14)***. These methods use information from profiles of related sequences to detect more distant relationships. Arguably, the most widely used of these methods is PSI-BLAST, which carries out iterative searches for a target protein on a data set of sequences using position-specific score matrices derived from BLAST profiles. The coverage and error rate of PSI-BLAST in remote homology detection in genome annotation was benchmarked by Müller et al. ***(53)***. Using a "model genome" derived from structural classification of proteins

(SCOP) *(54)* classified sequences, Müller et al. claimed that PSI-BLAST was able to recognize homologs for 40% of domains with <20% identity *(53)*.

In 1998, Park and co-workers benchmarked ISS, PSI-BLAST, and their hidden Markov model-based method SAM-T98 against pairwise methods. They found that up to three times as many remote homologs could be detected using the profile-based methods than could be from using the traditional pairwise methods. FFAS is another profile-based method, which differs fundamentally from PSI-BLAST in that uses profiles on both sides of the alignment, that is, a profile is generated for the target sequence, which is aligned to template profiles of proteins from the PDB *(14)*. This profile–profile approach has proved to be a major advance, and many of the current top-performing structure prediction methods have since incorporated similar scoring schemes—see Ohlson et al. *(55)* for a review of methods.

Sequence searching using profile methods can be used to detect protein sequences with very remote common ancestry, that is, distant homologs with similar function. However, these methods perform poorly at recognizing non-homologous proteins with similar folds *(14)*. For detecting analogous proteins (proteins with similar folds but no sequence detectable common ancestry *(26)*), methods which make use of additional structural information are the only clear option *(56,14)*.

3.2. Hybrid Methods

The development of hybrid methods, which combine sequence profile searching with methods derived from fold recognition, has been designed to quickly, reliably, and automatically align sequences to analogous structures on a genomic scale. Although these methods do rely on finding some sequence homology to target sequences, in doing so they allow inferences to be made about possible protein function.

The GenTHREADER method, developed by Jones in 1999, was one of the earliest hybrid approaches to fully automated fold recognition *(56)*. The original GenTHREADER protocol incorporated sequence alignment profiles, which were evaluated using energy potentials derived from THREADER *(30)*. The resulting alignment scores, pairwise energy scores, solvation energy scores, and length information were used as inputs to neural network, which was trained to recognize whether proteins shared the same fold according to the Class, Architecture, Topology, Homology (CATH) *(57)* definition of fold. GenTHREADER has since been updated to include additional structural information, which has resulted in the detection of more remote homologs and a higher overall quality of sequence to structure alignments *(58)*. The current GenTHREADER neural network architecture is shown in **Fig. 2**.

A number of alternative hybrid approaches designed to enhance sequence searching by incorporating structural information were also developed at the start of the millennium. INBGU, developed by Fischer in 2000 *(61)*, used a combination of sequence profiles and PHD *(62)* predicted secondary structure versus observed secondary structure in an extension of the fold recognition method Sequence Derived Properties (SDP) *(63)*. Kelley et al. *(64)* developed 3D position-specific scoring matrix (PSSM), which also incorporated predicted secondary structure. 3D-PSSM used PSI-BLAST to generate 1D profiles, which were then augmented using combination of solvation potentials, observed secondary structure, and SAP structural alignments to generate 3D profiles for each sequence in the library of templates. For each query sequence, an 1D-PSSM was produced using PSI-BLAST, and the secondary structure was predicted using PSIPRED *(13)*. Using this information, the query was then aligned to each template PSSM. The method FUGUE by Shi et al. *(65)* also made use of structural alignments, secondary structure, and solvent accessibility information. Using this information, environment-specific score matrices and structure-dependent score gap penalties are derived and are then used to align target sequence profiles against the library of structural profiles. Each of these methods have undergone a number of incremental improvements since there initial development and have given rise to a plethora of new hybrid techniques.

3.3. Fully Automated Servers

Current methods, such as GenTHREADER, produce single output scores, which can be reliably interpreted as a measure of confidence in the prediction because of their consistency. The implementation of these tools as Web servers have allowed state-of-the-art, fully automatic methods to be easily utilized by non-expert users.

Intuitive Web interfaces are available for most automated servers, allowing users to submit their target proteins of interest using a Web browser from anywhere in the world. The results of each prediction are emailed back to users shortly after the sequence is submitted. Emails will contain the prediction results in plain text format, or often, hypertext links will be provided to Web pages containing more intuitive graphical data. For example, the Bioinformatics Unit at UCL offers a number of leading protein structure prediction methods as free services to academic users *(66)*. **Figure 3** shows the Web interface for the PSIPRED server where users can perform GenTHREADER predictions on a given sequence, among other options *(60)*.

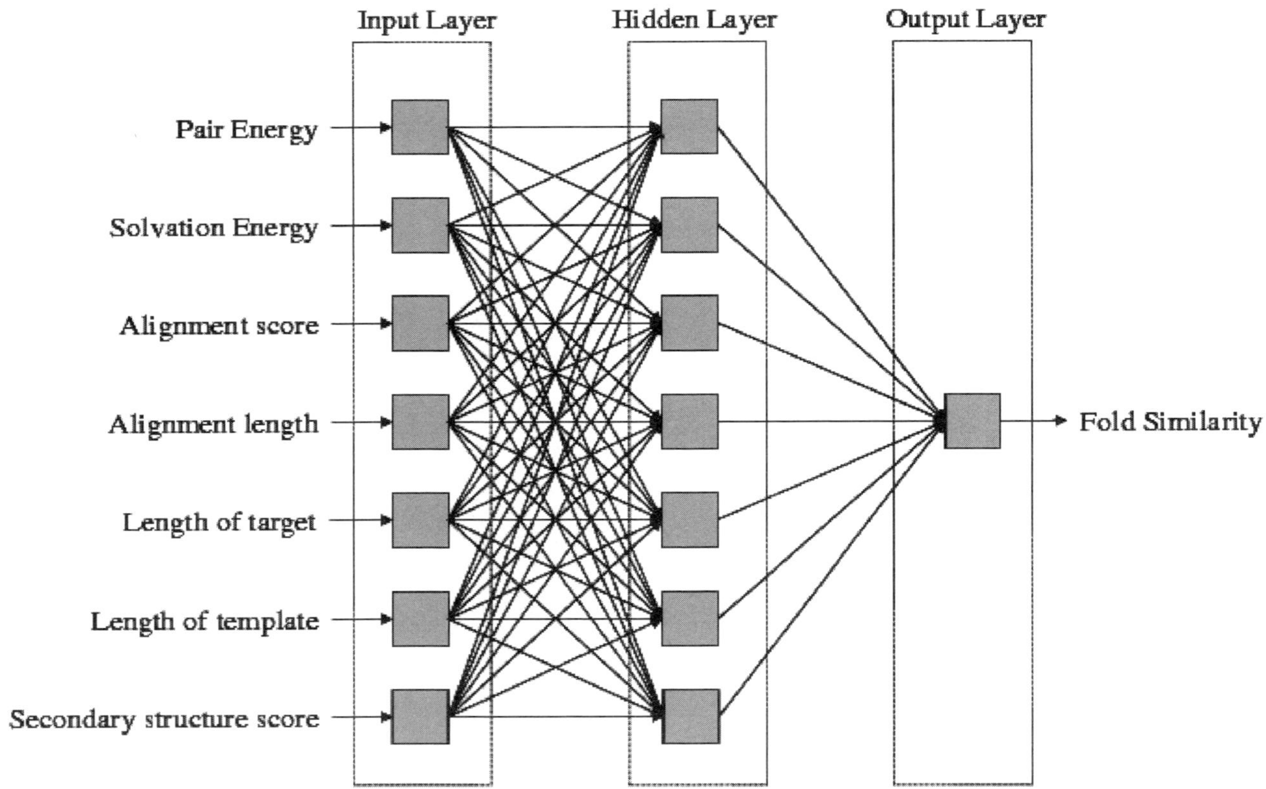
Input Layer
Hidden Layer
Output Layer
Pair Energy
Solvation Energy
Alignment score
Alignment length
Length of target
Length of template
Secondary structure score
Fold Similarity

The growth in number of such servers and the acknowledged requirement for the full automation of structure prediction initialized the critical assessment of fully automated structure prediction (CAFASP), which was first held in conjunction with CASP 3 *(67)*. Even in the relatively short time between CAFASP1 and CAFASP2 *(68)*, an increase in performance of fully automatic fold recognition was observed. Indeed by CAFASP3, fully automated servers were outperforming many human experts *(69)*. In addition to the biennial assessments, continual assessments of automatic structure prediction methods have been initiated, namely, LiveBench *(70)* and EVA *(71)*.

During CASP5/CAFASP3, it became apparent that improved template selection and increased accuracy of sequence to structure alignments could be achieved from a consensus of prediction methods. Many of the top prediction groups made extensive use of so-called "meta-servers" such as 3D-SHOTGUN *(72)* (http://bioinfo.pl/meta/), Pcons, and Pmodeller *(73)*. Meta-servers work by simultaneously querying many individual independent prediction servers and then automatically collating the results to form a consensus prediction. Despite the outstanding success of meta-servers, they were criticized for stifling the innovation of novel independent methods. Nevertheless for both expert and non-expert users alike, consensus predictions are extremely valuable if no obvious solution can be found using individual methods independently.

4. Databases Serving Structural Annotations for Entire Proteomes

Another important development has been the introduction of databases serving the results of proteome-wide sequence-structure alignments. These dedicated structural annotation resources are freely available to academics and provide intuitive Web interfaces with various search options. Databases such as 3D-GENOMICS *(74)* and Gene3D *(75)* primarily use PSI-BLAST and other sequence-based search methods to assign folds to related sequences. The Genomic Threading Database *(59)* (*see* **Fig. 4** differs from 3D-GENOMICS and Gene3D in that GenTHREADER *(58)* is used as the key part of the annotation

Fig. 2. The neural network architecture of the GenTHREADER method—one of the earliest hybrid approaches to rapid, fully automated fold recognition *(56,58)*. The current version combines powerful profile–profile sequence searches with traditional threading potentials and structural information. GenTHREADER was designed to be a used on whole proteome sequences for use in the construction of the Genomic Threading Database *(59)*. Individual sequences can be submitted for GenTHREADER predictions through the PSIPRED server *(60)*.

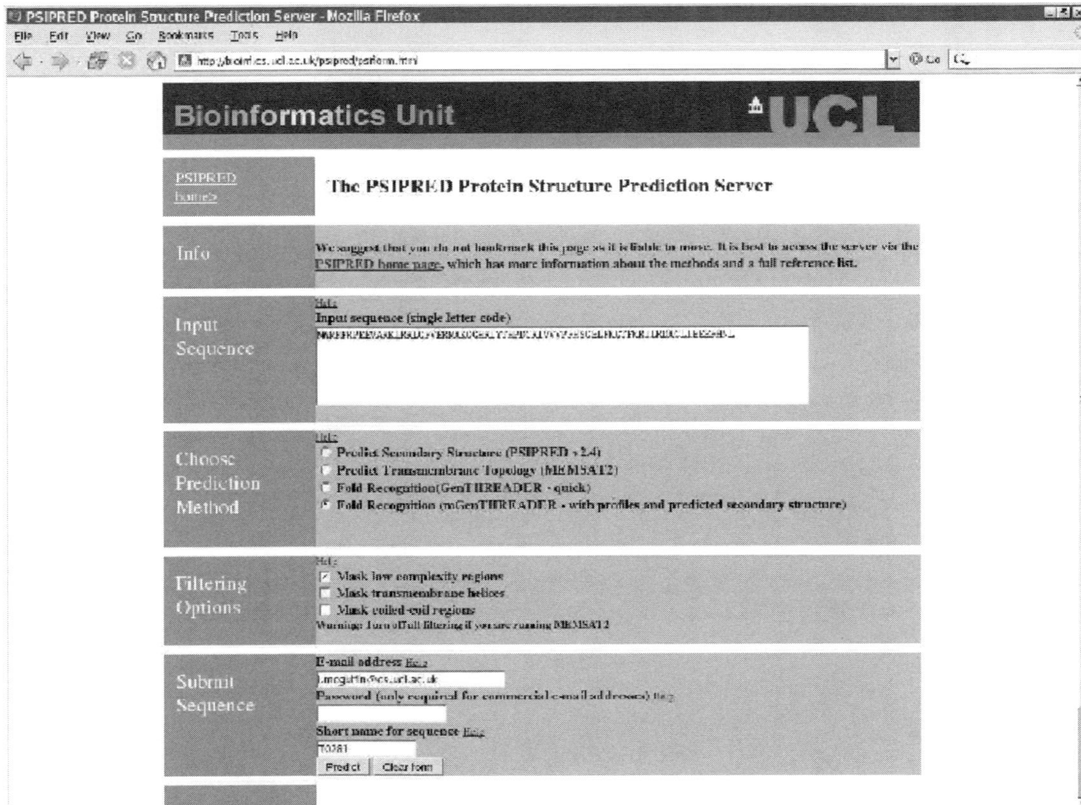

Fig. 3. Screen shot of the PSIPRED server submission form accessible through http://www.psipred.net *(60)*. Individual sequences may be submitted for fully automated fold recognition using the GenTHREADER method.

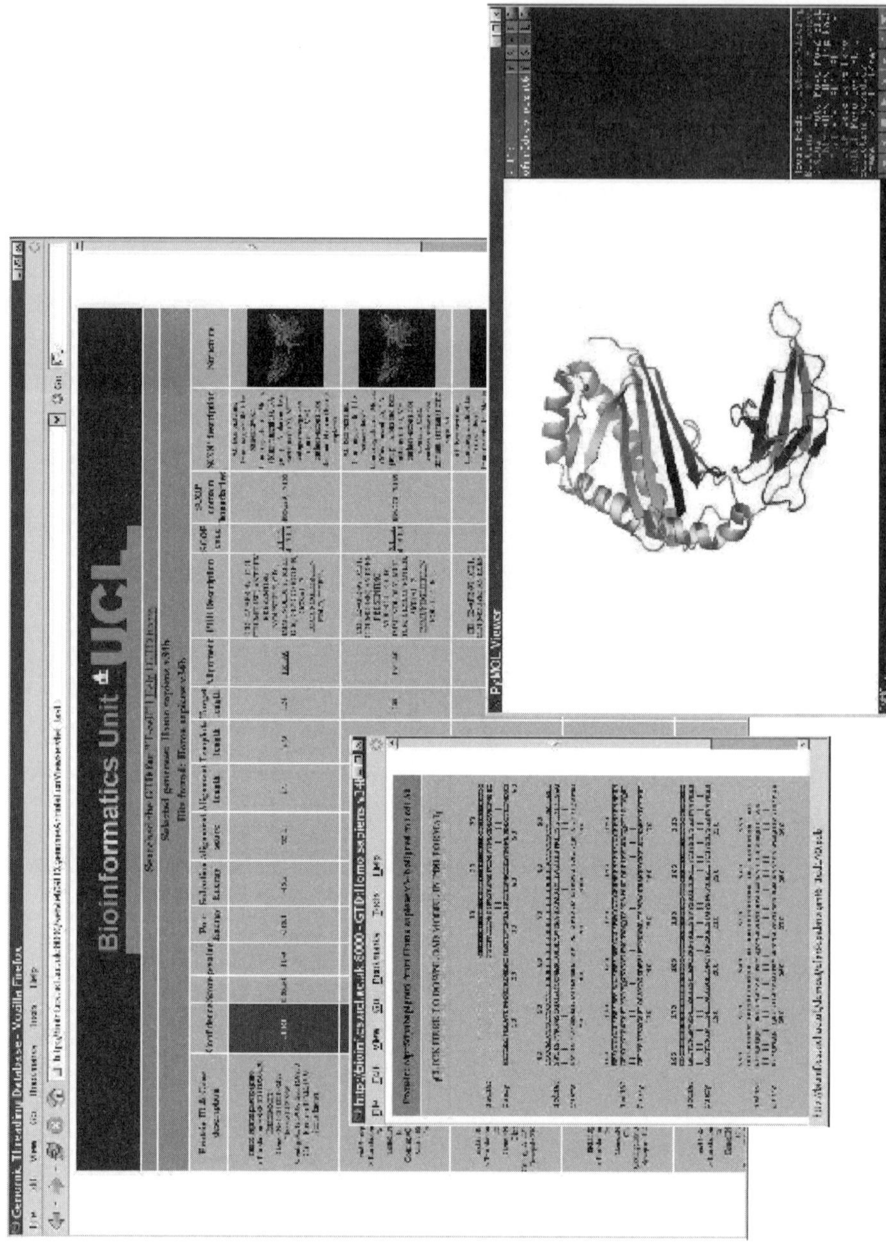

system with a view to providing more sensitive and selective detection of remote homology. It is anticipated that each of these structural databases will employ more rigorous profile–profile methods in the near future, with the advent of Grid technology.

As each database uses a different prediction strategy, it is possible to combine the results from several different annotation databases together to form a consensus of methods. The e-Protein project (http://www.e-protein.org) was setup to bring together structural and functional annotation databases from University College London, Imperial College London, and the European Bioinformatics Institute, through a single interface using the Distributed Annotation System ***(76)***. Another aspect of the project was to develop prototypes of Grid technology to distribute the computational load across processing clusters at each site. This has lead to a rapid increase in the speed at which proteome-wide predictions can be made ***(100)***.

5. How to Align Your Sequence to a Known Structure

The following is a practical guide through the steps of aligning a given target sequence to a template structure and producing a 3D model of your protein. The flow chart in **Fig. 5** outlines the basic steps described below.

5.1. Check Structural Annotation Databases for a Model

Perhaps the first thing to do is to check whether a model already exists for your target protein within one of the structural annotations databases. Databases such as the GTD and 3D-Genomics provide users with models for most sequences in PDB format, which can be downloaded and viewed using your favorite molecular model viewer. If, however, the annotation is very out of date, of low confidence, or if you cannot find a match to your sequence, then you will need to build a new model from scratch.

5.2. Preparing Your Sequence

Preparing your target sequence is an essential step to building an accurate model of a globular protein. Your target sequence may contain low complexity or non-globular regions, which should be identified and filtered or masked

Fig. 4. The Genomic Threading Database: a comprehensive resource for structural annotations of the genomes from key organisms (http://bioinf.cs.ucl.ac.uk/GTD) ***(59)***. Pre-computed models for each globular protein within a proteome can be downloaded and viewed through a Web browse plugin. The PYMOL molecular viewer is used in the figure to view the model (http://www.pymol.org).

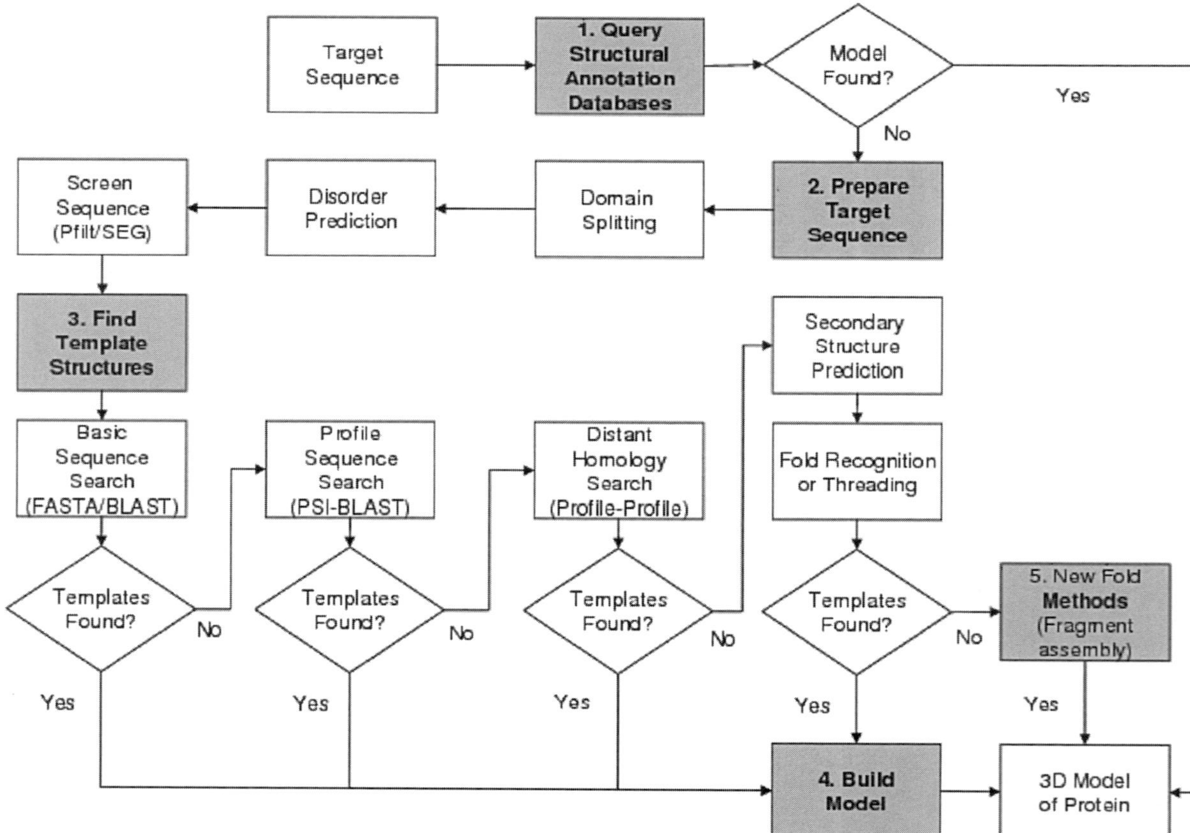

Fig. 5. Flow chart showing the steps for aligning a sequence to a known structure.

out before sequence searching. This will ensure that you will be searching for appropriate globular regions within your protein only.

The program *pfilt* (ftp://bioinf.cs.ucl.ac.uk/pub/pfilt/) *(77)* can be used to filter out (i.e., replace amino acid characters with Xs) those regions of low complexity, coiled-coil regions, and regions with extremely biased amino acid compositions such as transmembrane helices. Another popular sequence filter is used by the *seg* method *(78)*. Many prediction servers either include user selectable sequence filtering options or apply them automatically before performing a prediction.

If your sequence is particularly long, then it may be useful to identify separate globular domains and chop your protein before identifying likely fold templates. There are many domain prediction servers to choose from, some of which can be accessed through the Meta-DP server (http://meta-dp.bioinformatics.buffalo.edu/) *(79)*.

An additional consideration is to specifically identify potential regions of native disorder using the DISOPRED server *(80)*. Regions of disorder often occur in linker regions between domains and may also form complete separate functional domains.

5.3. Finding Template Structures

Once the sequence has been prepared, the next step is to attempt to find a template structure from which to build a model. The method used to identify a template is dependent on the level of sequence similarity of the target sequence to the database of known folds. It is sensible to begin with a basic sequence search in the first instance as each progressive level of searching requires more time to complete.

5.3.1. Simple Sequence Search

It is often best to start with a simple, rapid sequence search, using either BLAST or FASTA, against the sequences within the PDB *(81)*. A quick Web search will reveal a number of Web sites providing interfaces for performing FASTA and BLAST searches. Perhaps the most widely used Web forms are the BLASTP interface at the National Center for Biotechnology Information (NCBI) (http://www.ncbi.nlm.nih.gov/blast/) and FASTA33 interface at the European Bioinformatics Institute (EBI) (http://www.ebi.ac.uk/fasta33/), both of which have options to search your sequence against the PDB. Alternatively, both programs can be downloaded, installed, and run in-house. The FASTA package can be downloaded from ftp://ftp.virginia.edu/pub/fasta; this package contains the latest version of FASTA plus SSEARCH, which may be useful for optimally realigning your sequence to the chosen template (see **Subheading 5.4**). BLAST can be downloaded from the NCBI as part of the toolkit (ftp://ftp.ncbi.nih.gov/toolbox/) or as a standalone package (http://www.ncbi.nlm.nih.gov/blast/download.shtml).

A good match to a PDB sequence is indicated by a low E-value. Generally, if your top hit has an E-value of > 0.001, it is worth carrying out a more sensitive sequence search. However, if a homolog can be found with E < 0.001 using BLAST or FASTA searches only, then you may move on to **Subheading 5.4**.

5.3.2. Complex Sequence Searches

If a very quick search does not reveal any obvious sequence homologs to known structures, then a more sensitive sequence search will be necessary. In this case, PSI-BLAST is the next step, which performs a more distant search by iteratively building alignment profiles and using them on subsequent sequence searches. Thus, with each iteration, more and more distant homologs are found, the profiles get larger, and the search time gets longer. It is therefore advisable to set a few iterations, maybe 3, in the first instance and then save a checkpoint file. It is then possible to increase the number of iterations from then on using the checkpoint file if required. It has to be said that using PSI-BLAST most effectively is something of a black art, and the documentation concerning the 30 or so configurable parameters may appear to be impenetrable to beginners. Jones and Swindells ***(77)*** have written an informative account on "Getting the most from PSI-BLAST" to shed some light on the key steps. Again, there are a plethora of Web interfaces available for PSI-BLAST, although the most up-to-date versions of the program and databases can be found at the NCBI.

Although PSI-BLAST is the most popularly used sequence-searching technique, there are many other methods available, which may provide more sensitive searches. The SAM method by Karplus et al. ***(82)*** uses an iterative Hidden Markov Model approach to sequence searching and has become quite widely used in comparative modeling because of its increased sensitivity over PSI-BLAST. Both standalone and Web servers are freely available to academics through the Karplus Group homepages (http://www.soe.ucsc.edu/research/compbio/sam.html).

For sequence only searching, most of the state-of-the-art methods carry out some variation of profile–profile alignments. The FFAS method ***(14)*** was one of the first to develop the idea of using profile–profile alignments and remains one of the best approaches. Although there is currently no downloadable version available, the FFAS server *(83)* is freely available for academic use through the Godzik group homepage (http://bioinformatics.ljcrf.edu/).

Most methods and servers provide users with a list of alternative hits to a given sequence and scores relating to the strength of the match of each hit. Users should refer to current documentation of individual methods to gauge the accuracy of their match. If sequence-based searching fails to identify a convincing template from which to model a fold, then a search for analogous fold templates must be carried out.

5.3.3. Hybrid Fold Recognition Searches

Hybrid techniques are often able to find analogous fold templates where using sequence information alone will fail. Approaches such as 3DPSSM *(64)*, mGenTHREADER *(58)*, FUGUE *(65)*, INBGU *(61)*, and SPARKS *(84)* all incorporate structural information in their scoring functions in some way and are freely available through Web servers. Most of these methods carry out comparisons of the predicted secondary structure of the target sequence to the known secondary structures of template folds to detect evolutionary more distant relationships. Many of these methods now also incorporate profile–profile alignments as the initial step; so, it is also worth considering hybrid methods for identifying close homologous templates. The best of the sequence-based methods and hybrid techniques are evaluated continuously by the LiveBench server *(70)*, and it is worthwhile checking the site for the latest ranking of methods.

If analogous templates cannot found by independent hybrid methods, then publicly available traditional threading methods such as THREADER *(30)* and PROSPECTOR *(85)* may still be worth investigating. If neither hybrid methods nor traditional threading techniques can identify a likely template, or if various weakly detected fold templates are listed as top hits, then a Meta-server search may be required.

5.3.4. Meta-Server Search

Meta-servers such as 3DJury *(72)* provide users with a list of templates obtained from various independent methods. The top template chosen is often the most commonly identified structure by all methods; however, several configurable options are available to users, which allow results to be combined in different ways. 3DJury will also provide users with a list of the sequence to structure alignments and corresponding 3D models for each of the independent methods.

5.4. Templates Found—Build or Refine Models

If any of the techniques listed above fail to identify an adequate template, then a new fold prediction method may be required to build a model (*see* **Subheading 5.5.**). However, for the vast majority of target sequences, a template can be found. Often for very close homologs, the sequence to structure alignments provided by the servers are adequate enough to build a model, and many servers will provide users with the 3D coordinates of a modeled fold. However, users may wish to further refine their models to build loops, model side chains more accurately, or "fill in" any gaps in the sequence to structure alignment.

There are a few publicly accessible servers and methods available for building refined models given a sequence to structure alignment. Perhaps the most popular server freely available for academic use is SWISS-MODEL (http://swissmodel.expasy.org/) ***(86)***. Alternatively, the MODELLER software is also very popular and available to download for various platforms to both academic and commercial users (http://salilab.org/modeller/) ***(15)***.

Both SWISS-MODEL and MODELLER provide the option of constructing high-quality models from multiple alignments. Multiple alignments can be built using a number of programs, although the SWISS-MODEL server recommends the T-COFFEE method ***(87)***.

MODELLER also provides the option of building models from multiple templates. Multiple template modeling is often beneficial because many templates provide consensus information on the parts of the structure least likely to change with variations in sequence. However, multiple template modeling may not be so beneficial in cases where the templates are very distant from one another.

A number of alternative programs for building models may also be worth investigating such as NEST ***(88)***, 3D-JIGSAW ***(24)***, and SegMod/ENCAD ***(89)***. Side chain conformations can be more effectively predicted using the specialized SCWRL method ***(90)***. Wallner and Elofsson ***(91)*** provide a rigorous comparison of the performance and reliability of model building programs mentioned above.

Once a model has been built, it is necessary to estimate its reliability. The stereochemistry of your model can be evaluated using a program such as PROCHECK ***(22)*** or WHAT-CHECK ***(92)***. These programs basically check the extent to which your model deviates from real X-ray structures based on a number of observed measures.

Various model quality assessment programs (MQAPs) are available which attempt to discriminate between native-like models and decoy structures. VERIFY3D ***(93)*** and PROSAII ***(21)*** have been in popular use for some

(Dayhoff, M.O., ed). Silver Springs, National Biomedical Research Foundation, pp. 345–352.

7. Gonnet, G.H., Cohen, M.A. and Brenner, S.A. (1992) Exhaustive matching of the entire protein database. *Science*, **256**, 1443–1445.
8. Jones, D.T., Taylor, W.R. and Thornton, J.M. (1992) The rapid generation of mutation data matrices from protein sequences. *CABIOS*, **8**, 275–282.
9. Henikoff, S., Henikoff, J.G. (1992) Amino acid substitution matrices from protein blocks. *Proc. Natl. Acad. Sci. U. S. A.*, **89**, 10915–10919.
10. Overington, J., Donnelly, D., Johnson, M.S., Sali, A. and Blundell, T.L. (1992) Environment-specific amino-acid substitution tables: tertiary templates and prediction of protein folds. *Protein Sci.*, **1**, 216–226.
11. Kann, M., Qian, B. and Goldstein, R.A. (2000) Optimization of a new score function for the detection of remote homologues. *Proteins*, **41**, 498–503.
12. Altschul, S.F., Madden, T.L., Schaffer, A.A., Zhang, J., Zhang, Z., Miller, W. and Lipman, D.J. (1997) Gapped BLAST and PSI-BLAST: a new generation of protein database search programs. *Nucleic Acids Res.*, **25**, 3389–3402.
13. Jones, D.T. (1999) Protein secondary structure prediction based on position-specific scoring matrices. *J. Mol. Biol.*, **292**, 195–202.
14. Rychlewski, L., Jaroszewski, L., Li, W. and Godzik (2000) A Comparison of sequence profiles. Strategies for structural predictions using sequence information. *Protein Sci.*, **9**, 232–241.
15. Marti-Renom, M.A., Stuart, A.C., Fiser, A., Sanchez, R., Melo, F. and Sali, A. (2000) Comparative protein structure modeling of genes and genomes. *Ann. Rev. Biophys. Biomolec. Struct.*, **29**, 291–325.
16. Sanchez R and Šali A. (2000) Comparative protein structure modeling. In: *Methods in Molecular Biology vol. 143: Protein Structure Prediction: Methods and Protocols* (Webster, D.M., ed). Humana Press, New Jersey, pp. 97–129.
17. Jones, D.T. (2000) A practical guide to protein structure prediction. In: *Methods in Molecular Biology vol. 143: Protein Structure Prediction: Methods and Protocols* (Webster, D.M., ed). Humana Press, New Jersey, pp. 131–154.
18. Blundell, T.L., Sibanda, B.L., Sternberg, M.J.E. and Thornton, J.M. (1987) Knowledge based prediction of protein structures and the design of novel molecules. *Nature*, **326**, 347–352.
19. Levitt, M. (1992) Accurate modeling of protein conformation by automatic segment matching. *J. Mol. Biol.*, **226**, 507–533.
20. Šali, A. and Blundell, T.L. (1993) Comparative modelling by satisfaction of spatial restraints. *J. Mol. Biol.*, **234**, 779–815.
21. Sippl, M.J. (1993) Recognition of errors in three-dimensional structures of proteins. *Proteins*, **17**, 355–362.
22. Laskowski, R.A., McArthur, M.W., Moss, D.J. and Thornton, J.M. (1993) PROCHECK: a program to check the stereochemical quality of protein structures. *J. Appl. Cryst.*, **26**, 283–291.

23. Peitsch, M.C. (1996) ProMod and Swiss-model: Internet-based tools for automated comparative protein modeling. *Biochem. Soc. T.*, **24**, 274–279.
24. Bates, P.A. and Sternberg, M.J.E. (1999) Model building by comparison at CASP3: using expert knowledge and computer automation. *Proteins*, **37**, 47–54.
25. Rost B. (1999) Twilight zone of protein sequence alignments. *Protein Eng.*, **12**, 85–94.
26. Orengo, C.A., Jones, D.T. and Thornton, J.M. (1994) Protein superfamilies and domain superfolds. *Nature*, **372**, 631–634.
27. Ponder, J.W. and Richards, F.M. (1987) Tertiary templates for proteins – use of packing criteria in the enumeration of allowed sequences for different structural classes. *J. Mol. Biol.*, **193**, 775–791.
28. Bowie, J.U., Clarke, N.D., Pabo, C.O. and Sauer, R.T. (1990) Identification of protein folds – matching hydrophobicity patterns of sequence sets with solvent accessibility patterns of known structures. *Proteins*, **7**, 257–264.
29. Bowie, J.U., Lüthy, R. and Eisenberg, D. (1991) A method to identify protein sequences that fold into a known 3-dimensional structure. *Science*, **253**, 164–170.
30. Jones, D.T., Taylor, W.R. and Thornton, J.M. (1992) A new approach to protein fold recognition. *Nature*, **358**, 86–89.
31. Taylor, W.R. and Orengo, C.A. (1989) Protein structure alignment. *J. Mol. Biol.*, **280**, 1–22.
32. Sippl, M.J. (1990) Calculation of conformational ensembles from potentials of mean force – an approach to the knowledge-based prediction of local structures in globular-proteins. *J. Mol. Biol.*, **213**, 859–883.
33. Murzin, AG. (1999) Structure classification-based assessment of CASP3 predictions for the fold recognition targets. *Proteins*, **37**(S3), 88–103.
34. Godzik, A., Kolinski, A. and Skolnick, J. (1992) Topology fingerprint approach to the inverse protein folding problem. *J. Mol. Biol.*, **227**, 227–238.
35. Bryant, S.H. (1996) Evaluation of threading specificity and accuracy. *Proteins*, **26**, 172–185.
36. Thiele, R., Zimmer, R. and Lengauer, T. (1999) Protein threading by recursive dynamic programming. *J. Mol. Biol.*, **290**, 757–779.
37. Karplus, K., Karchin, R., Barrett, C., Tu, S., Cline, M., Diekhans, M., Grate, L., Casper, J. and Hughey, R. (2001) What is the value added by human intervention in protein structure prediction? *Proteins*, **45**(S5), 86–91
38. Levinthal, C. (1968) Are there pathways for protein folding? *J. Chim. Phys.*, **65**, 44–45.
39. Zagrovic, B., Sorin, E.J. and Pande, V. (2001) Beta-hairpin folding simulations in atomistic detail using an implicit solvent model. *J. Mol. Biol.*, **313**, 151–169.
40. Allen F. et al. (2001) Blue gene: a vision for protein science using a petaflop supercomputer. *IBM Sys. J.*, **40**, 310.
41. Pande, V.S., Grosberg, A.Y., Tanaka, T. and Rokhsar, D.S. (1998) Pathways for protein folding: is a 'new view' needed? *Curr. Opin. Struct. Biol.*, **8,** 68–79.

42. Jones, D.T. (1997) Successful *ab initio* prediction of the tertiary structure of NK-lysin using multiple sequences and recognized supersecondary structural motifs. *Proteins*, **29**(S1), 185–191.
43. Jones, D.T. (2001) Predicting novel protein folds by using FRAGFOLD. *Proteins*, **45**(S5), 127–132.
44. Jones, D.T. and McGuffin, L.J. (2003) Assembling novel protein folds from super-secondary structural fragments. *Proteins*, **53** (S6), 480–485.
45. Simons, K.T., Kooperberg, C., Huang, E. and Baker, D. (1997) Assembly of protein tertiary structures from fragments with similar local sequences using simulated annealing and Bayesian scoring functions. *J. Mol. Biol.*, **268**, 209–225.
46. Bonneau, R., Tsai, J., Ruczinski, I., Chivian, D., Rohl, C., Strauss, C.E.M. and Baker, D. (2001) Rosetta in CASP4: progress in *ab initio* protein structure prediction. *Proteins*, **45**(S5),119–126.
47. Bradley, P., Chivian, D., Meiler, J., Misura, K.M., Rohl, C.A., Schief, W.R., Wedemeyer, W.J., Schueler-Furman, O., Murphy, P., Schonbrun, J., Strauss, C.E. and Baker D. (2003) Rosetta predictions in CASP5: successes, failures, and prospects for complete automation. *Proteins*, **53**(S6),457–468.
48. Jones, D.T., Miller, R.T. and Thornton, J.M. (1995) Successful protein fold recognition by optimal sequence threading validated by rigorous blind testing. *Proteins*, **23**, 387–379.
49. Brenner, S.E., Chothia, C. and Hubbard, T.J.P. (1998) Assessing sequence comparison methods with reliable structurally identified distant evolutionary relationships. *Proc. Natl. Acad. Sci. U. S. A.*, **95**, 6073–6078.
50. Altschul, S.F. and Gish, W. (1996) Local alignment statistics. *Methods Enzymol.* **266**, 460–480.
51. Park, J., Teichmann, S.A., Hubbard, T., et al. (1997) Intermediate sequences increase the detection of homology between sequences. *J. Mol. Biol.*, **273**, 349–354.
52. Park, J., Karplus, K., Barret, C., Hughey, R., Haussler, D., Hubbard, T. and Chothia, C. (1998) Sequence comparisons using multiple sequences detect three times as many remote homologues as pairwise methods. *J. Mol. Biol.*, **284**, 1201–1210.
53. Müller, A., MacCallum, R. and Sternberg, M.J.E. (1999) Benchmarking PSI-BLAST in genome annotation. *J. Mol. Biol.*, **293**, 1257–1271.
54. Murzin, A.G., Brenner, S.E., Hubbard, T. and Chothia C. (1995) SCOP: a structural classification of proteins database for the investigation of sequences and structures. *J. Mol. Biol.*, **247**, 536–540.
55. Ohlson, T., Wallner, B. and Elofsson, A. (2004) Profile-profile methods provide improved fold-recognition: a study of different profile-profile alignment methods. *Proteins*, **57**,188–197.
56. Jones, D.T. (1999b) GenTHREADER: an efficient and reliable protein fold recognition method for genomic sequences. *J. Mol. Biol.*, **287**, 797–815.

57. Orengo, C.A., Michie, A.D., Jones, S., Jones, D.T., Swindells, M.B. and Thornton, J.M. (1997) CATH – a hierarchic classification of protein domain structures. *Structure*, **5**, 1093–1108.
58. McGuffin, L.J. and Jones, D.T. (2003) Improvement of the GenTHREADER method for genomic fold recognition. *Bioinformatics*, **19**, 874–881.
59. McGuffin, L.J., Street, S., Bryson, K., Sorensen, S.A. and Jones, D.T. (2004) The genomic threading database: a comprehensive resource for structural annotations of the genomes from key organisms. *Nucleic Acids Res.*, **32**, D196–199.
60. McGuffin L.J., Bryson K. and Jones, D.T. (2000) The PSIPRED protein structure prediction server. *Bioinformatics* **16**, 404–405.
61. Fischer, D. (2000) Hybrid fold recognition: combining sequence derived properties with evolutionary information. In: *Proceedings of the Pacific Symposium on Biocomputing*. World Scientific Press, Hawii, pp. 119–130.
62. Rost, B. (1996) PHD: predicting one-dimensional protein structure by profile based neural networks. *Methods Enzymol*, **266**, 525–539.
63. Fischer, D. and Eisenberg, D. (1996) Protein fold recognition using sequence-derived predictions. *Protein Sci.*, **5**, 947–955.
64. Kelley, L.A., MacCallum, R.M. and Sternberg, M.J.E. (2000) Enhanced genome annotation using structural profiles in the program 3D-PSSM. *J. Mol. Biol.*, **299**, 499–520.
65. Shi, J., Blundell, T.L. and Mizuguchi, K. (2001) FUGUE: sequence-structure homology recognition using environment-specific substitution tables and structure-dependent gap penalties. *J. Mol. Biol.*, **310**, 243–257.
66. Bryson, K., McGuffin, L.J., Marsden, R.L., Ward, J.J., Sodhi, J.S. and Jones, D.T. (2005) Protein structure prediction servers at University College London. *Nucleic Acids Res.*, **33**, W36–38.
67. Fischer, D., Christian, B., Bryson, K., Elofsonn, A., Godzik, A., Jones, D., Karplus, K.J., Kelley, L.A., MacCallum, R.M., Pawowski, K., Rost, B., Rychlewski, L. and Sternberg, M. (1999) CAFASP-1: critical assessment of fully automated structure prediction methods. *Proteins*, **37**(S3), 209–217.
68. Fischer, D., Elofsson, A., Rychlewski, L., Pazos, F., Valencia, A., Rost, B., Ortiz, A.R. and Dunbrack, R.L., Jr. (2001) CAFASP2: the second critical assessment of fully automated structure prediction methods. *Proteins*, **45**(S5), 171–183.
69. Fischer, D., Rychlewski, L., Dunbrack, R.L., Jr., Ortiz, A.R. and Elofsson, A. (2003) CAFASP3: the third critical assessment of fully automated structure prediction methods. *Proteins*, **53**(S6), 503–516.
70. Bujnicki, J.M., Elofsson, A., Fischer, D. and Rychlewski, L. (2001) LiveBench-2: large-scale automated evaluation of protein structure prediction servers. *Proteins*, **45**(S5), 184–191.
71. Eyrich, V., Martí-Renom, M.A., Przybylski, D., Madhusudhan, M.S., Fiser, A., Pazos, F., Valencia, A., Sali, A. and Rost, B. (2001) EVA: continuous automatic

evaluation of protein structure prediction servers. *Bioinformatics*, **17**, 1242–1243.
72. Ginalski, K., Elofsson, A., Fischer, D. and Rychlewski, L. (2003) 3D-Jury: a simple approach to improve protein structure predictions. *Bioinformatics*, **19**, 1015–1018.
73. Wallner, B., Fang, H. and Elofsson, A. (2003) Automatic consensus-based fold recognition using Pcons, ProQ, and Pmodeller. *Proteins*, **53** (S6), 534–541.
74. Fleming, K., Muller, A., MacCallum, R.M. and Sternberg, M.J. (2004) 3D-GENOMICS: a database to compare structural and functional annotations of proteins between sequenced genomes. *Nucleic Acids Res.*, **32**, D245–250.
75. Buchan, D.W., Rison, S.C., Bray, J.E., Lee, D., Pearl, F., Thornton, J.M. and Orengo, C.A. (2003) Gene3D: structural assignments for the biologist and bioinformaticist alike. *Nucleic Acids Res.*, **31**, D469–473.
76. Dowell, R.D., Jokerst, R.M., Day, A., Eddy, S.R. and Stein, L. (2001) The distributed annotation system. *BMC Bioinformatics*, **2**, 7.
77. Jones, D.T. and Swindells, M.B. (2002) Getting the most from PSI-BLAST. *Trends Biochem. Soc.*, **27**, 161–164.
78. Wootton, J.C. and Federhen, S. (1993) Statistics of local complexity in amino acid sequences and sequence databases. *Comput. Chem.*, **17**, 149–163.
79. Saini, H.K. and Fischer, D. (2005) Meta-DP: domain prediction meta-server. *Bioinformatics*, **15**, 2917–2920.
80. Ward, J.J., McGuffin, L.J., Bryson K., Buxton, B.F. and Jones, D.T. (2004) The DISOPRED server for the prediction of protein disorder. *Bioinformatics*, **20**, 2138–2139.
81. Berman, H.M., Westbrook, J., Feng, Z., Gilliland, G., Bhat, T.N., Weissig, H., Shindyalov, I.N. and Bourne PE. (2000) The protein data bank. *Nucleic Acids Res.*, **28**, 235–242.
82. Karplus, K., Karchin, R., Draper, J., Casper, J., Mandel-Gutfreund, Y. and Diekhans, M. (2003) Combining local-structure, fold-recognition, and new fold methods for protein structure prediction. *Proteins*, **53** (S6), 491–496.
83. Jaroszewski, L., Rychlewski, L., Li, Z., Li, W. and Godzik, A. (2005) FFAS03: a server for profile-profile sequence alignments. *Nucleic Acids Res.*, **33**, W284–288.
84. Zhou, H. and Zhou, Y. (2004) Single-body residue-level knowledge-based energy score combined with sequence-profile and secondary structure information for fold recognition. *Proteins*, **55**, 1005–1013.
85. Skolnick, J., Kihara, D. and Zhang, Y. (2004) Development and large scale benchmark testing of the PROSPECTOR_3 threading algorithm. *Proteins*, **56**, 502–518.
86. Schwede, T., Kopp, J., Guex, N. and Peitsch, M.C. (2003) SWISS-MODEL: an automated protein homology-modeling server. *Nucleic Acids Res.*, **31**, 3381–3385.
87. Notredame, C., Higgins, D. and Heringa, J. (2000) T-coffee: a novel method for multiple sequence alignments. *J. Mol. Biol.*, **302**, 205–217.
88. Petrey, D., Xiang, Z., Tang, C.L., Xie, L., Gimpelev, M., Mitros, T., Soto, C.S., Goldsmith-Fischman, S., Kernytsky, A., Schlessinger, A., Koh, I.Y., Alexov, E.

and Honig, B. (2003) Using multiple structure alignments, fast model building, and energetic analysis in fold recognition and homology modeling. *Proteins*, **53**(S6), 430–435.

89. Cardozo, T., Totrov, M. and Abagyan, R. (1995) Homology modeling by the ICM method. *Proteins*, **23**, 403–414.
90. Canutescu, A.A., Shelenkov, A.A. and Dunbrack, R.L., Jr. (2003) A graph-theory algorithm for rapid protein side-chain prediction. *Protein Sci.*, **12**, 2001–2014.
91. Wallner, B. and Elofsson, A. (2005) All are not equal: a benchmark of different homology modeling programs. *Protein Sci.*, **14**, 1315–1327.
92. Hooft, R.W., Vriend, G., Sander, C. and Abola, E.E. (1996) Errors in protein structures. *Nature*, **381**, 272.
93. Eisenberg, D., Luthy, R. and Bowie, J.U. (1997) VERIFY3D: assessment of protein models with three-dimensional profiles. *Methods Enzymol.*, **277**, 396–404.
94. Pettitt, C.S., McGuffin, L.J. and Jones, D.T. (2005) Improving sequence-based fold recognition by using 3D model quality assessment. *Bioinformatics*, **21**, 3509–3515.
95. Kim, D.E., Chivian, D. and Baker, D. (2004) Protein structure prediction and analysis using the Robetta server. *Nucleic Acids Res.*, **32** (S2), W526–531.
96. Zhang, Y. and Skolnick, J. (2005) The protein structure prediction problem could be solved using the current PDB library. *Proc. Natl. Acad. Sci. U. S. A.*, **102**, 1029–1034.
97. Aloy, P., Bottcher, B., Ceulemans, H., Leutwein, C., Mellwig C., Fischer, S., Gavin, A.C., Bork, P., Superti-Furga, G., Serrano, L. and Russell, R.B. (2004) Structure-based assembly of protein complexes in yeast. *Science*, **303**, 2026–2029.
98. Lu, L., Arakaki, A.K., Lu, H. and Skolnick, J. (2003) Multimeric threading-based prediction of protein-protein interactions on a genomic scale: application to the Saccharomyces cerevisiae proteome. *Genome Res.*, **13**, 1146–1154.
99. Schueler-Furman, O., Wang, C. and Baker, D. (2005) Progress in protein-protein docking: atomic resolution predictions in the CAPRI experiment using RosettaDock with an improved treatment of side-chain flexibility. *Proteins*, **60**, 187–194.
100. McGuffin, L. J., Smith R. T., Bryson, K., Sorensen, S. A., & Jones, D. T. (2006) High throughput profile-profile based fold recognition for the entire Human proteome. *BMC Bioinformatics*, **7**, 288.

4

Protein Structure Prediction Using Threading

Jinbo Xu, Feng Jiao, and Libo Yu

Summary

This chapter discusses the protocol for computational protein structure prediction by protein threading. First, we present a general procedure and summarize some typical ideas for each step of protein threading. Then, we describe the design and implementation of RAPTOR, a protein structure prediction program based on threading. The major focuses are three key components of RAPTOR: a linear programming approach to protein threading, two machine learning approaches (SVM and Gradient Boosting) to fold recognition, and evaluation of the statistical significance of the prediction results. The first part of this chapter is a brief review of protein threading, and the second part contains original research results. Some key ideas and results have been previously published.

Key Words: Protein structure prediction; protein threading; linear programming; SVM; Gradient Boosting.

1. Introduction

Protein threading predicts the 3D structure for a new protein by aligning its primary sequence to proteins in the Protein Data Bank (PDB) to see if a similar structure can be found. This chapter uses "target" or "target protein" to refer to a new protein with structure to be predicted and "template" for a protein in the PDB. If a target protein can be aligned to a template in the PDB very well, then the target protein is assumed to have a similar structure as the template and the structure of the target can be constructed based on this alignment. The goodness of one target-template alignment is evaluated using a scoring function. Protein threading makes a structure prediction according

From: *Methods in Molecular Biology, vol. 413: Protein Structure Prediction, Second Edition*
Edited by: M. Zaki and C. Bystroff © Humana Press Inc., Totowa, NJ

to the following procedures. First, the target protein is aligned to each of the proteins in the PDB, and the optimal alignment between the target and each template is calculated according to a given scoring function. Secondly, the best one or several templates are chosen for the target based on their alignments to the target. The spatial positions of the aligned residues in the target can be copied from the chosen templates. Usually, the aligned residues lie in the core regions (i.e., the spatially conserved region) of the target protein. Finally, a loop modeling method is used to predict the coordinates of unaligned residues, and a side-chain packing program is used to predict coordinates for the side-chain atoms.

In the past 10 years, protein structure prediction based on protein threading has made significant progress due to both the enlargement of the PDB and the improvements in prediction protocols. In order to produce protein structures in high-throughput mode, NIH has launched a Protein Structure Initiative, which has produced several thousand non-redundant protein structures and continues to produce structures. The NIH initiative uses the strategy of producing protein structures to maximize the number of proteins in nature that are within modeling distance of proteins in the PDB. According to some statistics (see http://www.rcsb.org/pdb/contentGrowthChart.do?content=fold-scop and http://www.rcsb.org/pdb/contentGrowthChart.do?content=supfam-scop for detailed statistics), each year, approximately 90% of all new proteins deposited to the PDB have a structure that is similar to one or more proteins in the PDB. According to **refs *1* and *2*)**, more than 90% of all single-domain proteins with up to 200 residues can be superimposed with a protein in the PDB with an average cRMSD less than 5 Å and an average coverage of 70%. This means that in principle, the structures of most new proteins can be predicted if we can develop a perfect protein threading protocol.

Generally speaking, there are three types of protein structure prediction methods: homology modeling, protein threading, and ab initio folding. Protein threading can be used for protein structure prediction when (1) the target protein does not share a high sequence similarity with any protein in the PDB and (2) the target protein shares a similar structure with some proteins in the PDB. Homology modeling predicts the structure for a target by identifying some homologous proteins from the PDB. Two homologous proteins usually share similar sequences and similar structures. Therefore, homology modeling detects whether two proteins are homologous by aligning their sequences. Compared to homology modeling, which only considers sequence similarity between the target and the template, protein threading makes use of the structural information encoded in the template to improve prediction accuracy, including the use of secondary structure, solvent accessibility, and pairwise interactions. In order to generate a good sequence-template alignment, homology modeling usually requires that the target and the template share at least 25% sequence identity *(3)*. Protein threading can go beyond this limitation and sometimes can align the target and the template very well even when their sequence identity is well below 25%. Ab inito folding predicts the structure for a target without using any complete protein structure in the PDB as a template. The number of possible conformations increases dramatically with respect to the target protein size. For a target protein with n residues, its backbone conformation roughly depends on $2n$ - 2 torsion angles as each non-terminal residue is associated with two angles and each terminal residue is associated with one angle. Compared to the ab initio folding method that searches through the entire conformational space, protein threading reduces computational complexity by restricting the conformational space to only several thousand templates in the PDB.

Protein threading consists of the following five components: (1) a library of template structures, (2) representation of targets and templates, (3) objective function measuring the quality of sequence-template alignment, (4) an algorithm finding the best sequence-template alignment, and (5) one method selecting the best template based on all the sequence-template alignments. A library of template structures is a set of representative structures selected from the PDB. Usually, in order to save computing time, among all the highly similar protein structures, only one is kept in the template library. To construct a library of template structures, we can cluster all the proteins in the PDB into several thousand groups and then choose one representative from each group as a structural template. For example, we can use the weekly updated PDB cluster results at ftp://ftp.rcsb.org/pub/pdb/derived_data/NR/ as the source of the template library. We can also build a template library using a set of representative proteins from the SCOP database *(4)*.

2. Representation of Targets and Templates

Besides its primary sequence, for a target protein, usually we also use its sequence profile and predicted secondary structure to improve the prediction accuracy of protein threading. A sequence profile can be generated using PSI-BLAST *(5)* or ClustalW *(6)*, based on a multiple sequence alignment of some proteins homologous to the target protein. A sequence profile encodes the evolutionary information of the target protein and the sequence variability among all the proteins homologous to the target. Two homologous proteins evolved from the same ancestor are more likely to have a similar sequence profile. The sequence profile of a target protein is a matrix with 20 rows and n

columns where *n* is the target protein size. Each column of this matrix corresponds to the occurring frequency of 20 different amino acids at a specific position of the target protein. The secondary structure of a target protein can be predicted using a secondary structure prediction program, which will give a confidence score (or probability) indicating the likelihood of each secondary structure type. Usually three secondary structure types are used, so the predicted secondary structure can be represented by a matrix with 3 rows and *n* columns. Each element in the matrix is the predicted probability of being a specific secondary structure type. Two widely used protein secondary structure prediction programs are PSIPRED ***(7)*** and PHD ***(8)***. Most of current threading programs like FUGUE ***(9)***, 3D-PSSM ***(10)***, PROSPECT ***(11)*** and RAPTOR ***(12)***, and others ***(13–16)*** incorporate both the sequence profile and secondary structure into their scoring functions. Besides sequence profile and predicted secondary structure, some protein threading programs also use predicted solvent accessibility to enhance the prediction accuracy ***(17)***.

The representation of a template structure is more complicated. A simple method is to use a 1D model to represent a template structure. Just like the model of a target protein, each template position is associated with a position-specific profile, a secondary structure type, and a solvent accessibility type. Because a template structure contains more information, there are several different methods to generate a profile for a template. A simple method is to use PSI-BLAST or ClustalW to generate a sequence-based profile for a template, just like PROSPECT-II ***(11,18)*** and RAPTOR ***(12)***. A structure-based profile constructed from a multiple structure alignment of some proteins with similar structures is also investigated ***(19)***. An issue with this kind of method is that some templates do not have enough similar structures in the PDB so that its structure-based profile might not have a good generalization performance. 3D-PSSM ***(10)*** first uses PSI-BLAST to generate a sequence profile and then obtains a 3D-based profile by merging the sequence profiles of many proteins with a similar fold as the template. SPARK ***(20)*** generates a structure-based profile for each nine-residue structural fragment and then assembles all the profiles together to obtain a profile for the whole template.

The major difference between homology modeling and protein threading is that besides sequence information, protein threading can make use of structural information such as secondary structure and solvent accessibility to improve both alignment accuracy and fold recognition rate. The secondary structure type at each template position can be calculated using DSSP ***(21)***. Many threading programs use three secondary structure types for the template: α-helix, β-strand, and loop. Solvent accessibility also plays an important role in many protein threading programs ***(12,17,22)***. Typically, the solvent accessibility at each template position is clustered into three types (buried, intermediate, and exposed) or just simply binary types (buried or exposed). The potential of one amino acid type being in a specific solvent accessibility type can be calculated by statistically analyzing existing proteins in the PDB.

To model the interaction relationship among template residues, we can use a 2D contact graph to represent a template structure. Besides the information contained in the 1D model, the 2D model also takes into account the pairwise contacts between two spatially close residues. An even more complicated model is to consider interactions among multiple residues ***(23)***. The 2D model representation of a template structure can be abstracted as a graph. In the graph, each vertex represents a residue in the template protein and each edge denotes a contact between two spatially close residues. The threading programs that use pairwise interactions include PROSPECT ***(22)***, PROSPECTOR ***(24)***, and RAPTOR ***(12)***. There are several different methods to construct the contact graph for a template. One is distance-based and the other is based upon Delaunay tessellation of protein structures ***(25)***. RAPTOR, PROSPECT, and PROSEPCTOR use a distance-based method to construct a contact graph. PROSPECT also explored distance-dependent pairwise interaction parameters. RAPTOR is exploring a contact definition method described in **ref. *26***. The preliminary result indicates that this is a promising method.

Due to different representations of a protein template structure, different alignment algorithms are needed to find the best sequence-template alignment. If only the 1D model is used, then a dynamic programming algorithm can be used to align the target to the template. If the 2D model is used, then a more involved algorithm is needed to find the optimal sequence-template alignment. Some threading programs ***(16,19)*** use a Hidden Markov Model (HMM) to represent a protein template or a multiple structure alignment of many proteins homologous to the template. This method can be treated as a variant of the 1D model as the HMM model can only capture the dependence relationship between two sequentially adjacent residues.

3. Threading Energy Function

The energy function should be able to quantitatively measure the quality of a given sequence-template alignment. Generally speaking, the energy function consists of sequence similarity score, environmental fitness score, structure consistency score, and gap penalty. Sequence similarity score measures the sequence similarity between the target and the template. If both the target and the template have their profiles generated, then we can calculate the sequence similarity score by comparing these two profiles. The environmental fitness score measures how well it is to align a target residue into the local environment at one specific template position. The structure consistency score contains two components: local structure consistency (i.e., secondary structure compatibility) and global structure consistency (i.e., the pairwise contact consistency or multiple-body contact consistency). We can use a weight factor to control the relative importance of various energy items in the scoring function. The weight factors can be adjusted for various purposes. For example, if only the sequence similarity is considered, then protein threading becomes homology modeling. If only the structural consistency is considered, then protein threading can be used to predict structures for the targets with only distant homologs in the PDB. Usually, these weight factors are trained to achieve the optimal alignment accuracy and fold recognition rate. Not many protein threading programs use pairwise interaction explicitly. The threading programs using pairwise interaction include RAPTOR *(12)*, PROSPECTOR *(24)*, and PROSPECT-I *(22)*. These programs also use other energy items such as sequence similarity and gap penalty, while Madej et al. *(27)* tried one method using only pairwise interaction.

Formally, the threading scoring function can be deduced using the Bayesian rule *(28)*. Let $P(T|S)$ denote the probability of the target S being in the same fold as the template T. Let $A = [A(1), A(2), \ldots, A(n)]$ denote an alignment between the target and the template where the target position j is aligned to the template position $A(j)$. If residue j is not aligned to any template residue, then $A(j)$ is empty. Let $P(T|S, A)$ denote the probability that the alignment between S and T is A. Then, $P(T|S)$ is equal to the maximum of $P(T|S, A)$ over all possible alignments. Applying Bayesian rules, we have $P(T|S, A) = \frac{P(T,S|A)}{P(S)} = \frac{P(S|T,A)P(T)}{P(S)}$. If we assume that $P(T)$ is a uniform distribution, given a specific target, then $P(T|S) \propto \max_A P(S|T, A)$.

Assume the target sequence S to be $a_1, a_2, \ldots, a_n$ and the template sequence to be $t_1, t_2, \ldots, t_m$. Then, $P(S|T, A)$ can be expanded as follows.

$$
\begin{aligned}
P(S|T, A) = {} & P(a_1, a_2, \ldots, a_n | t_1, t_2, \ldots, t_m, A) \\
= {} & \prod_i P\left(a_i | t_{A(i)}\right) \prod_{i<j} \frac{P\left(a_i, a_j | t_{A(i)}, t_{A(j)}\right)}{P\left(a_i | t_{A(i)}\right) P\left(a_j | t_{A(j)}\right)} \\
& \times \prod_{i<j<k} \frac{P\left(a_i, a_j, a_k | t_{A(i)}, t_{A(j)}, t_{A(k)}\right) P\left(a_i | t_{A(i)}\right) P\left(a_j | t_{A(j)}\right) P\left(a_k | t_{A(k)}\right)}{P\left(a_i, a_j | t_{A(i)}, t_{A(j)}\right) P\left(a_i, a_k | t_{A(i)}, t_{A(k)}\right) P\left(a_j, a_k | t_{A(j)}, t_{A(k)}\right)} \qquad (1)
\end{aligned}
$$

The first term of the right-hand side of **Eq. 1** is the probability of one particular residue a_i being aligned at position $A(i)$ regardless of the alignment of other residues. The first term generally refers to the probability of the template residue at position $A(i)$ mutating to the sequence residue a_i, and the probability of the sequence residue a_i occurring at the local structural environment of position $A(i)$. The local structural environment refers to secondary structure and solvent accessibility. The second term is the probability of two residues a_i and a_j simultaneously being aligned to two specific template positions $A(i)$ and $A(j)$. This item measures the pairwise intera ction of any two residues. Usually, we only consider those cases where $A(i)$ and $A(j)$ are spatially close. The remaining items refer to the probability of the multiple target residues simultaneously occurring at multiple specific template positions. Apart from first two terms, the other term in the right-hand side of **Eq. 1** is often ignored because it is not easy to obtain an accurate estimation of the parameters due to insufficient experimental data and it is also computationally difficult to optimize the objective function. If the template structure is represented by the 1D model, then we should also ignore the second term of the right-hand side of **Eq. 1**. As $P(S|T, A)$ is the product of several items, we can use its negative logarithm form $f(S|T, A) = -\log P(S|T, A)$ for the sake of convenience and computation.

Given a specific target and a template, the general form of the energy function is as follows.

$$f(S|T,A) = \sum_j f_1[j, A(j)] + \sum_{j_1, j_2} f_2[j_1, j_2, A(j_1), A(j_2)] + \ldots + \sum_{j_1, j_2, \ldots, j_M} f_M[j_1, j_2, \ldots, j_M, A(j_1), A(j_2), \ldots, A(j_M)], \quad (2)$$

where $f_1[j, A(j)]$ is the singleton score when the amino acid in target position j is placed to the template position $A(j)$; $f_2[j_1, j_2, A(j_1), A(j_2)]$ represents the pairwise score when $A(j_1)$ and $A(j_2)$ are spatially nearby and the residues at position j_l, $(l = 1, 2)$, are placed to the template positions $A(j_l)$ at the same time, and $f_3, f_4, \ldots,$ denote the multi-body interaction scores that are often ignored in practice. If the template structure is represented by the 2D model, then only the first two terms in the right-hand side of **Eq. 2** are kept.

Equation 2 is only general forms of an energy function for protein threading. The accuracy of the parameters used in the energy function is important for the prediction accuracy of protein threading. There are many different methods to estimate the parameters. The knowledge-based method measures a parameter by statistically analyzing a subset of non-redundant proteins in the PDB. A general form in statistically estimating a parameter ρ is $-\log \frac{f_{\text{obs}}(\rho=\nu)}{f_{\text{ref}}(\rho=\nu)}$, where $f_{obs}(\rho = \nu)$ is the observed frequency of ρ being at value ν and $f_{\text{ref}}(\rho = \nu)$ is the reference frequency or expected frequency of $\rho = \nu$. Zhang et al. *(29)*

proposed a physics-based method to estimate the reference frequency of a parameter. Besides the statistics-based methods to evaluate the parameters for the scoring function, some optimization methods are also used to design the scoring function. For example, Meller and Elber *(30)* used a linear programming method to optimize the parameters for protein threading.

4. Computational Complexity of Sequence-Template Alignment Problem

If the template structure is represented as a 1D model, then a dynamic programming algorithm can be used to find the optimal sequence-template alignment within low-degree polynomial time, regardless of the threading scoring function. However, if the pairwise interactions or multi-body interactions are taken into consideration, then it is NP-hard to find the best sequence-template alignment *(31,32)*, which means it is unlikely to have a polynomial-time algorithm to find the best alignment.

Various algorithms have been proposed for the optimal sequence-template alignment problem in the case where the template structure is represented by the 2D model. Many available threading programs *(24,33,34)* do not rigorously treat the second term of the right-hand side of **Eq. 2** as it is time-consuming to search for the optimal alignment. If **Eq. 2** is used as the objective function and gaps are allowed in the alignment, then the threading problem is NP-hard. Akutsu and Miyano *(32)* conducted a comprehensive study concerning the computational complexity of the protein threading problem. They have demonstrated that the protein threading problem is MAX SNP-hard, which means that no approximation algorithm can guarantee to generate a solution with any constant factor guarantee of accuracy within polynomial time unless $NP = P$ (it is generally believed that $NP \neq P$). Besides this result, they also proposed several approximation algorithms for this problem. The above computational complexity results suggest that the protein threading problem is computationally very difficult. But the theoretical results are somewhat misleading and the approximation algorithms are not very useful in developing practical protein threading programs. Empirically, it is acceptable if we can have an algorithm to find the best alignment within a reasonable amount of time even if the time complexity of the algorithm is exponential.

5. Threading Algorithms to Date

A dynamic programming algorithm can only find the optimal sequence-template alignment if the template is represented by a 1D model. This section summarizes several algorithms that can deal with the case where the template is represented by a 2D model. Lathrop and Smith *(35–37)* developed a branch-and-bound algorithm to solve the space is split into many small subspaces by partitioning the alignment domain of a single template position into several intervals. The lower and upper bound of the objective function in each subspace is estimated based on the sequence-template alignment generated without considering pairwise interactions. During the search process, some subspaces can be discarded based on the estimated lower and upper bounds. The algorithm terminates until the pruned search space contains only the best sequence-template alignment. This algorithm runs fast when the similarity between the target and the template is high.

Xu et al. designed a divide-and-conquer algorithm that is used in their structure prediction computer program, PROSPECT *(22,38)*, based on an observation that if the contact cutoff distance is not big, then the residue interaction pattern of many templates can be represented as a sparse graph. This algorithm splits a template into two subsegments such that there are few inter-segment contacts, recursively align each subsegment to the target respectively, and, finally, merge the alignments of two subsegments to form a complete alignment. PROSPECT runs very fast for approximately three-quarters of the templates. However, it runs very slowly or runs out of memory on a 32-bit platform on the remaining one-quarter templates with many inter-residue contacts. In the same spirit of cutting a protein template into some small components and then aligning each component almost independently to the target, Xu et al. *(39)* also proposed a tree decomposition-based approach to protein threading. This approach guarantees to generate an exact solution to the threading problem within a subexponential time. The difference between the divide-and-conquer-based and the tree decomposition-based algorithms is that the latter can cut a protein template into smaller components.

Both the branch-and-bound algorithm and the divide-and-conquer algorithm can find the globally optimal solution to the sequence-template alignment problem. However, both algorithms are still computationally expensive and not suitable for genome-scale protein structure prediction. In **Subheading 7.**, we will describe a linear programming approach to the optimal sequence-template alignment problem. Empirically, the LP approach can find the optimal alignment for 99% of threading instances within polynomial time *(12)*.

There are also many approximate or heuristic algorithms proposed for protein threading. Madej et al. *(27)* developed a Gibbs sampling technique to search for the optimal sequence-template alignment. Godzik et al. *(40)* and Jones et al. *(34)* proposed an interaction-frozen approximation algorithm to find a good sequence-template alignment iteratively. In each iteration, this algorithm assumes that one end of a contact is fixed when calculating pairwise interaction score. Thiele and Zimmer *(41)* developed a recursive dynamic programming. Recently, Balev *(42)* proposed a Larangian relaxation algorithm to solve this problem. The advantage is that this algorithm can estimate the gap between the optimal value and the best-so-far objective function value. In addition, this algorithm can also run very efficiently in practice when the target and the template are similar although it cannot guarantee the optimal solution.

6. Fold Recognition

Fold recognition identifies the best template for a given target based on all the generated sequence-template alignments. The sequence-template alignment score cannot be directly used to rank the templates due to the bias introduced by residue composition and the number of alternative sequence-template alignments for a given pair of target and template *(43)*. Both Z-score *(27,43)* and machine learning methods *(12,13)* are used to do fold recognition. Most of the current structure prediction programs use Z-score *(15,18,24)* to recognize the best-fit templates, whereas several programs such as GenTHREADER *(13)* and PROSPECT-I *(44)* use a neural network model to rank the templates. The neural network method formulates the fold recognition problem as a classification problem. We will discuss later (see Section 7.4) that formulating the template selection problem as a classification problem is not good enough for the purpose of building a structural model for the target with the best accuracy. The machine learning methods extract some features from a sequence-template alignment to describe the quality of this alignment in many different aspects and then try to predict if this pair of target and template is in the same fold or to predict the overall quality of the alignment. A machine learning model directly predicting the quality of a 3D structural model built from a sequence-template alignment can also be used to conduct fold recognition *(45)*. According to **ref.** ***46***, the machine learning methods are better than Z-score in terms of both sensitivity and specificity. In fact, Z-score cannot cancel out all the bias introduced by the protein sizes. A large target protein tends to have a large Z-score.

The Z-score was proposed to cancel out the bias caused by sequence residue composition and by the number of alternative sequence-template alignments. A typical procedure to calculate Z-score *(43)* is as follows: (1) shuffle the residues of the target randomly, (2) find the optimal alignment between the shuffled target and the template and calculate the alignment score, and (3) repeat the above two steps as many as 100 times or until the distribution of the generated alignment scores converges. Z-score is the alignment score in standard deviation units relative to the mean alignment score. Suppose that the mean and standard deviation of the shuffled alignment score distribution is u and σ. If the alignment score between the un-shuffled target and the template is S, then Z-score can be calculated by $\frac{u-S}{\sigma}$. The higher Z-score is, the better the alignment.

The Z-score method has the following two drawbacks. First, it takes a lot of extra time to calculate Z-score for a pair of target and template. In order to calculate Z-score for each pair, the target has to be shuffled and threaded many times to the template. This hinders the use of Z-score methods in genome-scale structure prediction. In contrast, machine learning methods require only one-time threading for a given target and template. Secondly, the Z-score is hard to interpret, especially when the scoring function is the weighted sum of various energy items such as mutation score, environmental fitness score, pairwise score, secondary structure score, gap penalty, and score induced from NMR data. For example, when the target is shuffled, shall we shuffle the position specific profile information and the predicted secondary structure type at each sequence residue? If we choose to shuffle the secondary structure, then the shuffled secondary structure arrangement does not look like a protein's structure arrangement as the regular secondary structure types (i.e., α-helix and β-strand) disperse randomly in the target. Otherwise, if we choose to predict the secondary structure again, the whole process will take a very long time.

7. RAPTOR: Optimal Protein Threading by Linear Programming

This section describes several key components of protein threading program RAPTOR: scoring function, a linear programming approach to finding the optimal sequence-template alignment, two machine learning approaches to fold recognition, and statistical significance of prediction results. Some of the work described in this section have been published in several papers ***(12,46,47)***.

7.1. Scoring Function

RAPTOR uses a threading scoring function consisting of mutation score E_m, environmental fitness score E_s, secondary structure compatibility score E_{ss}, pairwise interaction score E_p and gap penalty E_g. The scoring function takes into consideration the evolutionary information of both the target and the template. PSI-BLAST is used to generate a sequence profile for both the template and the target. The sequence profile of the template is represented by a position-specific score matrix (PSSM), where PSSM (i, a) denotes the score of the residue at template position i mutating to residue a. It is defined as the log-odds of the occurring probability of residue a at position i. The sequence profile of the target is represented as a position-specific frequency matrix (PSFM), where

PSFM (j, b) denotes the occurring frequency of residue b at target position j. Let $A(i)$ denote the target position that is aligned by template position i. If the template position i is not aligned to any target position, then $A(i)$ is invalid.

7.1.1. Mutation Score

RAPTOR calculates the mutation score at each template position i using $\sum_a \mathrm{PSFM}[A(i), a] \times \mathrm{PSSM}(i, a)$. The total mutation score of a given alignment A can be calculated by $E_{\mathrm{m}} = \sum_i \sum_a \mathrm{PSFM}[A(i), a] \times \mathrm{PSSM}(i, a)$.

7.1.2. Environmental Fitness Score

RAPTOR uses two local structural features to describe the structural environment env_i at template position i: secondary structure type (ss) and solvent accessibility (sa). RAPTOR uses three types of secondary structure, α-helix, β-sheet, and loop, and three levels of solvent accessibility, buried (inaccessible), intermediate, and accessible. The boundaries between the different solvent accessibility levels are determined by the equal-frequency discretization method. The calculated boundaries are at 7 and 37%. The combination of these two local structure features yields nine different local structural environments. Let $F(\mathrm{env}, a)$ denote the environment fitness potential for a particular combination of amino acid type a and environment descriptor env. $F(\mathrm{env}, a)$ can be calculated as follows.

$$F(\mathrm{env}, a) = F(\mathrm{ss}, \mathrm{sa}, a) = -K_B T \log \frac{N(\mathrm{ss}, \mathrm{sa}, a)}{N_{\mathrm{E}}(\mathrm{ss}, \mathrm{sa}, a)}$$

where K_B is the Boltzmann's constant, T is the temperature, $N(\mathrm{ss}, \mathrm{sa}, a)$ is the number of amino acid type a occurring in secondary structure type *ss* and with solvent accessibility *sa*, and $N_E(\mathrm{ss}, \mathrm{sa}, a)$ is the expected value of $N(\mathrm{ss}, \mathrm{sa}, a)$, calculated as

$$N_{\mathrm{E}}(\mathrm{ss}, \mathrm{sa}, a) = \frac{N(\mathrm{ss}, \mathrm{sa})\, N(a)}{N}$$

where $N(\mathrm{ss}, \mathrm{sa})$ is the number of residues in secondary structure type *ss* and with solvent accessibility *sa*, $N(a)$ is the number of amino acids of type a and N is the number of amino acids. The total fitness score can be calculated as follows.

$$E_{\mathrm{s}} = \sum_i \sum_a \mathrm{PSFM}[A(i), a] \times F(\mathrm{env}_i, a)$$

7.1.3. Pairwise Contact Score

If the two ends of a contact in the template are aligned to target positions j_1 and j_2 respectively, then the pairwise score between j_1 and j_2 can be calculated by $\text{Pair}(j_1, j_2) = \sum_a \left[\text{PSFM}(j_1, a) \times \sum_b \text{PSFM}(j_2, b) P(a, b) \right]$, where $P(a, b)$ denotes the pairwise interaction potential between two residues a and b and is taken from **ref.** ***(38)***.

7.1.4. Secondary Structure Score

Let $SS(i, j)$ denote the difference between the template secondary structure at position i and the predicted target secondary structure at position j. Suppose that at the target position j the predicted confidence scores for α-helix, β-strand, and loop are $x(j)$, $y(j)$, and $z(j)$, respectively. If the secondary structure at template position i is α-helix, then $\text{SS}(i, j)$ is defined as $x(j) - z(j)$. If the secondary structure at template position i is β-strand, then $\text{SS}(i, j)$ is defined as $y(j) - z(j)$. Otherwise it is 0.

7.1.5. Gap Penalty

It is very unlikely that the target and the template have an exact match. Therefore, some gaps should be allowed in order to guarantee the quality of the alignment. However, if there are too many gaps, especially gap openings, in the sequence-structure alignment, then it might indicate that the target does not have a similar structure as the template. RAPTOR uses a gap penalty function $b + ge$ to penalize the number of gap openings and gap extensions. Meanwhile, b is a gap open penalty, e is a gap extension penalty, and g is the gap length.

7.1.6. Contact Capacity Score

Contact capacity score is an optional energy item. Contact capacity potential accounts for the hydrophobic contribution of free energy. Contact capacity characterizes the capability of a residue making a certain number of contacts with any other residues in a single protein. The threading program 123D has explored this feature in its energy function ***(48)***. Let $\text{CC}(a, k)$ denote the potential of amino acid type a having k contacts. It can be calculated by $\text{CC}(a, k) = -\log \frac{N(a,k)}{N(k)N(a)/N}$, where $N(a, k)$ is the number of residues of type a and with k contacts, $N(k)$ is the number of residues having k contacts, $N(a)$ is the number of residues of type a, and N the total number of residues. The total contact capacity score can be calculated by

$E_c = \sum_i \sum_a \text{PSFM}[A(i), a] \times \text{CC}[a, CN(i)]$, where CN (i) denotes the number of contacts at template position i.

7.2. Threading Assumptions and Model

RAPTOR formulates the protein threading problem using the following assumptions *(27,38)*.

1. RAPTOR parses each structural template as a linear series of cores with connecting loops between adjacent cores. Cores are the most conserved regions in a protein. RAPTOR does not allow gaps in core regions as the chance of insertions or deletions within cores is very small.
2. RAPTOR considers only contacts (interactions) between two core residues. Generally, it is believed that the interactions involving the loop residues can be ignored since their contribution to fold recognition is relatively insignificant. An interaction exists between two residues if the spatial distance between their C_α atoms is below 7 Å, and they are at least four positions apart along the primary sequence of the template. An interaction exists between two cores if there is at least one inter-residue contact between them.

RAPTOR uses a contact graph to represent a template structure. A vertex in the graph represents a template residue and an edge represents an inter-residue contact. Based on the assumption that no gaps are allowed within a core, RAPTOR further simplifies the template contact graph by modeling each core as a vertex and adding one edge between two cores if there is at least one inter-residue contact between them (*see* **Fig. 1** for an example of a template contact graph). For simplicity, when we say that core c_i is aligned to target position s_l, we mean that this core is aligned to the segment starting from position s_l. Let $D[i]$ denote all the valid target positions that c_i can be aligned to and $R[i, j, l]$ all the valid alignment positions of c_j given that c_i is aligned to s_l. For any two alignments (c_i, s_l), (c_j, s_k), $k \in R[i, j, l]$, if and only if $l \in R[j, i, k]$.

7.3. Integer Programming Formulation

7.3.1. Introduction to Linear and Integer Programming

Linear programming and integer programming is a subfield of mathematical programming. A mathematical program tries to identify an extreme (i.e., minimum or maximum) point of a function $f(x_1, x_2, \ldots, x_n)$ in a feasible region formed by a set of constraints, for example, $g(x_1, x_2, \ldots, x_n) \geq b$. When both the objective function f and the constraints are linear, this mathematical program becomes a linear program (LP). When $x_1, \ldots, x_n$ are required to be integers, it becomes an integer linear program. Linear programming and integer

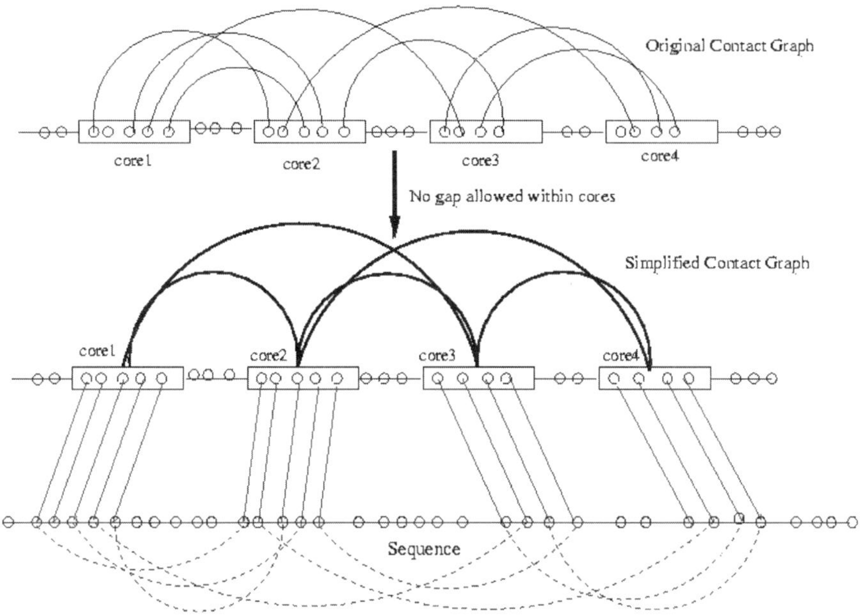

Fig. 1. A template contact graph and a sequence-template alignment. A small circle represents one residue and one solid arc indicates an interaction between two residues. A dashed arc shows that if two target residues are aligned to two template residues with a contact, then the interaction score of these two target residues must be counted in the scoring function. The interaction score between two target segments is the sum of the interaction scores of two target residues which are aligned to two interacting template residues.

programming have been extensively used to solve many optimization problems derived from various application areas such as finance, economics, and planning ***(49,50)***. A general form of a LP with m constraints and n variables is

$$\min \left\{ \boldsymbol{c}^T \boldsymbol{x} : A\boldsymbol{x} \geq \boldsymbol{b}, \boldsymbol{x} \geq 0, \boldsymbol{x} \in R^n, A \in R^{m \times n}, \boldsymbol{b} \in R^m, \boldsymbol{c} \in R^n \right\},$$

where $\boldsymbol{x}$ is the vector of variables, $\boldsymbol{c}$ the cost vector, A the constraint matrix, and b the vector of allowed resources.

LPs can be solved within polynomial time whereas IPs are NP-hard ***(51–53)***. Two major methods for solving an LP are the Simplex method discovered by G. Dantzig in 1947 ***(54)*** and the interior-point method invented by Karmarkar in 1984 ***(55)***. However, it is difficult to say in practice which method is better *(52)*. The current state-of-the-art LP software packages allow us to deal with problems where the constraint matrix A has millions of non-zero elements.

7.3.2. Linear Programming Formulation

RAPTOR employs two kinds of binary variables to formulate the threading problem as a LP. Let $x_{i,l}$ be a binary variable such that $x_{i,l} = 1$ if and only if core c_i is aligned to position s_l. Similarly, for any two interacting cores c_{i_1} andc_{i_2}, let $y_{(i_1,l_1),(i_2,l_2)} = 1$ indicate that core c_{i_1} is aligned to position s_{l_1} and simultaneously core c_{i_1} is aligned to position s_{l_2}. Therefore, $y_{(i_1,l_1),(i_2,l_2)} = 1$ if and only if $x_{i_1,l_1} = 1$ and $x_{i_2,l_2} = 1$. The objective function of the protein threading problem can be formulated as follows.

$$f = W_{\mathrm{m}}E_{\mathrm{m}} + W_{\mathrm{s}}E_{\mathrm{s}} + W_{\mathrm{p}}E_{\mathrm{p}} + W_{\mathrm{g}}E_{\mathrm{g}} + W_{\mathrm{ss}}E_{\mathrm{ss}}, \text{ where}$$

$$E_{\mathrm{m}} = \sum_{i=1}^{M} \sum_{l \in D[i]} \left[x_{i,l} \sum_{r=0}^{len_i - 1} \mathrm{Mutation}(\mathrm{head_i} + r, l + r) \right];$$

$$E_{\mathrm{s}} = \sum_{i=1}^{M} \sum_{l \in D[i]} \left[x_{i,l} \sum_{r=0}^{len_i - 1} \mathrm{Fitness}(\mathrm{head}_i + r, j + r) \right];$$

$$E_{\mathrm{ss}} = \sum_{i=1}^{M} \sum_{l \in D[i]} \left[x_{i,l} \sum_{r=0}^{len_i - 1} \mathrm{SS}(\mathrm{head}_i + r, j + r) \right];$$

$$E_{\mathrm{p}} = \sum_{(c_i, c_j) \in E(G)} \sum_{l \in D[i]} \sum_{k \in R[i,j,l]} y_{(i,l),(j,k)} P(i, j, l, k);$$

$$P(i, j, l, k) = \sum_{u=0}^{len_i - 1} \sum_{v=0}^{len_j - 1} \delta\left(t_{\mathrm{head}_i + u'} t_{\mathrm{head}_j} + v\right) \mathrm{Pair}(l + u, k + v);$$

$$E_{\mathrm{g}} = \sum_{i=1}^{M} \sum_{l \in D[i]} \sum_{k \in R[i,i+1,l]} y_{(i,l),(i+1,k)} \mathrm{Gap}(i, l, k).$$

In this objective function, the singleton score items E_{m}, E_{s}, and E_{ss} are the expansion of the first term of the right-hand side of **Eq. 2**; the pairwise score items E_{p} and E_{g} are the expansion of the second term of this equation; $\mathrm{head_i}$ is the head position of core i; $\delta(t_u, t_v) = 1$ if there is contact between two positions u and v in the template; otherwise it is 0. $\mathrm{Gap}(i, l, k)$ is the alignment score of aligning the template segment between core i and core $i + 1$ to the target segment from position l to k. This score contains gap penalty. As inter-residue contacts involving loop residues are ignored, $\mathrm{Gap}(i, l, k)$ can be computed by a dynamic programming algorithm, when i, ln, and k are given.

Let $G = (V, E)$ denote the simplified template contact graph. The binary variables are subject to the following constraints:

$$\sum_{j \in D[i]} x_{i,j} = 1, i = 1, 2, \ldots, M; \tag{3}$$

$$\sum_{k \in R[i,j,l]} y_{(i,j)(j,k)} = X_{i,l'} \left(c_i, c_j\right) \in E(G); \tag{4}$$

$$\sum_{l \in R[j,i,k]} y_{(i,l)(j,k)} = x_{j,k'} \left(c_i, c_j\right) \in E(G); \tag{5}$$

$$x_{i,j} \in \{0, 1\}; \tag{6}$$

$$y_{(i,l)(j,k)} \in \{0, 1\}. \tag{7}$$

Equation 3 indicates that one core can be aligned to a unique target position. **Equations 4** and **5** imply that one y variable is equal to 1 if and only if both of its two x variables are equal to 1. **Equations 6** and **7** restrict x and y to be either 0 or 1.

To solve the above IPs, we first relax the integral constraints to linear constraints. Then we can solve the LP using either the Simplex method or the interior-point method. Finally, we can obtain the solution to the IP using branch-and-bound. Experimental results show that empirically this formulation can solve 99% of the real-world threading instances within polynomial time.

7.4. Fold Recognition

RAPTOR uses two popular machine learning methods, Support Vector Machines and Gradient Boosting, to conduct fold recognition. We can formulate the fold recognition problem as a classification problem or a regression problem. When formulating the problem as a classification problem, we treat a target-template pair as a positive example if the target and the template are in the same SCOP fold class *(4)* and as a negative example otherwise. When formulating the problem as a regression problem, we want to predict the alignment accuracy of a specific sequence-template alignment based on some features extracted from the alignment. The alignment accuracy is the number of correctly aligned positions by the sequence-template alignment algorithm. We judge if one aligned position is correct or not by comparing this alignment with the alignment generated using a structure alignment program SARF *(56)*.

The regression formulation has some advantages over the classification formulation. The similarity of two proteins can be at fold level, superfamily level, or family level. A single binary classifier cannot effectively differentiate one similarity level from another. According to simple statistics, the more similar two proteins, the better alignment the two proteins can have. Two proteins similar at a family level will have a better alignment than two at a superfamily level, which in turn better than two at a fold level. By using regression method with the alignment accuracy as the objective function, we can differentiate these three similarity levels in a single regression model. The most important problem is that even if a classifier can predict two proteins to be similar, it is possible that the alignment accuracy between them is not good. Instead, what we need is one template with the best alignment to the target. Classification-based methods can only recognize those templates with a similar fold as the target but cannot tell which template has the best alignment to the target. A template with a similar fold as the target cannot guarantee a good alignment to the target. The preferred result is that the better alignment the template has to the target, the better the rank of the template.

The predicted alignment accuracy by RAPTOR has a correlation coefficient 0.71 with the real alignment accuracy ***(46)***. For a given target, the templates can be ranked by the predicted sequence-template alignment accuracy. Experimental results show that the predicted alignment accuracy has a much better sensitivity and specificity than Z-score method and a much better computational efficiency. The regression-based method is also better than the classification-based method in terms of sensitivity.

7.4.1. Experimental Data

In order to train the machine learning models, we randomly chose 300 structures from the FSSP list as templates ***(57,58)*** and 200 proteins as targets from the Holm and Sander's test set ***(58)***. We generated a set of 60,000 training data by threading each of the 200 targets to each of the 300 templates. We also used Fischer et al.'s benchmark ***(59)*** as the test set to fix the parameters in our training models. Finally, we used the Lindahl's benchmark ***(60)*** as the test set to measure the generalization performance of the machine learning models. The Lindahl's benchmark contains 976 proteins, any two of which share at most 40% sequence identity. By threading each one against all the others, we obtain a set of 976×975 threading pairs. As the training set is chosen randomly from a set of non-redundant proteins, the overlap between the training set and Lindahl's benchmark is fairly small, which is no more than 0.4% of the whole test set. To make sure the complete separation of training and test sets, these overlap pairs are excluded from the test data.

7.4.2. Feature Extraction

To use the machine learning methods, RAPTOR extracts the following features from each sequence-template alignment.

1. Target size, which is the number of residues in the target.
2. Template size, which is the number of residues in the template.
3. Alignment length, which is the number of aligned residues. Usually, two proteins from the same fold class should share a large portion of similar substructure. If the alignment length is considerably smaller than their sizes, then it indicates that this alignment is not good.
4. Sequence identity. Although a low sequence identity does not imply that two proteins are not similar, a high sequence identity can indicate that two proteins should be considered as similar *(3)*.
5. Number of contacts with both ends being aligned to the target. There is a contact between two residues if their spatial distance is within a given cutoff. Usually, a larger protein should have more contacts.
6. Number of contacts with only one end being aligned to the target. If this number is big, then it might indicate that the target is aligned to an incomplete domain of the template, which is not good as the target is supposed to fold as an independent unit.
7. Total alignment score.
8. Mutation score, which measures the sequence similarity between the target and the template.
9. Environment fitness score, which measures how well to put a residue into a specific environment.
10. Gap penalty. Some gaps are allowed in the sequence-template alignment.
11. However, if there are too many gaps in the alignment, it might indicate that the quality of the alignment is bad and the target and the template might not have the same fold.
12. Secondary structure compatibility score, which measures the difference between the template secondary structure type and the predicted target secondary structure.
13. Pairwise interaction score, which characterizes the capability of a residue making a contact with another residue.
14. The Z-score of the total alignment score and the Z-score of a single score item such as mutation score, environment fitness score, secondary structure score, and pairwise interaction score. Z-score is only used for the comparison purpose. Our experiments indicate that using Z-score in our machine learning methods does not improve prediction accuracy at all.

7.4.3. Introduction to Support Vector Machines

Support Vector Machines and Kernel methods were developed in the late 1970s by Vapnik *(61)*. However, it is only in recent years that the Kernel methods have been gaining more attentions. The most commonly used SVM is the non-linear SVM. We will start with the simple linear SVM because the non-linear SVM is just a kernelized linear SVM. We briefly introduce linear and non-linear SVM regression ***(62)***. SVM classification can be treated as a special case of SVM regression.

7.4.3.1. Linear SVM Regression

Given a set of training data $\{x_i, y_i\}$, $i = 1, 2, \ldots, l$, $y_i \in R$, and $x_i \in R^m$, we call x_i the input data point, and y_i the observed response given an input x_i. Our goal is to find a function $f(x)$ that has at most ε deviation from the observed response. Suppose that the relationship between x and y is linear. That is, there is a vector $w \in R^m$ such that $f(x) = wx + b$. There might be multiple w satisfying this equation, so we require that w has the smallest Euclidean norm to guarantee a unique w. Therefore, we can write this problem as the following optimization problem:

$$\min \frac{1}{2} ||w||^2$$

subject to

$$y_i - wx_i - b \leq \varepsilon$$
$$wx_i + b - y_i \geq \varepsilon$$

It is almost impossible to guarantee such a $f(x)$ exists. In order to have a feasible solution, we allow for some errors. That is, we introduce slack variable ζ_i and ζ_i^* $(i = 1, 2, \ldots, l)$ to achieve the following optimization problem:

$$\min \frac{1}{2} ||w||^2 + C \sum_{i=1}^{l} (\zeta_i, \zeta_i^*)$$

subject to

$$y_i - wx_i - b \leq \varepsilon + \zeta_i$$
$$wx_i + b - y_i \geq \varepsilon + \zeta_i^*$$
$$\zeta_i, \zeta_i^* \geq 0$$

where C is the penalty factor.

By introducing Lagarangian multiplier λ_i and λ_i^* $(i = 1, 2, \ldots, l)$ for the constraints, we have the following dual problem:

$$\max L_D = -\frac{1}{2}\sum_{i,j=1}^{l}(\lambda_i - \lambda_i^*)(\lambda_j - \lambda_j^*)(x_i x_j) - \varepsilon\sum_{i=1}^{l}(\lambda_i + \lambda_i^*) + \sum_{i=1}^{l} y_i(\lambda_i - \lambda_i^*)$$

subject to

$$\sum_{i=1}^{l}(\lambda_i - \lambda_i^*) = 0$$

$$\lambda_i, \lambda_i^* \in [0, C].$$

After solving λ_i and λ_i^*, we have:

$$f(x) = \sum_{i=1}^{l}(\lambda_i - \lambda_i^*)(x_i x) + b.$$

7.4.3.2. Nonlinear SVM Regression

Now we generalize the linear SVM to accommodate the case where the observed outputs are not a linear function of the input data. A very straightforward idea is to map the data points into a higher dimension space and then do linear regression in the higher dimension space. The only difference lies in that in the objective function L_D, we replace $(x_i x_j)$ with $\phi(x_i)\phi(x_j)$ where ϕ is the mapping function. Theoretically, there is no problem if we know the mapping function ϕ. However, there is a computational challenge if we calculate $\phi(x_i)$ directly, when its dimension is very large, say millions of dimensions or infinite. Notice that in L_D only the products $\phi(x_i)\phi(x_j)$ but not any $\phi(x_i)$ are needed. In order to circumvent this difficulty, the mapping function ϕ is chosen such that the inner product of any two points in the new space can be represented as a function of the original two points. That is, there is a function K such that $\phi(x_i)\phi(x_j) = K(x_i, x_j)$. Then, we do not need to directly calculate $\phi(x_i)$ and $\phi(x_i)\phi(x_j)$ because we only need to compute $K(x_i, x_j)$. Function K is also called a kernel function.

7.4.4. SVM Approach to Fold Recognition

We tried several different kernel functions in training the SVM models and decided that the RBF kernel is the best, no matter whether SVM regression or classification is used. For a given target, all the templates are ranked by the outputs of the SVM models. If SVM regression is used, then the SVM outputs can be interpreted as the predicted alignment accuracy. Otherwise, the SVM outputs indicate the likelihood of the target and the template being structurally similar.

As mentioned before, the ultimate goal is to rank all the sequence-template alignments for a given target such that the first-ranked sequence-template alignment has the best alignment accuracy. In order to compare SVM classification and regression in terms of alignment accuracy, we calculate average alignment accuracy of the first-ranked sequence-template alignments over three different similarity levels: family level, superfamily level, and fold level. The average alignment accuracy on the Lindahl's benchmark test set is listed in **Table 1**. This table clearly indicates that SVM regression can improve the average alignment accuracy by 30% at fold level, 25% at superfamily level, and 10% at family level. It is seen that both SVM classification and regression methods fail to predict good alignments for some targets. Therefore, the average alignment accuracy shown in this table is fairly low.

7.4.5. Introduction to Gradient Boosting Algorithm

Given an input vector variable x, a response variable y, and some samples $\{y_i, x_i\}_{i=1}^{N}$, we want a function $F^*(x)$ that can predict y from x such that over the joint distribution of $\{y, x\}$, the expected value of a particular loss function $L[y, F(x)]$ is minimized ***(63)***. The loss function is used to measure the deviation between the real y value and its predicted value.

$$F^*(x) = \arg\min_{F(x)} E_{y,x} L[y, F(x)] = \arg\min_{F(x)} E_x \left\{ E_y L[y, F(x)] \mid x \right\} \tag{8}$$

Normally, $F(x)$ is a member of a parameterized class of functions $F(x; P)$, where P is a set of parameters. We use the form of the "additive"' expansions to design the function as follows:

$$F(x; P) = \sum_{m=0}^{M} \beta_m h(s; \alpha_m), \tag{9}$$

Table 1
Performance Comparison Between SVM Classification and Regression Methods

Method	Fold level	Superfamily level	Family level
SVM classification	13.8	20.6	49.8
SVM regression	17.3	25.5	55.9

The numbers in the table are average alignment accuracy of the first-ranked alignment produced by the two methods. When the superfamily level is considered, then all the target-template pairs similar in the family level are removed from the test data. Similarly, when the fold level is considered, all the target-template pairs similar in the family level and superfamily level are removed from the test data.

where $P = \{\beta_m, \alpha_m\}_{m=0}^{M}$. The functions $h(x;\alpha)$ are usually simple functions of x with parameter α (The parameter α could be a vector.) The parameters in **Eq. 9** can be trained using greedy approach. That is, $\{\beta_m, \alpha_m\}$ is trained after $\{\beta_i, \alpha_i\}$ $(i = 0,1,\ldots,m-1)$ are trained. Friedman ***(64)*** proposed a Gradient Boosting algorithm to solve the optimization problem described in **Eq. 9**. By employing the least square loss function $[L(y,F)] = (y-F)^2/2$, we have a least-square boosting algorithm. Suppose that $\tilde{y}_i = y_i - F_{m-1}(x_i)$. Then, β_m can be calculated as follows:

$$(\beta_m, \alpha_m) = \arg\min_{\beta,\alpha} \sum_{i=1}^{N} [\tilde{y}_i - \beta h(x_i;\alpha)]^2, \beta_m = N \times \tilde{y}_i \Big/ \sum_{i=1}^{N} h(x_i;\alpha_m). \tag{10}$$

The simple function $h(x,\alpha)$ can have any form that can be conveniently optimized over α. We chose a linear regression function $h(x,\alpha) = ax + b$ for the prediction of alignment accuracy where x represent the features extracted from the alignment and $\alpha = (a,b)$. Using this simple linear function, the parameters a and b can be solved easily by the following equation:

$$\begin{aligned} a &= \frac{l_{xy}}{l_{xx}}, b = \tilde{y} - ax \\ l_{xx} &= n \times \sum_{i=1}^{n} x_i^2 - \left(\sum_{i=1}^{n} x_i\right)^2. \\ l_{xy} &= n \times \sum_{i=1}^{n} x_i\tilde{y}_i - \left(\sum_{i=1}^{n} x_i\right)\left(\sum_{i=1}^{n} \tilde{y}_i\right) \end{aligned} \tag{11}$$

7.4.6. Gradient Boosting Approach to Fold Recognition

Here, we treat the alignment accuracy as the response variable and the features extracted from a sequence-template alignment as the input variables. In training the least-square boosting regression model to predict alignment accuracy from the extracted features, at each round, the boosting training algorithm chooses a single feature and obtain a linear function $h(x,\alpha) = ax + b$ with the minimum least-square error where x represents the chosen feature. The training will terminate if no further improvement on accuracy is achieved. In the end, all the generated linear functions are added to form the final regression function. The underlying reasons of choosing a single feature at each round are (1) we would like to see how important each feature is for fold recognition and (2) we notice that alignment accuracy is proportional to some features. For example, the higher the alignment accuracy, the lower the mutation score, fitness score, and pairwise score. Specifically, we employ the following procedures to train the boosting regression model.

1. Calculate the difference between the real alignment accuracy and the predicted alignment accuracy. We call this difference as alignment accuracy residual. Assume the initial predicted alignment accuracy to be the average alignment accuracy of all the training alignments.
2. Choose a single feature that correlates most with the alignment accuracy residual. The parameters α and β are calculated by using **Eqs 10** and **11**. Then, the alignment accuracy residual is predicted by using this chosen feature and the parameters.
3. Update the predicted alignment accuracy by adding the predicted alignment accuracy residual to previous predicted alignment accuracy. Repeat the above two steps until the predicted alignment accuracy converges.

7.4.7. RAPTOR Fold Recognition Performance

7.4.7.1. Sensitivity

Table 2 compares the performance of several popular machine learning methods. In this experiment, we used RAPTOR to generate all the sequence-template alignments. For each method, we tuned the parameters on the training set and tested the model on the test set. Besides SVM classification, SVM

Table 2
Performance Comparison Among Seven Machine Learning Methods

Methods	Family level (%)		Superfamily level (%)		Fold level (%)	
	Top 1	Top 5	Top 1	Top 5	Top 1	Top 5
Boosting (regression)	86.5	89.2	60.2	74.4	38.8	61.7
SVM (regression)	85.0	89.1	55.4	71.8	38.6	60.6
SVM (classification)	82.6	83.6	45.7	58.8	30.4	52.6
Ada_Boost	82.8	84.1	50.7	61.1	32.2	53.3
Neural network	81.1	83.2	47.4	58.3	30.1	54.8
Bayes classifier	69.9	72.5	29.2	42.6	13.6	40.0
Naive Bayes classifier	68.0	70.8	31.0	41.7	15.1	37.4

These methods use the same set of sequence-template alignments generated by RAPTOR. "Top N" (N = 1, 5) means that the best template among the top N templates output by the machine learning method is used to judge if the prediction is correct or not. If this best template and the target are similar in one level, then we say that this target is predicted correctly in this similarity level. The results are obtained using the Lindahl's benchmark.

regression, and the boosting algorithm, we also tested the following machine learning methods.

1. AdaBoost: The standard AdaBoost algorithm ***(63)*** for classification is similar to the boosting algorithm described in this chapter except that AdaBoost does classification instead of regression and uses the exponential instead of least-squares loss function. The AdaBoost algorithm achieves a comparable result to SVM classification but is worse than the boosting regression algorithm and SVM regression.
2. Neural network: Neural network is one of the most popular methods used in machine learning ***(65)***. We use the Matlab neural network tools to implement this method. The performance of neural network is similar as SVM classification and AdaBoost.
3. Bayesian classifier: A Bayesian classifier is a probability-based classifier which assigns a sample to a class based on its probability of belonging to the class ***(65)***.
4. Naive Bayesian classifier: The Naive Bayesian classifier is similar to the Bayesian classifier except that it assumes that the features of each class are independent, which greatly simplifies computation ***(65)***. We can see both Bayesian classifier and Naive Bayesian classifier obtain a poor performance.

Our experimental results show clearly that (1) the regression-based approaches demonstrate better performance than the classification-based approaches, (2) the boosting regression algorithm performs slightly better than SVM regression and significantly better than the other methods, and (3) the computational efficiency of the boosting regression algorithm is much better than SVM regression, SVM classification, and neural network.

One of the advantages of the boosting algorithm over SVM regression is its ability to identify important features, as at each round the boosting algorithm only chooses a single feature to approximate the alignment accuracy residual. The top five features chosen by the boosting algorithm are sequence identity, total alignment score, fitness score, mutation score, and pairwise interaction score.

It seems surprising that the widely used Z-score is not chosen as one of the most important features. This indicates that the Z-score may not be the most important feature and redundant. To confirm our hypothesis, we re-trained our model using all the features except all the Z-scores. The results show that for the boosting algorithm there is almost no difference between using Z-score as an additional feature and without using it. This means that we can greatly improve the computational efficiency of protein threading without sacrificing accuracy, by completely avoiding the calculation of the expensive Z-score.

7.4.7.2. Specificity

We further examined the specificity of the boosting regression algorithm on the Lindahl's benchmark. All threading pairs are ranked by confidence score (i.e., the predicted alignment accuracy or the classification if a SVM classifier is used). At the superfamily level, the boosting algorithm is consistently better than SVM regression and classification within the whole spectrum of sensitivity. At both the family level and fold level, the boosting algorithm is a little better when the specificity is high whereas worse when the specificity is low. At the family level, the boosting algorithm achieves a sensitivity of 55.0 and 64.0% at 99 and 50% specificities, respectively, whereas SVM regression achieves a sensitivity of 44.2 and 71.3%, and SVM classification achieves a sensitivity of 27.0 and 70.9% respectively. At the superfamily level, the boosting algorithm has a sensitivity of 8.2 and 20.8% at 99 and 50% specificities, respectively. In contrast, SVM regression has a sensitivity of 3.6 and 17.8%, and SVM classification has a sensitivity of 2.0 and 16.1% respectively. At the fold level, there is no big difference among the three methods.

7.5. E Value of RAPTOR

To evaluate the statistical significance of a particular prediction generated by RAPTOR, we developed a method to calculate the E value of a specific sequence-template alignment. Given a sequence-template alignment with n correctly aligned positions, its P value is the probability that by chance a target protein of the same length has an alignment with more than n correctly aligned positions. Given a probability density function (PDF) of the number of aligned positions, the P value of the alignment is the area under the PDF curve from n to the positive infinity.

The alignment is considered to be significant when its P value is sufficiently small. Its E value is the expected number of false alignments that has more than n correctly aligned positions, which is the product of its P value and the size of the template database. With E value, the quality of alignments is more intuitive, without consideration of protein sizes, which is more convenient for users. With E value, by setting some thresholds, it is much easier for users, especially biologists, to understand the outputs of RAPTOR. To evaluate E value of any sequence-template alignment, we should have the distribution of the numbers of correctly aligned positions. A large amount of data is required to empirically approximate this distribution. We used PDB25 to obtain the distributions of the number of correctly aligned positions. PDB25 is a set of approximately 900 proteins, any two of which share no more than 25% sequence identity. Each protein in PDB25 is aligned to each of the other structures in PDB25 using SARF ***(56)***, a protein structure alignment program. Each protein in PDB25 has a distribution of the number of correctly aligned positions.

We discretize protein size into some intervals and calculate a distribution of the number of correctly aligned positions for each interval. The distribution can be represented as a histogram, a normal distribution, or an extreme value distribution (EVD), whichever has a good fit to the empirical distribution. The protein size is divided into small intervals with length 13 such that each interval has at least one PDB25 protein falling in it. Altogether we have 47 intervals, with 47 distributions accordingly. The normal distribution curve fits the data better than the EVD curve. As a result, we only used the histogram and normal distribution to calculate the E value.

The basic procedure of calculating the E value of a new alignment is as follows. First of all, RAPTOR uses a machine learning model to predict the alignment accuracy of the alignment, which is the predicted number of correctly aligned positions in the alignment. Then RAPTOR calculates P value and E value of the alignment based on the distribution associated with the target protein size.

8. Discussions

This chapter presented a general procedure of protein threading for structure prediction including template library construction, target and template representation, threading algorithm, and template selection. In this chapter, we also described some design and implementation details of our protein threading program RAPTOR. By employing the well-studied linear programming technique, RAPTOR can solve the optimal sequence-template alignment problem very efficiently. By employing the advanced machine learning algorithms, RAPTOR can avoid calculating Z-score without losing prediction accuracy.

One of the key issues with threading technique is that currently threading scoring function is residue based. By ignoring the atomic details, we can do threading quickly. However, an atom-level scoring function usually is more sensitive than a residue-level energy function. Current protein structure prediction program often does threading and side-chain packing in two separate steps. It might improve prediction accuracy if we can make use of an atom-level scoring function and conduct threading and side-chain packing simultaneously. Another issue with protein threading is that when the similarity between the target and the template is low, the minimized threading energy function does not correspond to the best alignment between the target and the template. A better way is to generate multiple sequence-template alignments with similar scores instead of a single alignment. Then, we can construct a complete structural model based on each alignment and use an atom-level energy function to choose the best structural model.

When the similarity between the target and the template is low, protein threading cannot generate a very good structural model for the target regardless of the threading protocol. Only a partial structure of the target can be predicted well. To generate a good complete structural model for the target, we need to conduct structural refinement after protein threading. As demonstrated in Zhang and Skolnick paper ***(66)***, the structural models generated by threading can be significantly improved if a proper refinement procedure is applied. Another possible solution is to conduct protein threading and fragment assembly simultaneously. Instead of aligning the whole target to the template, we only align a partial target to the template and then employ the fragment assembly technique ***(67)*** to generate a conformation for the unaligned regions. This method enables us to evaluate the quality of the complete structural model directly.

References

1. Kihara, D. and J. Skolnick, The PDB is a covering set of small protein structures. J Mol Biol, 2003. 334(4): p. 793–802.
2. Zhang, Y. and J. Skolnick, The protein structure prediction problem could be solved using the current PDB library. Proc Natl Acad Sci USA, 2005. 102(4): p. 1029–1034.
3. Rost, B., Twilight zone of protein sequence alignments. Protein Eng, 1999. 12: p. 85–94.
4. Murzin, A.G., et al., SCOP: a structural classification of proteins database for the investigation of sequences and structures. J Mol Biol, 1995. 247: p. 536–540.
5. Altschul, S.F., et al., Gapped BLAST and PSI-BLAST: a new generation of protein database search programs. Nucleic Acids Res, 1997. 25: p. 3389–3402.
6. Higgins, D., et al., CLUSTALW: improving the sensitivity of progressive multiple sequence alignment through sequence weighting, position-specific gap penalties and weight matrix choice. Nucleic Acids Res, 1994. 22: p. 4673–4680.
7. Jones, D.T., Protein secondary structure prediction based on position-specific scoring matrices. J Mol Biol, 1999. 292: p. 195–202.
8. Rost, B., C. Sander, and R. Schneider, PHD–an automatic mail server for protein secondary structure prediction. Comput Appl Biosci, 1994. 10(1): p. 53–60.
9. Shi, J., L.B. Tom, and M. Kenji, FUGUE: sequence-structure homology recognition using environment-specific substitution tables and structure-dependent gap penalties. J Mol Biol, 2001. 310: p. 243–257.
10. Kelley, L.A., R.M. MacCallum, and M.J. Sternberg, Enhanced genome annotation using structural profiles in the program 3D-PSSM. J Mol Biol, 2000. 299(2): p. 499–520.
11. Kim, D., et al., PROSPECT II: protein structure prediction method for genome-scale applications. Bioinformatics, 2003. 16(9): p. 641–650.

12. Xu, J., et al., RAPTOR: optimal protein threading by linear programming. J Bioinform Comput Biol, 2003. 1(1): p. 95–9117.
13. Jones, D.T., GenTHREADER: an efficient and reliable protein fold recognition method for genomic sequences. J Mol Biol, 1999. 287: p. 797–815.
14. Fischer, D. Hybrid fold recognition: combining sequence derived properties with evolutionary Information. Proceedings of the 2000 Pacific Symposium Biocomputing. 2000, World.
15. Rychlewski, L., et al., Comparison of sequence profiles. Strategies for structural predictions using sequence information. Protein Sci, 2000. 9(2): p. 232–241.
16. Karplus, K., et al., Combining local-structure, fold-recognition, and new fold methods for protein structure prediction. Proteins, 2003. 53 Suppl 6: p. 491–496.
17. Rost, B., R. Schneider, and C. Sander, Protein fold recognition by prediction-based threading. J Mol Biol, 1997. 270(3): p. 471–480.
18. Kim, D., et al., PROSPECT II: protein structure prediction method for genome-scale applications. Bioinformatics, 2002. 16(9): p. 10.
19. Al-Lazikani, B., F. Sheinerman, and B. Honig, Combining multiple structure and sequence alignments to improve sequence detection and alignment: application to the SH2 domains of Janus kinases. Proc Natl Acad Sci, 2001. 98(26): p. 14796–14801.
20. H. Zhou and Y. Zhou, "SPARKS 2 and SP3 servers in CASP 6.", Proteins, S7, p. 152–156, 2005.
21. Kabsch, W. and C. Sander, Dictionary of protein secondary structure: protein recognition of hydrogen-bonded and geometrical features. Biopolymers, 1983. 22: p. 2577–2637.
22. Xu, Y. and D. Xu, Protein threading using PROSPECT: design and evaluation. Proteins, 2000. 40: p. 343–354.
23. Singh, R. K., A. Tropsha, and Vaisman, II, Delaunay tessellation of proteins: four body nearest-neighbor propensities of amino acid residues. J Comput Biol, 1996. 3(2): p. 213–221.
24. Skolnick, J. and D. Kihara, Defrosting the frozen approximation: PROSPECTOR–a new approach to threading. Proteins, 2001. 42(3): p. 319–331.
25. Zheng, W., et al., A new approach to protein fold recognition based on Delaunay tessellation of protein structure, Pacific Symposium in Biocomputing. 1997. p. 486–497.
26. McConkey, B.J., V. Sobolev, and M. Edelman, Quantification of protein surfaces, volumes and atom-atom contacts using a constrained Voronoi procedure. Bioinformatics, 2002. 18(10): p. 1365–1373.
27. Madej, T., J. F. Gibrat, and S. H. Bryant, Threading a database of protein cores. Proteins, 1995. 23.
28. Lathrop, R., et al., A Bayes-optimal probability theory that unifies protein sequence-structure recognition and alignment. Bull Math Biol, 1998. 60: p. 1039–1071.

29. Zhang, C., et al., An accurate, residue-level, pair potential of mean force for folding and binding based on the distance-scaled, ideal-gas reference state. Protein Sci, 2004. 13(2): p. 400–411.
30. Meller, J. and R. Elber, Linear programming optimization and a double statistical filter for protein threading protocols. Proteins, 2001. 45(3): p. 241–261.
31. Lathrop, R.H., The protein threading problem with sequence amino acid interaction preferences is NP-complete. Protein Eng, 1994. 7: p. 1059–1068.
32. Akutsu, T. and S. Miyano, On the approximation of protein threading. Theor Comput Sci, 1999. 210: p. 261–275.
33. Bryant, S. H. and C. E. Lawrence, An empirical energy function for threading protein sequence through folding motif. Proteins, 1993. 16: p. 92–112.
34. Jones, D. T., W. R. Taylor, and J. M. Thornton, A new approach to protein fold recognition. Nature, 1992. 358: p. 86–98.
35. Lathrop, R. H. and T. F. Smith. A branch-and-bound algorithm for optimal protein threading with pairwise (contact potential) amino acid interactions. Proceedings of the 27th Hawaii International Conference on System Sciences. 1994: IEEE.
36. Lathrop, R. H. and T. F. Smith, Global optimum protein threading with gapped alignment and empirical pair score functions. J Mol Biol, 1996. 255: p. 641–665.
37. Lathrop, R. H., An anytime local-to-global optimization algorithm for protein threading in theta (m2n2) space. J Comput Biol, 1999. 6(3–4): p. 405–418.
38. Xu, Y., D. Xu, and E. C. Uberbacher, An efficient computational method for globally optimal threadings. J Comput Biol, 1998. 5(3): p. 597–614.
39. Xu, J., F. Jiao, and B. Berger, A tree-decomposition approach to protein structure prediction. Proc IEEE Comput Syst Bioinform Conf, 2005. p. 247–256.
40. Godzik, A., A. Kolinski, and J. Skolnick, Topology fingerprint approach to the inverse protein folding problem. J Mol Biol, 1992. 227(1): p. 227–238.
41. Thiele, R., R. Zimmer, and T. Lenguaer, Protein threading by recursive dynamic programming. J Mol Biol, 1999. 290: p. 757–779.
42. S. Balev, Solving the Protein Threading Problem by Lagrangian Relaxation, In Proceedings of 6th Workshop on Algorithms in Bioinformatics (WABI 2004), LNBI 3240, p. 182–193, 2004.
43. Bryant, S. H. and S. F. Altschul, Statistics of sequence-structure threading. Curr Opin Struct Biol, 1995. 5: p. 236–244.
44. Xu, Y., D. Xu, and V. Olman, A practical method for interpretation of threading scores: an application of neural networks. Statistica Sinica Special Issue on Bioinformatics, 2002. 12: p. 159–177.
45. Wallner, B. and A. Elofsson, Can correct protein models be identified? Protein Sci, 2003. 12(5): p. 1073–1086.
46. Xu, J., Fold recognition by predicted alignment accuracy. IEEE/ACM Trans Comput Biol Bioinformatics, 2005. 2(2): p. 157–165.
47. Xu, J., et al., Protein threading by linear programming, Pacific Symposium in Biocomputing. 2003. p. 264–275.

48. Alexandrov, N. N., R. Nussinov, and R. M. Zimmer, Fast protein fold recognition via sequence to structure alignment and contact capacity potentials, Pacific Symposium in Biocomputing. 1996. Hawaii, USA. p. 53–72.
49. Shepp, L., Linear Programming in Tomography, Probability and Finance, DIMACS TR97-67, 1997, Rutgers University, NJ, USA.
50. Dorfman, R., P.A. Samuelson and R.M. Solow, Linear Programming and Economic Analysis, 1987, Mc-Graw Hill Co., New York.
51. Schrijver, A., Theory of Linear and Integer Programming. 1998. John Wiley & Sons, New York.
52. Beasley, J.E., Advances in Linear and Integer Programming. 1996. Oxford University Press, University of Oxford, United Kingdom.
53. Vanderbei, R. J., Integer Programming. 2001. Springer, New York. p. 307–313.
54. Dantzig, G.B., Linear Programming and Extensions. 1963. Princeton University Press, Princeton, N. J.
55. Karmarkar, N., A new polynomial-time algorithm for linear programming. Combinatorica, 1984. 4: p. 373–395.
56. Alexandrov, N. N., SARFing the PDB. Protein Eng, 1996. 9: p. 727–732.
57. Holm, L. and C. Sander, Mapping the protein universe. Science, 1996. 273: p. 595–602.
58. Holm, L. and C. Sander. Decision support system for the evolutionary classification of protein structures. Proceedings of the Fifth International Conference on Intelligent Systems for Molecular Biology. 1997.
59. Fischer, D., et al. Assessing the performance of fold recognition methods by means of a comprehensive benchmark. Proceedings of the 1996 Pacific Symposium on Biocomputing. 1996. World.
60. Lindahl, E. and A. Elofsson, Identification of related proteins on family, superfamily and fold level. J Mol Biol, 2000. 295: p. 613–625.
61. Vapnik, V. N., The Nature of Statistical Learning Theory. 1995. Springer, New York.
62. Burges, C. J. C., A tutorial on support vector machines for pattern recognition. Data Mining and Knowledge Discovery, 1998. 2(2), 121–167.
63. Freund, Y. and R. E. Schapire. A decision-theoretic generalization of on-line learning and an application to boosting. European Conference on Computational Learning Theory. 1995.
64. Friedman, J. H., Greedy function approximation: a gradient boosting machine. Annals of Statistics, 2001. 29(5), 1189–1232.
65. D. Michie, D.J. Spiegelhalter, C.C. Taylor, Machine Learning, Neural and Statistical Classification (edit collection). 1994. Elllis Horwood, London.
66. Zhang, Y. and J. Skolnick, Automated structure prediction of weakly homologous proteins on a genomic scale. Proc Natl Acad Sci, 2004. 101(20): p. 7594–7599.
67. Simons, K., et al., Ab initio protein structure prediction of CASP III targets using ROSETTA. Proteins, 1999. S3: p. 171–176.

III

Structure Alignment and Indexing

5

Algorithms for Multiple Protein Structure Alignment and Structure-Derived Multiple Sequence Alignment

Maxim Shatsky, Ruth Nussinov, and Haim J. Wolfson

Summary

Primary amino acid content and the geometry of the folded protein 3D structure are major parameters of protein function. During the course of evolution the protein 3D structure is more preserved than its primary sequence. Thus, analysis of protein structures is expected to lead to a deep insight into protein function. Recognition of a structural core common to a set of protein structures serves as a basic tool for the studies of protein evolution and classification, analysis of similar structural motifs and functional binding sites, and for homology modeling and threading.

In this chapter, we discuss several biologically related computational aspects of the multiple structure alignment and propose a method that provides solutions to these problems. Finally, we address the problem of structure-based multiple sequence alignment and propose an optimization method that unifies primary sequence and 3D structure information.

Key Words: Multiple structure alignment; partial alignment; structure base sequence alignment; structure-sequence conservation.

1. Introduction

The increasing number of determined protein structures opens new horizons for studies of protein function. There are numerous examples of similar functioning proteins, for example, isomerases, cytokines, myoglobins, immunoglobulins, and transferases, with similar 3D structure but less than 25% sequence identity. Therefore, in order to study relationships between such proteins sequence analysis alone is not sufficient. While methods for sequence analysis have significantly advanced in the past years, methods for structural

From: *Methods in Molecular Biology, vol. 413: Protein Structure Prediction, Second Edition*
Edited by: M. Zaki and C. Bystroff © Humana Press Inc., Totowa, NJ

analysis are still at an earlier, exploratory stage. Here, we address one of the most basic structure-related problems, the problem of multiple structure alignment and structure-derived multiple sequence alignment.

A number of methods have been proposed to solve the problem of structural alignment between a pair of proteins, for example, VAST *(1)*, Geometric Hashing (GH) *(2)*, CE *(3)*, DALI *(4)*, FlexProt *(5)*, and others *(6)*. Obviously, multiple structure alignment can provide much more information. Recognition of a structural core common to a set of protein structures has many applications in the studies of protein evolution and classification *(4,7)*, analysis of similar functional binding sites and protein–protein interfaces *(8–10)*, homology modeling and threading *(11,12)*, and so on. However, despite this need, the multiple structure alignment problem has not been extensively studied, and, consequently, there are very few available methods that solve this task.

Let us formulate a list of some principal requirements for a multiple structure alignment method. Subjects like protein structure representation and structural similarity scoring functions are extensively discussed in several reviews *(6,13,14)*; therefore, our aim is to emphasize the topics that are specifically related to the multiple alignment problem and topics that were paid less attention in previous reviews and they are the following:

- Partial alignment.
- Subset alignment.
- Flexible alignment.
- Sequential alignment.
- Sequence order independent alignment.
- Time efficiency.

1.1. Partial Alignment

There might be only a sub-structure (motif and domain) that is similar between a set of molecules, for example, in a set of multi-domain proteins having one or several common domains. In case of protein domain swapping *(15)*, a protein chain of a monomer can be partially aligned with two protein chains of a dimer. Another example is alignment between multi-protein complexes having some structurally similar combination of molecules. Thus, a detection of all common motifs, domains, or multi-protein combinations may be required for a multiple structure alignment method. We consider a *local* alignment as a special case of *partial* alignment. For example, a partial alignment may consist of several locally matched structural elements that can be aligned under the same Euclidean transformation.

1.2. Subset Alignment

An important aspect of any multiple, sequence, or structure alignment is a detection of a subset of molecules that are more similar than the whole input set. For example, consider an input set of 10 proteins from one family and 5 proteins from another family. Assume that the proteins in each family are structurally similar, but there is little similarity between any two proteins from the first and second family. A multiple alignment between these 15 molecules would probably detect at most one common secondary structure element. Therefore, it is very important for a multiple alignment method to be able to automatically distinguish between two such subsets.

To demonstrate the significance of the *partial* and *subset* alignment ability, consider a schematic example in **Fig. 1A**. Three proteins share a common small pattern, and each pair of the proteins share additional, larger, patterns. The

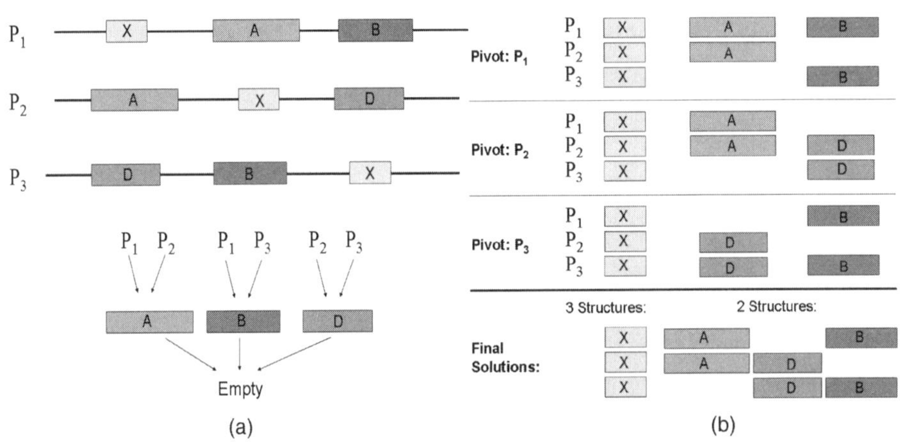

Fig. 1. (**A**) A schematic example of three proteins that share a common pattern X. Applying a pairwise alignment method that detects the most similar common pattern will result in pattern A for proteins P_1 and P_2, pattern B for P_1 and P_3, and pattern D for P_2 and P_3. Therefore, no common pattern can be derived from the patterns A, B, and D. One possible solution is to store two (or more) high scoring solutions for each pairwise comparison. However, in this case, the number of iterations to compare all pairwise results to detect the best combination of multiple alignments becomes exponential. (**B**) The MultiProt method aims to compute a large number of different local multiple alignments. The depicted alignments are detected while selecting each structure as a pivot, for example, patterns X, A, and B are detected when protein P_1 is selected as a pivot. Finally, pattern X will appear in the multiple alignment of three molecules. Patterns A, B, and D will appear in the set of alignments consisting of two molecules.

desired goal of a multiple structure alignment method is to detect all four patterns. An additional example with real protein structures is given below in **Subheading 4.3**. It should be clear that the number of all possible solutions that may be also biologically meaningful could be exponential in the number of input molecules. For example, consider proteins that contain a large number of α-helices. Each pair of α-helices could be structurally matched (at least partially, if they are different in their lengths). Any multiple combination of α-helices from different proteins results in some multiple alignment. Obviously, the number of such multiple alignments is exponential. Therefore, even if an algorithm is capable of detecting all such combinations, it is not practical to report them.

1.3. Flexible Alignment

Proteins are flexible molecules, which may appear in different conformations. Hinge motion may divide a protein structure into several almost rigid parts, which move one relative to the other. In such a case, a *rigid* structural alignment may detect only separate *partially* matched regions. Therefore, only a manual inspection of the final solution may reveal the whole picture as in the example given below in **Subheading 4.3**. Obviously, a method that is able to automatically detect a multiple flexible alignment is more beneficial.

1.4. Sequential and Sequence Order Independent Alignment

Sequence alignment methods naturally produce alignments that follow the protein sequence order, that is, aligned amino acids indices are always in increasing order. However, protein evolution imposes less constraints on the sequential order than on the structural properties. Consequently, proteins may have a similar function but topologically different 3D structure. One such example is the calcium/phospholipid-binding domain (CaLB, C2 domain) which consists of a β-sandwich (eight strands in two sheets). The proteins synaptogamin I (pdb:1rsy) and cytosolic phospholipase A2 (pdb:1rlw) both have the C2 domain but with different topology *(16)*. The phenomenon of similar secondary structure arrangements but with different topology has been the case of a previous study [*(17)* and references therein]. Therefore, in order to fully discover the structural similarities, sequence order independent alignments should be considered. In **Subheading 4.3**., we consider one such example.

1.5. Time Efficiency

Optimal pairwise structural alignment can be solved in polynomial time; however, it is still computationally expensive and currently not practical for an implementation *(18)*. Approximation techniques can significantly reduce time complexity with relatively small degradation in solution accuracy ***(19,20)***. However, even for three structures the multiple alignment problem is NP-hard ***(21)*** (i.e., practically solvable only in exponential time). While the worst case scenario may be computationally infeasible for the detection of an exact solution, considering specific geometrical properties of the protein molecules can, in practice, significantly reduce the computational cost. Examples of such properties, which are far from resembling a random point distribution, include sequentiality of the protein backbone, secondary structure element composition, and protein compactness. Therefore, "smart" heuristic methods that utilize such properties, in practice, may give results that are sufficient for a biological research. An excellent example of a heuristic method for multiple sequence alignment is MUSCLE ***(22)***. Still further research, theoretical and practical, is required for the multiple structural alignment problem.

Below we briefly review available methods for the multiple structure alignment task and try to correlate them with the list of requirements defined above.

A *center-star* approach is one of the efficient ways to compute a multiple sequence alignment. Analogously, it can be applied for multiple structure alignment. A *center* structure is selected which is most similar to the rest of the molecules. Then, iteratively, all other structures are joined into a multiple alignment based on their *pairwise* alignments with the *center* structure ***(11,23)***. Alternatively, one can apply a *tree-progressive* approach, where a multiple alignment is created according to some distance tree ***(24,25)***. Therefore, a *tree-progressive* alignment first aligns similar proteins, then proceeds to more distant relationships. An advantage of such an approach is its ability to detect sub-set alignments of structurally different families.

In order to tackle the flexible alignment problem, the POSA method ***(26)*** utilizes a partial order graph representation of multiple alignments. The advantage of this method is in automatic detection of larger structurally similar regions that cannot be detected without considering hinge motions of the protein backbone. The multiple alignments are computed from pairwise alignments using a *tree-progressive* approach.

The *center-star* and the *tree-progressive* approaches are essentially based on some pairwise alignment method that is iteratively applied for the construction of a multiple alignment. Therefore, such technique is less suitable for detection of small structurally similar motifs as at each stage of the iterative alignment only one, the best, solution is selected. **Figure 1A** shows a simple example where a straightforward application of a pairwise alignment method will fail to recognize a pattern common to more than two sequences/structures.

The MALECON *(27)* and MUSTANG *(28)* methods aim to avoid the shortcomings of the iterative pairwise approaches using two different techniques. MALECON *(27)* considers all possible combinations of the input molecules. When the number of input proteins is large, such a combinatorial approach becomes exponential; therefore, the method considers at least all possible protein triplets while other proteins are progressively added to the aligned triplets. MUSTANG *(28)* uses a *tree-progressive* approach; however, it reduces possible artifacts of the iterative multiple alignment by applying a refinement of the residue correspondence scores based on transitive relations between the aligned structure pairs. Consequently, the advantage of both methods is in detection of subset alignments. However, because only one solution is considered for any given combination of proteins, some smaller local alignments can be missed. Both methods produce sequential alignments.

The *MUSTA* algorithm *(29,30)* computes a common geometric core that appears simultaneously in all the input molecules, thus avoiding the shortcomings of the iterative pairwise approaches. The method applies the *Geometric Hashing* technique *(27)* which allows detection of the sequence order independent alignments. This technique was successfully applied in a number of pairwise structure alignment methods *(32,33)*. Because the method requires that all input molecules participate in the multiple structural alignment, the drawback of this method is inability to distinguish outliers. It is sufficient that one structure is very distinct from the others to result in an empty alignment. Consequently, this method cannot detect subset alignments. Second, its efficiency limits practical application for only 10–15 molecules.

Another approach, SPratt2 *(34)*, aims to detect small common, local structural motifs of size 3–20 amino acids. The method describes each residue as a short string of its spatial neighbors. Then, an efficient sequence pattern discovery technique is applied to detect sets of residues with common environmental descriptors. The computed alignments are sequential. The method is efficient and allows subset alignments.

The MASS *(16,35)* method utilizes the secondary structure information (SSE) to reduce the computational cost of initial common core detection. Therefore, it requires that at least two pairs of SSE be multiply aligned. The method is capable of detecting partial, subset, sequential, and non-sequential alignments.

In order to produce structure-based multiple sequence alignment, the recently developed method 3DCoffee *(36)* incorporates spatial weights into the multiple sequence alignment method TCoffee. The spatial weight of an amino acid pair is defined as a positive large constant number, when this pair is structurally aligned according to some *pairwise* structural alignment method. Therefore, the method does not distinguish between amino acids that are structurally aligned at different distances. Because the method applies information only from the best pairwise structure alignment, the weighting decision may be inaccurate for the multiple alignment problem.

Here, we discuss a method that aims to solve the multiple structural alignment problem with the support of the most above-defined requirements. One of the advantages of our method is its ability of subset and partial alignment. This is illustrated in **Fig. 1A** and **B**. Consider the set of proteins from **Fig. 1A**. The goal of our method is to detect local multiple alignments of all four patterns. This is achieved by performing all possible local multiple alignments of ungapped fragments. The final solutions are constructed from these locally aligned multiple fragments. This makes our approach different from most existing methods, which generally derive a multiple alignment from the high scoring pairwise superpositions. If a pattern appears more than once in some protein, our method recognizes only one combination of this pattern from all possible appearances; however, all sets of possible combinations are reported by the program. Our method is extremely efficient and is suitable for simultaneous comparison of up to tens of proteins.

Below, we start with the brief description of the multiple structural alignment method, MultiProt *(37)*. Then, we discuss the problem of structure-based multiple sequence alignment and propose an optimization method that unifies primary sequence and 3D structure information. Finally, we present some experimental results that include comparison with the HOMSTRAD *(38)* benchmark of manually curated multiple structure-based sequence alignments. We argue that our automated approach produces slightly more accurate alignments.

2. MultiProt—an Algorithm for Multiple Protein Structure Alignment

The input is k protein structures, $\{P_i\}_{i=1}^{k}$, each represented as a sequence of the centers of the C_α atoms. In addition, the input contains a parameter ε, which is the distance threshold between the matched C_α atoms. The goal of the algorithm is to compute, for each $r = 2, \ldots, k$, the largest multiple alignments consisting of exactly r structures. Practically, the number of multiple alignment solutions computed for each r is a user-defined parameter.

Here, we briefly explain the main idea of the method *(37)*. First, we pick a *pivot* structure and require that it is included in all multiple alignments. In order to prevent dependency on a *pivot* structure, all input structures are iteratively selected to be a *pivot* one. We call two sequential (without gaps) fragments of the same length to be ε-congruent if there exists a Euclidean 3D transformation that superimposes both fragments with root mean square deviation (rmsd) less than ε.

The MultiProt algorithm consists of three major stages. In the first stage, all ε-congruent fragment pairs are efficiently detected between the *pivot* and all other structures. Secondly, we compute all possible combinations of ε-congruent multiple (sub)-fragments. This stage is analogous to the detection of all non-gaped local multiple alignments. To prevent an exponential number of multiple local alignments, we do not compute them explicitly but rather store all possible alignments by means of combination sets. Such set consists of one fragment from the *pivot* structure and its ε-congruent fragments from other structures; therefore, such set may include several fragments from some molecule.Thirdly, for each local multiple alignment set, we heuristically select [the problem of selecting the optimal combination is NP-hard ***(39)***] a combination of fragments, one fragment from each structure. Once a unique combination is selected, we compute a global multiple correspondence between the C_α atoms. At this stage, we have a choice (user defined parameter) whether to compute a sequential alignment or a non-sequential one.

The main idea of the MultiProt approach is its ability to efficiently compute a large number of local non-gapped multiple structure alignments. Essentially, a local multiple alignment is computed for each possible fragment of the input molecules. Such local alignments serve as a basis for the extension to the larger partial multiple alignments. In addition, in order to detect subset alignments, the solutions are scored separately according to protein composition, that is, a scoring of alignment between proteins $\{a, b, c\}$ does not effect a ranking of an alignment between $\{a, b, d\}$ (the application for this requirement is demonstrated in **Subheading 4.3**).

To compute a significance of a multiple structural alignment, we apply a simple estimation by means of p-value. Naturally, the p-value depends on the number of input proteins, k, and their sizes. Therefore, we computed the multiple alignment size distribution for different values of k (practically only for $k = 2, \ldots, 10$). We selected a representative set of 5674 protein structures from the SCOP database (1.65) ***(40)*** which have less than 40% of pairwise sequence identity [this data set is provided by ASTRAL ***(41)***]. This resulted in 2304 protein domains (according to the SCOP classification). For each domain we arbitrarily selected only one structure. For each k, the number of structures, we applied MultiProt on k randomly selected structural domains. In total, we performed 10,000 such random alignments for each k. We computed these distributions separately for sequential and non-sequential alignments. Clearly, larger structures will likely produce larger alignments. Therefore, given some multiple alignment size and the minimal structure size, s_{min}, of the aligned molecules, we estimate its significance only from distributions of multiple alignments with minimal molecule size within 20% of s_{min}.

MultiProt is time efficient. On a standard PC with Pentium(R) 4, 2.00 GHz, on proteins with average size of 179 amino acids (from the above data set), the average running time for 2, ..., 10 structures is 0.5 s, 4 s, 10.5 s, 21 s, 38 s, 1 min, 1 min 30 s, 2 min 10 s and 3 min 3 s, respectively.

3. Structure-Derived Multiple Sequence Alignment

Sequence alignment methods may produce inaccurate alignments due to low sequence identity. For proteins with solved 3D structure, a structural superposition provides a basis for a more robust assessment of evolutionary relationships between amino acids. Yet, a structural 3D superposition does not uniquely define an alignment between protein sequences. Consider the following scenario, where we are given a multiple structure superposition between three proteins $\{a_i\},\{b_j\}$, and $\{c_k\}$. Assume that the following groups of amino acids have been superimposed close in 3D space: (a_1, b_1), (a_2, b_2, c_1), (a_3, b_3, c_2), and (a_3, b_3, c_3). Therefore, there can be several multiple sequence alignments that are consistent with the structural superposition, for example, the following three combinations (for the purpose of sequence alignment we assume that all sequence and structural alignments are according to protein sequence order):

I.				II.				III.			
a_1	a_2	a_3	–	a_1	a_2	–	a_3	a_1	a_2	–	a_3
b_1	b_2	b_3	–	b_1	b_2	b_3	–	b_1	b_2	–	b_3
–	c_1	c_2	c_3	–	c_1	c_2	c_3	–	c_1	c_2	c_3

Which alignment is preferable? All three satisfy the geometrical constraints. Obviously, in this case, we would prefer an alignment with less gaps that also places more similar amino acid types, according to some substitution matrix, in the same column. Therefore, we face an optimization problem that is similar to the multiple sequence alignment problem but has additional spatial constraints. Here, we propose to perform a multiple sequence alignment that unifies structural information, derived from a multiple structure alignment, with amino acid substitution matrices ***(42)***. Because protein structure is generally more conserved than sequence, we propose to perform an optimization of the multiple alignment, first, according to structure and then according to amino acid types combined with 3D information. Namely, we propose the following scheme: Given a set of protein structures, first, perform a multiple structural alignment [e.g., apply the MultiProt *(37)* method]. Second, based on the multiple structure superposition, perform a multiple sequence alignment optimizing a sequence structure unified scoring function.

The scoring function is the likelihood of amino acid a to substitute b assuming that $a(b)$ is located in a secondary structure of type SSE(a), SSE(b), and the 3D distance, according to a multiple structural alignment, between C_α atoms of a and b is $d\,(a, b)$. SSE(a) is either a helix, a strand or unspecified.

$$\text{LikelihoodRatio}\,(a,b) = \frac{P\,(a,b)\,P_{3D}[d\,(a,b)\,|\text{SSE}\,(a), \text{SSE}\,(b)]}{P'(a,b)P'_{3D}\,[d\,(a,b)\,|\text{SSE}\,(a), \text{SSE}\,(b)]},$$

where $P(a,b)/P'\,(a,b)$ is a commonly used likelihood ratio of amino acid substitutions, $P\,(a,b)$ is the probability for a to substitute b, and $P'\,(a,b)$ is the randomly expected probability [our default values are taken from the Blosum62 matrix ***(43)***], $P_{3D}\,[d\,(a,b)]$ is the observed probability of distance $d\,(a,b)$ in a set of structural alignments of closely related proteins (conditioned by the type of secondary structures), and $P'_{3D}\,[d\,(a,b)]$ is the randomly expected probability of 3D distances ***(42)***.

Finally, the score for the *3D substitution* matrix is defined as log-odds,

$$\text{Score}\,(a,b) = 2\log_2\left[\frac{P\,(a,b)}{P'\,(a,b)}\right] + 2\log_2\left\{\frac{P_{3D}\,[d\,(a,b)\,|\text{SSE}\,(a), \text{SSE}\,(b)]}{P_{3D}{}'\,[d\,(a,b)\,|\text{SSE}\,(a), \text{SSE}\,(b)]}\right\}.$$

Therefore, the values of the first term are taken from a standard substitution matrix and the newly computed values of the second term are given in **Table 1**.

Table 1
3D Substitution Scores

Distance	[0,1]	[1,2]	[2,3]	[3,4]	[4,5]	[5,6]	[6,7]	[7,8]	[8,9]	[9, ∞]
H H	5.23	4.23	3.82	2.62	−0.19	−1.93	−3.06	−3.37	−3.28	−3.13
H S	−5.01	−5.45	−4.90	−5.23	−5.79	−6.26	−6.40	−6.17	−6.09	−5.94
H U	2.36	1.07	0.92	0.15	−1.17	−2.26	−3.15	−3.58	−3.98	−4.23
S S	9.09	5.80	3.99	3.11	−2.07	−3.67	−4.33	−4.61	−4.97	−4.11
S U	4.73	2.37	1.63	0.81	−2.28	−3.54	−4.26	−4.56	−4.56	−4.72
U U	8.66	5.26	3.97	3.04	1.19	−0.30	−1.58	−2.42	−3.21	−3.63

[a] H, helix, S, strand, U, undefined.

These scores are applied in the multiple sequence alignment algorithm. To solve the multiple alignment, we apply an iterative profile-profile alignment procedure. First, each structure is initialized as a singleton profile. At each step of iteration, the two most similar profiles are joined into one until only one profile, which includes all the input structures, is left.

Optionally, the method allows to produce *distance constrained alignments*. This is achieved by requiring that all pairwise distances between the C_α atoms of amino acids from the same column are less than some predefined parameter. Such requirement is trivially incorporated into the scoring function of the profile-profile alignment procedure. The constrained multiple alignments allow identification and clustering of structurally similar regions. One such application is shown in **Subheading 4.6**.

The program output format of multiple alignments is ClustalX or PIR. Optionally, for each column of multiple alignment a structure-sequence conservation score can be reported. It combines the amino acid types and the structural superposition of these amino acids. For reasons of practical convenience, the scores are scaled into the range [0,9] and are displayed under each multiple alignment column. Therefore, a visual examination of these scores reveals whether a region is conserved or not (*see* examples in **Subheading 4.6**). In addition, for each input protein file, the amino acid temperature factor field can be set to the corresponding conservation score. This allows convenient 3D visualization in color of the amino acid conservation scores.

4. Experimental Examples

4.1. Pairwise Alignment Cases

First, we test the MultiProt method with non-trivial pairwise alignment cases. We repeated the experiment presented by Shindyalov and Bourne *(3)*. The experiment presents a set of 10 protein pairs and pairwise alignments performed by different (pairwise) methods. The results are presented in **Table 2**. Two kinds of MultiProt results are given: alignments that preserve protein backbone order and sequence order independent alignments. As can be observed from the table, our pairwise results are very competitive. The maximal running time (pair 1crl:534, 1ede:310) is less than 4 s.

4.2. Sequence Order Independent Structure Alignment

A four-helix arrangement appears in a large number of proteins. SCOP includes at least 40 folds with a four-helix bundle. Holm and Sander *(44)* show an alignment of the Rop protein (1rop) with cytochrome b56 (256b). Both

Table 2
Pairwise Structural Alignment test

Molecule 1 (size)	Molecule 2 (size)	VAST S_{al}/rms	Dali S_{al}/rms	CE S_{al}/rms	GH S_{al}/rms	$MultiProt_1$ S_{al}/rms	$MultiProt_2$ S_{al}/rms
1fxi:A(96)	1ubq(76)	48/2.1	–	–	51/1.6	44/1.7	50/1.8
1 ten(89)	3hhr:B(195)	78/1.6	86/1.9	87/1.9	81/1.7	81/1.3	82/1.3
3hla:B(99)	2rhe(114)	–	63/2.5	85/3.5	62/1.8	60/1.8	67/1.9
2aza:A(129)	1paz(120)	4/2.2	–	85/2.9	74/1.9	75/2.0	85/2.5
1cew:I(108)	1mol:A(94)	71/1.9	81/2.3	69/1.9	66/1.6	76/1.8	75/1.9
1cid(177)	2rhe(114)	85/2.2	95/3.3	94/2.7	70/1.5	84/1.8	88/1.9
1crl(534)	1ede(310)	–	211/3.4	187/3.2	80/1.9	161/2.3	232/2.4
2sim(381)	1nsb:A(390)	284/3.8	286/3.8	264/3.0	197/2.0	233/2.3	268/2.3
1bge:B(159)	2gmf:A(121)	74/2.5	98/3.5	94/4.1	72/1.8	78/2.5	88/2.2
1tie(166)	4fgf(124)	82/1.7	108/2.0	116/2.9	87/1.7	95/2.1	99/2.3

[a] RMS, root mean square; S_{al}, number of aligned atoms. The protein pairs are classified as 'difficult' for structural anaysis *(48)*. The alignments are performed by VAST *(1)*, Dali *(44)*, CE *(3)*, Geometric Hashing (GH) method *(2)* (http://bioinfo 3d.cs.ac.il/c_alpha_match/), and MultiProt. The information in this table, except for the GH method and MultiProt results, is taken from Shindyalov and Bourne *(3)* $MultiProt_1$ results do preserve the sequence order, while $MultiProt_2$ are sequence order independent.

proteins have a four-helix bundle, but the topological arrangement is different, that is, when the two structures are aligned, at least one helix pair is aligned in an opposite sequential order. Here, we show a multiple structural alignment of four proteins (1f4n, 2cbl:A, 1b3q, and 1rhg:A) which share a four-helix bundle (*see* **Fig. 2**). **Figure 2C** shows the direction of the protein sequences according to a structural alignment when all four helices are aligned. As one can see the direction is different for the last two helices. Thus, none of the commonly used sequence alignment methods can align simultaneously the four α-helices. **Figure 2B** shows a multiple structural alignment with the four helices aligned. The running time is 14 s.

4.3. Detection of Partial and Subset Alignments

Here, we demonstrate the ability of MultiProt to detect partial and subset multiple alignments. We consider five multi-domain molecules 1adj:A, 1hc7:A, 1qf6:A, 12as:A, and 1v95:A. Some domains are structurally similar. Our goal in studying such a set of proteins is to identify the two common domains (*see* **Fig. 3**): Class II aminoacyl-tRNA synthetase (aaRS)-like, catalyic domain and Anticodon-binding domain of Class II aaRS [the classification is according to

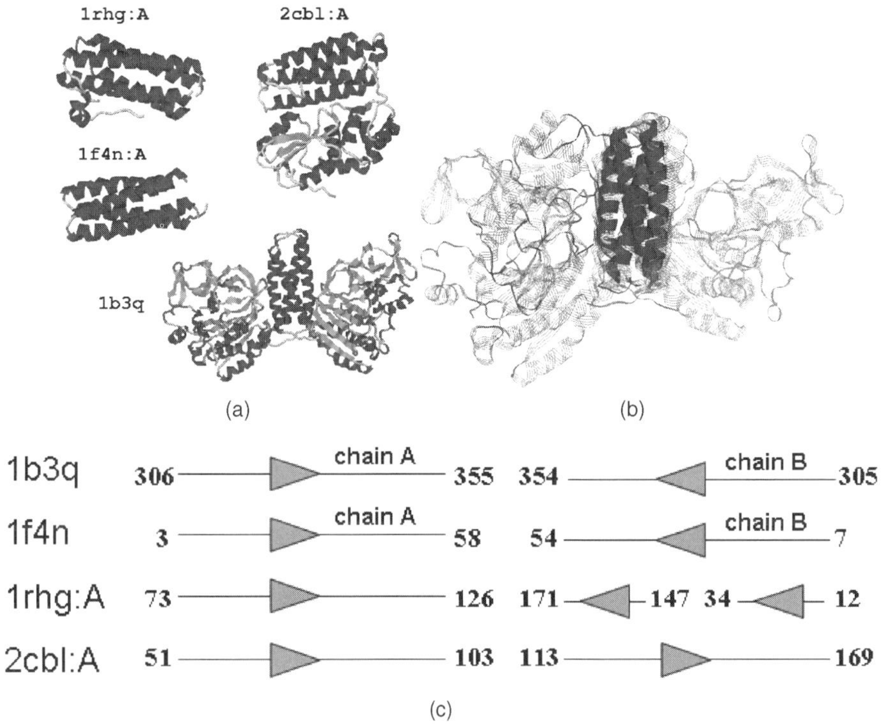

Fig. 2. Sequence order independent structure alignment. (**A**) Four proteins, 1f4n, 2cbl:A, 1b3q, and 1rhg:A, containing a four-helix bundle. (**B**) Multiple alignment of four-helix bundle produced by MultiProt. (**C**) Schematic representation of the sequence alignment derived from the multiple structural superposition. This common structural motif cannot be detected by standard sequence alignment methods due to different topological arrangement and different chain composition.

SCOP *(40)*]. The multiple alignment for all five structures resulted in a common structural core of size 39 amino acids, consisting mainly of β-sheet and α-helix. Despite the fact that these two domains are differently classified, there is some *partial* non-random (p-value < 0.001) structural similarity. The solutions containing four structures revealed two high scoring multiple alignments with different protein composition. These multiple alignments are alignments of the first domain (1adj:A, 1hc7:A, 1qf6:A, and 12as:A) and of the second domain (1adj:A, 1hc7:A, 1qf6:A, and 1v95:A). Therefore, in this example, MultiProt successfully carries out the task of subset and partial multiple alignment. The running time is 2 min.

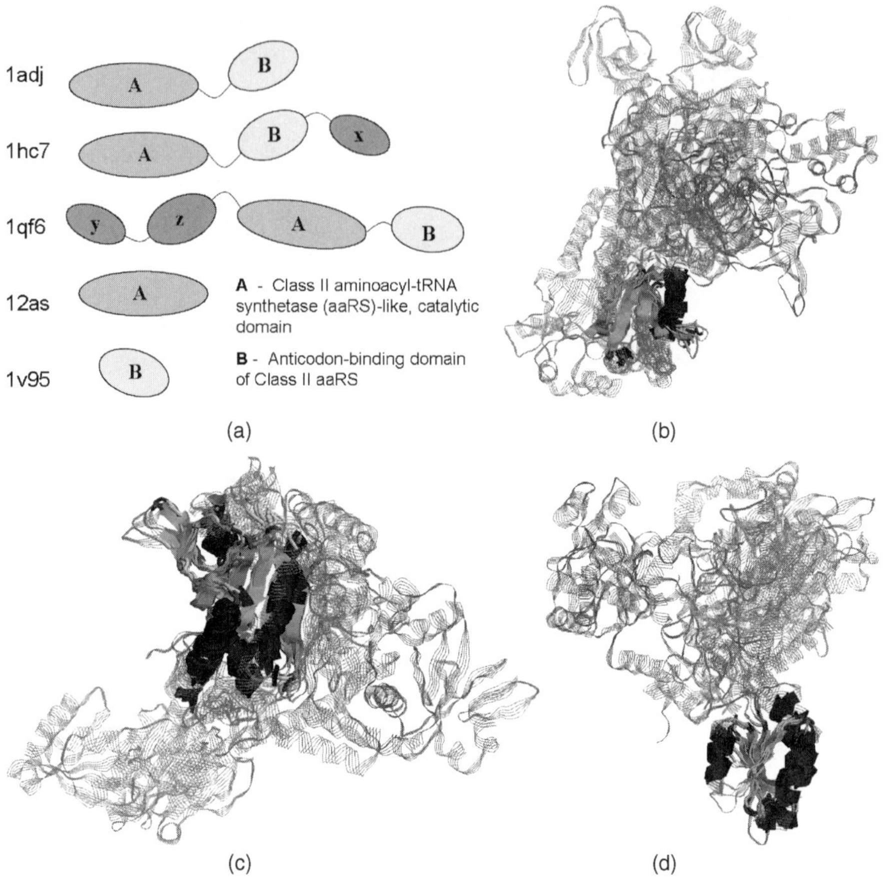

Fig. 3. Partial and subset alignments. (**A**) Simplified schematic view of protein domains of 1adj:A, 1hc7:A, 1qf6:A, 12as:A, and 1v95:A. (**B**) All five proteins are aligned. The common core is 39 amino acids (p-value < 0.001). (**C**) Domain A is aligned between the first four proteins. The common core has 125 amino acids, this is highest ranked solution for four structures. (**D**) Domain B is aligned between 1adj:A, 1hc7:A, 1qf6:A, and 1v95:A. The common core has 76 amino acids, this is the second ranked solution with molecule ID composition different from the larger alignment of domain A.

4.4. Comparison Against the HOMSTRAD Database

HOMSTRAD is a benchmark database of manually curated multiple alignments of 1032 homologous protein families with available 3D structures *(**38**)*. Our aim is to compare the quality of multiple alignments produced by MultiProt and STACCATO against the HOMSTRAD database. However, an objective comparison is not trivial as we do not know neither (1) the ultimate scoring function for multiple alignment nor (2) the correct alignment (which will require to know, for instance, the exact phylogenetic tree of the protein family). To overcome this problem, we decided to select two kinds of measures. The first measure is a commonly used sequence similarity score, namely, a sum-of-pairs score according to the BLOSUM62 matrix, which we denote by *Seq* score. The second measure computes the fitness of structural alignment according to a multiple sequence alignment. We denote the *Str* score by the number of columns in a multiple alignment that contains at least half non-gap positions and their 3D rmsd is less than 3 Å. Both scores, *Seq* and *Str*, are normalized by the length of the multiple alignment. Given two multiple alignments, we conjecture that an alignment with higher values of both scores is closer to the optimum. However, if only one score is higher at the expense of the other, it is not clear which alignment is better.

Table 3 demonstrates a comparison of MultiProt and STACCATO with the HOMSTRAD data set. We applied different gap opening penalties in order to detect the optimal value. The values in the table represent the average (over 1032 alignments) of the improvement/degradation of the *Seq* and *Str* scores. For example, for gap opening penalty –7, the number of multiple alignments

Table 3
Comparison Against the HOMSTRAD Database

Gap opening penalty	Seq %	Str %	Number of gaps
–3	+9.5 (+9.4)	+0.4 (+8.3)	+24
–7	+7.5 (+7.4)	+0.3 (+8.2)	–7
–10	+6.4 (+6.1)	–0.04 (+8.0)	–16
–13	+5.4 (+5.1)	–0.4 (+7.6)	–20
–15	+4.8 (+4.6)	–0.7 (+7.3)	–23
–30	+1.3 (+1)	–3 (+5.4)	–33

[a] Two experiments have been conducted. In the first one STACCATO has been applied on the multiple structure alignments as found in HOMSTRAD. In the second experiment, STACCATO has been applied on the multiple structure alignments computer by the MultiProt method. The numeric values represent the difference in scores of STACCATO and HOMSTRAD alignments measured in percents relative to the HOMSTRAD score (positive values mean an improvement over the HOMSTRAD alignments). The results of the second experiment are presented in the parentheses. The last column represents a relative difference in the number of gap openings (negative values mean less gaps are opened in the STACCATO alignments).

where our approach improved at least one score while the other score was at least as good as HOMSTRAD score is 587 (57% of the 1032 cases). The number of alignments where our approach degraded both *Seq* score and *Str* score is only 16. There are 1000 cases where our approach improved either *Seq* score or *Str* score (at the expense of the other).

The default value of gap opening penalty is selected to be –10, for which MultiProt and STACCATO arguably give more accurate alignments than HOMSTRAD. Outside the range of [–7,–10], either the number of gap openings is increased or the *Seq* and *Str* scores are decreased.

4.5. Low Sequence Identity with High Structural Similarity

Here, we give a simple example that demonstrates that in order to achieve a correct sequence alignment it is essential to use structural information (if available). We selected three proteins from the Glutathione S-Transferase family with less than 15% of pairwise sequence identity, which is extremely low. These are 1gnw:A:86-211 (Class phi GST from Arabidopsis thaliana), 1g7o:A:76-215 (Glutaredoxin 2 from Escherichia coli), and 1gwc:A:87-224 (Class tau GST from Triticum tauschii l.).

Due to low sequence similarity, multiple sequence alignment methods may produce an inaccurate alignment. However, these proteins come from the same family and share high structural similarity; therefore, an accurate alignment can be computed from a multiple structure superposition (*see* **Fig. 4**).

The active site residues of the Glutathione S-Transferase family were analyzed by Zhang et al. *(45)*. They proposed a method, SAPS, to compute a more accurate sequence alignment based on multiple structure information.

```
(a) 1gnw   136> --VAEEEAKLAKVLDV-YEARLKEFKYLAGETFTLTDLHHIPAIQYLLGTP-
    1gwc   129> --KKQAFAAVG-VLEGALRECSKGGGFFGGDGVGLVDVALGGVLSWMKVTEA
    1g7o   138> AESDGLIKNISDDLRALDKLIVKPNA-VNGE-LSEDDIQLFPLLRNLTLVA-
                             *        *      *      *

                0C223312222222221222333555422223223332222223333223330

(b) 1gnw        YGLTT-DEAVVAEEEAKLA-KVLDVYEARLKEFKYLAGETFTLT--DLHH
    1gwc        GVLEG-ALRECSKGGGFFGGDGVGLVDVALGGVLSWMKVTEALS--GDKI
    1g7o        KEASAGNFADLLAHSDGLIKNISDDLRALDKLIVKPNAVNGELSEDDIQL
                                                          *
```

Fig. 4. (**A**) A fragment of alignment produced by MultiProt and STACCATO. The structural superposition conservation score, which is displayed under each column, shows that most of the fragment is structurally aligned within 2–3 Å. (**B**) Alignment computed by ClustalW *(47)*. Some of the discrepancies are marked with boxes.

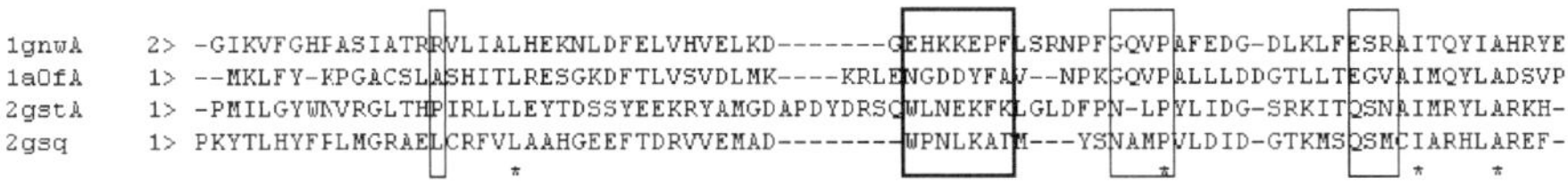

Fig. 5. Alignment of Glutathione S-Transferase proteins as computed by STACCATO. Active site residues are correctly aligned and are marked with boxes. The correct alignment of these regions was shown by the SAPS method ***(45)***. Bold box marks structurally more variable region which contains additional active site residues which were not aligned by SAPS.

However, SAPS has a restriction that only non-gap fragments are aligned. Therefore, our proposed method has an advantage of computing complete alignments including optimization of gap regions (*see* **Fig. 5**).

4.6. Loop Movement in Tyrosine Kinase

Tyrosine kinase represents a large family of evolutionarily conserved enzymes that play a critical role in cellular signaling pathways ***(46)***. Here, our aim is to analyze a multiple alignment of the activation loop. We selected two

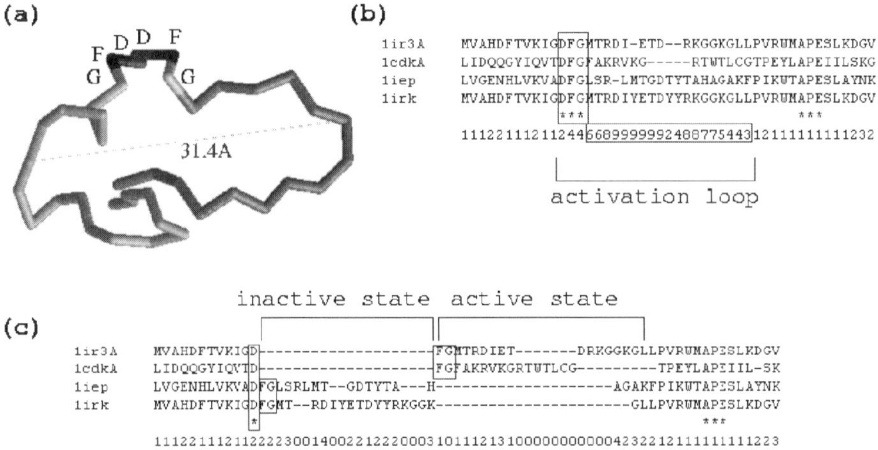

Fig. 6. (**A**) Activation loop of tyrosine kinase in the active and the inactive state. The DFG motif is located at the beginning of the loop. Structural disposition of some residues may be as large as 31 Å. (**B**) Alignment produced by STACCATO. Visual inspection of the structural superposition conservation score, which is displayed under each column, suggests a significant structural variability of the region. (**C**) Alignment produced with distance constraint of 5 Å. Two separate structurally similar clusters are clearly revealed. Only aspartic acid from the DFG motif is aligned.

kinase proteins in the active state (1ir3:A and 1cdk:A) and two in the inactive state (1irk and 1iep). These are insulin receptors from human (1ir3:A and 1irk), a cAMP-dependent protein kinase catalytic subunit from pig (1cdk:A), and Abelsone tyrosine kinase from mouse (1iep).

At the beginning of the activation loop there is a well-conserved DFG motif, which is involved in Mg-ATP binding. During the activation, the loop undergoes a significant conformational change, when some amino acids change their position by as much as 31 Å (*see* **Fig. 6A**). What kind of analysis should be performed in order to detect and distinguish between two conformational states and detect residues participating in this reorganization? Clearly, the multiple sequence alignment does not recognize active and inactive states, as it multiply aligns the sequence of the whole activation loop. In our approach, as discussed above in **Subheading 3.**, we are able to apply *distance constrained alignment* which aligns only amino acids closely located in space. Applying a distance threshold of 5 Å, we are able to distinguish between two states of the activation loop as shown in **Fig. 6C**. Only the aspartic residue from the DFG motif is spatially conserved (within 5 Å distance threshold).

5. Conclusions

Here, we have discussed a powerful method, MultiProt, for multiple protein structure alignment. The main advantages of our method are (1) simultaneous structure superposition (no side effects of pairwise alignment methods), (2) solutions are detected for any number of molecules (subset alignments), (3) proteins can consist of several domains or even several chains (partial alignments), (4) the final alignments can optionally preserve the sequence order or be sequence order independent, and (v) time efficiency (tens and even hundreds of molecules). In case that protein structures differ modulo hinge motion, a complete alignment can be detected by a manual examination of several largest partial solutions. In addition, we have discussed the problem of structure-based multiple sequence alignment, which in the case of protein structure availability overcomes the problems inherent to sequence alignment methods. The discussed method, STACCATO, with the combination of MultiProt produces multiple alignments as good (and arguably slightly better) as multiple alignments from the manually curated HOMSTRAD data base.

Acknowledgments

The research of M. Shatsky is supported by a PhD fellowship in "Complexity Science" from the Yeshaya Horowitz association. This research was supported by the Israel Science Foundation (grant no. 281/05), the Binational US-Israel

Science Foundation (BSF) and by the Hermann Minkowski-Minerva Center for Geometry at Tel Aviv University. The research of R. Nussinov has been funded in whole or in part with Federal funds from the National Cancer Institute, National Institutes of Health, under contract number NO1-CO-12400. This research was supported (in part) by the Intramural Research Program of the NIH, National Cancer Institute, Center for Cancer Research. The content of this publication does not necessarily reflect the view or policies of the Department of Health and Human Services, nor does mention of trade names, commercial products, or organization imply endorsement by the U.S. Government.

References

1. Madej, T., Gibrat, J., and Bryant, S. Threading a database of protein cores. Proteins 23:356–369, 1995. Online available on http://www.ncbi.nlm.nih.gov/Structure/VAST/vast.shtml.
2. Bachar, O., Fischer, D., Nussinov, R., and Wolfson, H. A computer vision based technique for 3-D sequence independent structural comparison. Protein Eng 6:279–288, 1993.
3. Shindyalov, I. and Bourne, P. Protein structure alignment by incremental combinatorical extension (ce) of the optimal path. Protein Eng 11(9):739–747, 1998. Online available on http://cl.sdsc.edu/ce.html.
4. Dietmann, S., Park, J., Notredame, C., Heger, A., Lappe, M., and Holm, L. A fully automatic evolutionary classification of protein folds: dali domain dictionary version 3. Nucleic Acids Res 29(1):55–57, 2001. Online available on http://www.embl-ebi.ac.uk/dali/.
5. Shatsky, M., Nussinov, R., and Wolfson, H. FlexProt: alignment of flexible protein structures without a pre-definition of hinge regions. Journal of Computational Biology 11(1):83–106, 2004.
6. Eidhammer, I., Jonassen, I., and Taylor, W. Structure comparison and structure patterns. J Comput Biol 7:685–716, 2000.
7. Orengo, C.A., Michie, A.D., Jones, S., Jones D.T., Swindells, M. B., and Thornton, J. M. CATH – a hierarchic classification of protein domain structure. Structure 5(8):1093–1108, 1997.
8. Ma, B., Elkayam, T., Wolfson, H., and Nussinov, R. Protein-protein interactions: structurally conserved residues distinguish between binding sites and exposed protein surfaces. Proc Natl Acad Sci USA 100(10):5772–5777, 2003.
9. Chung, J., Wang, W., and Bourne, P. Exploiting sequence and structure homologs to identify protein-protein binding sites. Proteins 62(3):630–640, 2006.
10. Aytuna, A., Gursoy, A., and Keskin, O. Prediction of protein-protein interactions by combining structure and sequence conservation in protein interfaces. Bioinformatics 21(12):2850–2855, 2005.

11. Akutsu, T. and Sim, K. Protein threading based on multiple protein structure alignment. In *Genome Informatics (GIW'99)*, Asai, K. and Miyano, S. and Takagi, T (eds). Universal Academy Press, Tokyo, 23–29, 1999.
12. Goldsmith-Fischman, S. and Honig, B. Structural genomics: computational methods for structure analysis. Protein Sci 12(9):1813–1821, 2003.
13. Koehl, P. Protein structure similarities. Curr Opin Struct Biol 11:348–353, 2001.
14. Kolodny, R., Koehl, P., and Levitt, M. Comprehensive evaluation of protein structure alignment methods: scoring by geometric measures. J Mol Biol 346(4): 1173–88, 2005.
15. Bennett, M., Schlunegger, M., and Eisenberg, D. 3d domain swapping: a mechanism for oligomer assembly. Protein Sci 4:2455–2468, 1995.
16. Dror, O., Benyamini, H., Nussinov, R., and Wolfson, H. MASS: multiple structural alignment by secondary structures. Bioinformatics 19 Suppl. 1:i95–i104, 2003.
17. Yuan, X. and Bystroff, C. Non-sequential structure-based alignments reveal topology-independent core packing arrangements in proteins. Bioinformatics 21(7):1010–1019, 2005.
18. Ambuhl, C., Chakraborty, S., and Gartner, B. Computing largest common point sets under approximate congruence. In *Proceedings of the 8th Annual European Symposium on Algorithms*, 52–63, Springer-Verlag, Springer, Berlin, 2000.
19. Akutsu, T. Protein structure alignment using dynamic programming and iterative improvement. IEICE Trans Inf Syst E79-D:1629–1636, Springer Berlin, 1996.
20. Kolodny, R. and Linial, N. Approximate protein structural alignment in polynomial time. Proc Natl Acad Sci USA 101(33):12201–12206, 2004.
21. Shatsky, M., Shulman-Peleg, A., Nussinov, R., and Wolfson, H. The multiple common point set problem and its application to molecule binding pattern detection. J Comput Biol 13(2):407–428, 2006.
22. Edgar, R. Muscle: multiple sequence alignment with high accuracy and high throughput. Nucleic Acids Res 32(5):1792–1797, 2004.
23. Gerstein, M. and Levitt, M. Using iterative dynamic programming to obtain accurate pairwise and multiple alignments of protein structures. In *Proceedings of the Fourth International Conference on Intelligent Systems in Molecular Biology*, 59–67, Menlo Park, CA, AAAI Press, Heidleberg, Germany, 1996.
24. Russell, R. and Barton, G. Multiple protein sequence alignment from tertiary structure comparison: assignment of global and residue confidence levels. Proteins 14:309–323, 1992.
25. Taylor, W. R., Flores, T., and Orengo, C. Multiple protein structure alignment. Protein Sci 3:1858–1870, 1994.
26. Ye, Y. and Godzik, A. Multiple flexible structure alignment using partial order graphs. Bioinformatics 21(10):2362–2369, 2005.
27. Ochagavia, M. E. and Wodak S. Progressive combinatorial algorithm for multiple structural alignments: application to distantly related proteins. Proteins 55(2):436–454, 2004.

28. Konagurthu, A., Whisstock, J., Stuckey, P., and Lesk, A. Mustang: a multiple structural alignment algorithm. Proteins 64(3):559–574, 2006.
29. Leibowitz, N., Nussinov, R., and Wolfson, H. MUSTA-a general, efficient, automated method for multiple structure alignment and detection of common motifs: application to proteins. J Comput Biol 8:93–121, 2001.
30. Leibowitz, N., Fligelman, Z., Nussinov, R., and Wolfson, H. Automated multiple structure alignment and detection of a common substructural motif. Proteins 43:235–245, 2001.
31. Wolfson, H. J. and Rigoutsos, I. Geometric hashing: an overview. IEEE Comput Sci Eng 4(4):10–21, 1997.
32. Nussinov, R. and Wolfson, H. Efficient detection of three-dimensional structural motifs in biological macromolecules by computer vision techniques. Proc Natl Acad Sci USA 88:10495–10499, 1991.
33. Shatsky, M., Fligelman, Z., Nussinov, R., and Wolfson, H. Alignment of flexible protein structures. In *8th International Conference on Intelligent Systems for Molecular Biology*, 329–343, AAAI press, Heidleberg, Germany, 2000.
34. Jonassen, I., Eidhammer, I., Conklin, D., and Taylor, W. Structure motif discovery and mining the pdb. Bioinformatics 18(2):362–367, 2002.
35. Dror, O., Benyamini, H., Nussinov, R., and Wolfson, H. Multiple structural alignment by secondary structures: – algorithm and applications. Protein Sci 12:2492–2507, 2003.
36. O'Sullivan, O., Suhre, K., Abergel, C., Higgins, D., and Notredame, C. 3Dcoffee: combining protein sequences and structures within multiple sequence alignments. J Mol Biol 340(2):385–395, 2004.
37. Shatsky, M., Nussinov, R., and Wolfson, H. A method for simultaneous alignment of multiple protein structures. Proteins 56(1):143–156, 2004.
38. Mizuguchi, K., Deane, C., Blundell, T., and Overington, J. Homstrad: a database of protein structure alignments for homologous families. Protein Sci 7:2469–2471, 1998.
39. Akutsu, T. and Halldorson, M. M. On the approximation of largest common subtrees and largest common point sets. Theor Comput Sci 233:33–50, 2000.
40. Murzin, A., Brenner, S., Hubbard, T., and Chothia, C. SCOP: a structural classification of proteins database for the investigation of sequences and structures. J Mol Biol 247:536–540, 1995.
41. Chandonia, J., Hon, G., Walker, N., Lo Conte, L., Koehl, P., Levitt, M., and Brenner, S. The astral compendium in 2004. Nucleic Acids Res 32:D189–D192, 2004.
42. Shatsky, M., Nussinov, R., and Wolfson, H. T. Optimization of multiple sequence alignment based on multiple structure alignment. Proteins 62(1):209–217, 2006.
43. Henikoff, S. and Henikoff, J. Amino acid substitution matrices from protein blocks. Proc Natl Acad Sci USA 89(22):10915–10919, 1992.

44. Holm, L. and Sander, C. Protein structure comparison by alignment of distance matrices. J Mol Biol 233:123–138, 1993.
45. Zhang, Z., Lindstam, M., Unge, J., Peterson, C., and Lu, G. Potential for dramatic improvement in sequence alignment against structures of remote homologous proteins by extracting structural information from multiple structure alignment. J Mol Biol 332(1):127–142, 2003.
46. Hubbard, S. and Till, J. H. Protein tyrosine kinase structure and function. Ann Rev Biochem 69:373–398, 2000.
47. Higgins, D., Thompson, J., and Gibson, T. CLUSTAL W: improving the sensitivity of progressive multiple sequence alignment through sequence weighting, position-specific gap penalties and weight matrix choice. Nucleic Acids Res 22:4673–4680, 1994.
48. Fischer, D., Elofsson, A., Rice, D., and Eisenberg, D. Assessing the performance of fold recognition methods by means of a comprehensive benchmark. In *Proceedings of Pacific Symposium on Biocomputing* (Hunter, L. and Klein, T., editors), World Scientific Press, Singapore, 300–318, 1996.

6

Indexing Protein Structures Using Suffix Trees

Feng Gao and Mohammed J. Zaki

Summary

Approaches for indexing proteins and fast and scalable searching for structures similar to a query structure have important applications such as protein structure and function prediction, protein classification and drug discovery. In this chapter, we describe a new method for extracting the local feature vectors of protein structures. Each residue is represented by a triangle, and the correlation between a set of residues is described by the distances between C_α atoms and the angles between the normals of planes in which the triangles lie. The normalized local feature vectors are indexed using a suffix tree. For all query segments, suffix trees can be used effectively to retrieve the maximal matches, which are then chained to obtain alignments with database proteins. Similar proteins are selected by their alignment score against the query. Our results show classification accuracy up to 97.8 and 99.4% at the superfamily and class level according to the SCOP classification and show that on average 7.49 out of 10 proteins from the same superfamily are obtained among the top 10 matches. These results outperform the best previous methods.

Key Words: Protein structure indexing; suffix trees; structural motifs; 3D database search; approximate matches.

1. Introduction

Traditionally, the problem of determining similar proteins was approached by finding the amount of similarity in their amino acid sequences. However, biologists have determined that even proteins that are remotely homologous in their sequence similarities can perform surprisingly very similar functions in living organisms *(1)*. This fact has been attributed to the dependency of

From: *Methods in Molecular Biology, vol. 413: Protein Structure Prediction, Second Edition*
Edited by: M. Zaki and C. Bystroff © Humana Press Inc., Totowa, NJ

the functional role of proteins on their actual three-dimensional (3D) structure. In view of this, it can be stated that two proteins with remote sequence homology can be functionally classified as similar if they exhibit structural homology.

Searching the growing database of protein structures for structural homologues is a difficult and time-consuming task. For example, we may want to retrieve all structures that contain sub-structures similar to the query, a specific 3D arrangement of surface residues, and so on. Searches such as these are the first step toward building a systems level model for protein interactions. In fact, high-throughput proteomics methods are already accumulating the protein interaction data that we would wish to model, but fast computational methods for structural database searching lag far behind; biologists are in need of a means to search the protein structure databases rapidly, similar to the way BLAST *(2)* rapidly searches the sequence databases.

1.1. Prior Research

Protein structural similarity determination can be classified into three main approaches: pair-wise alignment, multiple structure alignment and database indexing.

Pair-wise structure alignment methods can be classified into three classes *(3)*. The first class works at the residue level ***(4,5)***. The second class focuses on using secondary structure elements (SSEs) such as α-helices and β-strands to align two proteins approximately ***(6–8)***. The third approach is to use geometric hashing, which can be applied at both the residue ***(9)*** and SSE levels ***(10)***.

Previous work has also looked at multiple structure alignment. These methods are also based on geometric hashing ***(11)*** or SSE information ***(12)***. A recent method ***(13)*** aims to solve the multiple structural alignment problem through detection of partial solutions; it computes the best scoring structural alignments, which can be either sequential or sequence-order independent ***(14)***, if one seeks geometric patterns which do not follow the sequence order.

Due to their time complexity, the pair-wise and multiple structure alignment approaches are not suitable for searching for similarity over thousands of protein structures. Database indexing and scalable searching approaches satisfy this requirement. There are two classes of protein structure indexing approaches depending on the kinds of representation used to capture the local structural features. The first class focuses on indexing the local features at the residue level directly, and the other class uses SSEs to approximate those local features.

The CTSS *(15)* program approximates the protein C_α backbone with a smooth spline with minimum curvature. The method then stores the curvature, torsion angle and the secondary structure that each C_α atom in the backbone belongs to, in a hash-based index. ProGreSS *(16)* is a recent method, which extracts the features for both the structure and sequence, within a sliding window over the backbone. Its structure features are the same as the CTSS features (curvature, torsion angles and SSE information); its sequence features are derived using scoring matrices like PAM or BLOSUM.

The Local Feature Frequency (LFF) profile algorithm *(17)* first extracts representative local features from the distance matrices of all protein fold families by medoid analysis, where the distance matrix for a protein is the symmetric matrix giving all pair-wise distances between the C_α atoms. In the next step, each C_α distance matrix of a protein structure is encoded by labeling all its sub-matrices by the index of the nearest representative LF patterns. Each structure is finally represented using a vector of the frequency of the representative local features. The structural similarity between two proteins is computed as the Euclidean distance between their LFF profile vectors.

There are also some methods that index the protein structures using SSEs. For each protein, PSI *(18)* uses a R^*-tree to index a 9D feature vector, a representation of all triplet SSEs within a range. After retrieving the matching triplet pairs, a graph-based algorithm is used to compute the alignment of the matching SSE pairs. Another SSE-based method, ProtDex *(19)* obtains the sub-matrices of the SSE contact patterns from the distance matrix of a protein structure. The grand sum of the sub-matrices and the contact-pattern type are indexed by an inverted file index. By their nature, SSEs model the protein only approximately, and therefore, these SSE-based approaches are not very accurate and, furthermore, are not very useful for small query proteins with few SSEs.

For a given query, the most common similarity-scoring scheme is the number of votes accumulated from the matching residues *(9,15,16)*. CTSS and ProGreSS further define the p-value of a protein based on the number of votes, and smaller p-values imply better similarity. These scoring schemes, however, do not take into account the local similarity.

The work most related to our approach is PAST *(20)*, which also uses a suffix tree to index protein structures. While PAST shares with our approach the general idea of using a discretized alphabet to represent structural sequences, and indexing them using suffix trees, the actual details of the methods are very different. We use a different feature representation, and searches for chains of maximal matches, and most importantly is especially designed for approximate matches. Furthermore, after computing the similar structural segments, we chain them into longer approximate alignments, whereas PAST is designed for structural motif extraction.

1.2. Our Contributions

In this chapter, we present a fast, novel protein indexing method called PSIST (which stands for Protein Structure Indexing using Suffix Trees). As the name implies, our new approach transforms the local structural information of a protein into a "sequence" on which a suffix tree is built for fast matches. We first extract local structural feature vectors using a sliding window along the backbone. For a pair of residues, the distance between their C_α atoms and the angle between the planes formed by the C_α, N and C atoms of each residue are calculated. The feature vectors for a given window include all the distances and angles between the first residue and the rest of the residues within the window. Compared with the local features from a single residue, our feature vectors contain both the translational and rotational information. After normalizing the feature vectors, the protein structure is converted to a sequence, called the "structure-feature sequence or SF sequence," over the discretized symbols.

We use suffix trees to index the protein SF sequences. A suffix tree is a versatile data structure for substring problems *(21)*, and it has been used for various problems such as protein sequence indexing *(22,23)* and genome alignment *(24,25)*. Suffix trees can be constructed in $O(n)$ time and space *(26,27)*, where n is the sequence length. Thus, suffix trees are an effective choice for indexing our protein SF-sequences.

For a given query, all the maximal matches are retrieved from the suffix tree and chained using a greedy approach. The top proteins with the highest alignment scores are finally selected. Our results show classification accuracy up to 97.8% and 99.4% at the superfamily and class levels according to the SCOP classification, and show that on average 7.49 out of 10 proteins from the same superfamily are obtained among the top 10 matches. These results are better than the best previous methods.

2. Indexing Proteins

2.1. Local Feature Extraction

A protein is composed of an ordered sequence of residues linked by peptide bonds. Each residue has C_α, N and C atoms, which constitute the backbone of the protein. Although the backbone is linear topologically, it is very complex

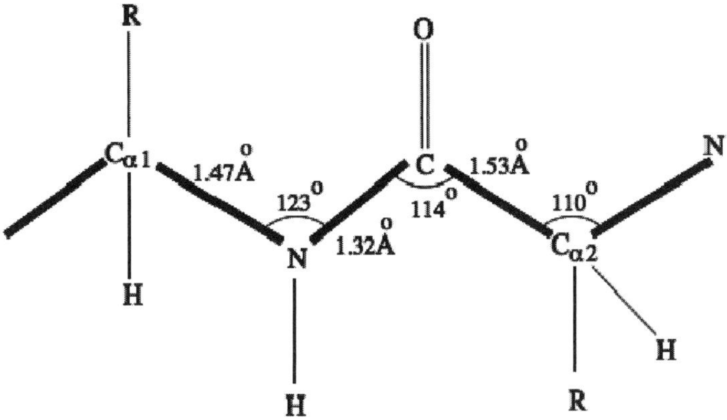

Fig. 1. Bond length and bond angles.

geometrically. The bond lengths and bond angles (*see* **Fig. 1**) and the torsion angles (ϕ, φ, ω; *see* **Fig. 2**) completely define the conformation and geometry of the protein.

The bond length is the distance between the bonded atoms, and the bond angle is the angle between any two covalent bonds that include a common atom (*see* **Fig. 1**). For instance, the bond length of N–C is 1.32 Å (Å denotes distance in angstroms), the bond angle between C_α–N and N–C is 123°. Torsion angles are used to describe conformations around rotatable bonds (*see* **Fig. 2**). Assume four consecutive atoms are connected by three bonds b_{i-1}, b_i and b_{i+1}.

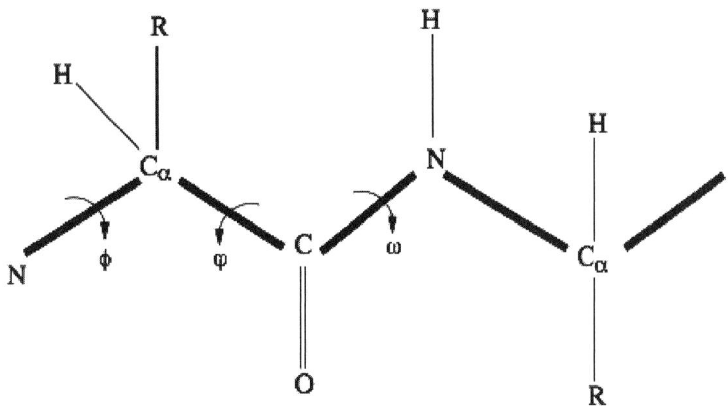

Fig. 2. Torsion angles.

The torsion angle of b_i is defined as the smallest angle between the projections of b_{i-1} and b_{i+1} on the plane perpendicular to bond b_i. In **Fig. 2**, ϕ, φ, ω are the torsion angles on the bonds $N - C_\alpha$, $C_\alpha - C$ and C–N, respectively.

To capture the local features more accurately, we need to extract the structural features from a set of local residues. To obtain the local feature vector, we first represent each residue individually and then consider the relationship between a pair of residues and a set of residues. For each residue, the length of $C_\alpha - N$ bond is 1.47 Å and that of the C_α–C bond is 1.53 Å, and the angle between $C_\alpha - N$ and $C_\alpha - C$ bonds is 110°. Thus, all the triangles formed by $N - C_\alpha - C$ atoms in each residue are equivalent, and each residue can be represented by a triangle of the same size. The relationship between a pair of residues in 3D space can be fully described by the rigid transformation between two residues, which is a vector of six dimensions, containing three translational and three rotational degrees of freedoms. To reduce the dimension of the vector, we use a distance and an angle to describe the transformation features between two residues.

We define the distance d between a pair of residues as the Euclidean distance between their C_α atoms. The angle θ between a pair of residues is defined as the angle between the planes that contain $N - C_\alpha - C$ triangles representing each residue (*see* **Fig. 3**). The distance and angle between a pair of residues are invariant to translation and rotation of the protein. The Euclidean distance between two C_α atoms is calculated using their 3D coordinates directly. The angle between the two planes defined by the $N - C_\alpha - C$ triangles is calculated

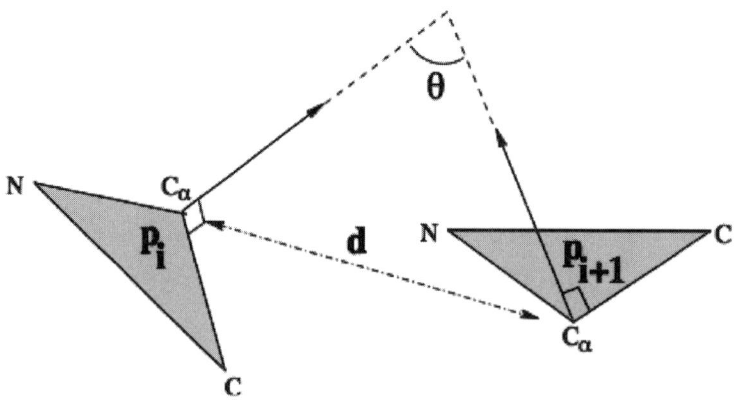

Fig. 3. The distance and angle between two residues.

between their normals with C_α as the origin. The normal of the plane defined by the triangle $N - C_\alpha - C$ is given as

$$\vec{n} = \frac{\overrightarrow{C_\alpha N} \times \overrightarrow{C_\alpha C}}{\left\| \overrightarrow{C_\alpha N} \times \overrightarrow{C_\alpha C} \right\|}.$$

The angle between the two normals $\vec{n_1}$ and $\vec{n_2}$ is then calculated as

$$\cos\theta = \vec{n_1} \bullet \vec{n_2},$$

where $\times$ and $\bullet$ denote the cross and dot product, respectively, between two vectors.

To describe the local features between a set of residues, we slide a window of length w along the backbone of the protein. The distances and angles between the first residue i and all the other residues j (with $j \in [i+1, i+w-1]$) within the window are computed and added to a feature vector. Each window is associated with one feature vector.

Let $P = \{p_1, p_2, \ldots, p_n\}$ represent a protein, where p_i is the i-th residue along the backbone. The feature vector of the protein is defined as $P^v = \{p_1^v, p_2^v, \ldots, p_{n-w-1}^v\}$, where w is the sliding window size, and p_i^v is a feature vector $\left[d\left(p_i, p_{i+1}\right), \cos\theta\left(p_i, p_{i+1}\right), \ldots, d\left(p_i, p_{i+w-1}\right), \cos\theta\left(p_i, p_{i+w-1}\right)\right]$, where $d\left(p_i, p_j\right)$ is the distance between the residues p_i and p_j, and $\cos\theta\left(p_i, p_j\right)$ gives the angle between the residues p_i and p_j. With window size w, the dimension of each feature vector p_i^v is $2 * (w-1)$.

2.2. Normalization

Our feature vector is a combination of distances and angles, which have different measures. A normalization procedure is performed after the feature vectors are extracted. The angle θ is in the range $[0, \pi]$, so the range of $\cos\theta \in [-1, 1]$.

For normalizing the distances, we need to know the upper bound on the distance between the i-th and $(i+w-1)$-th residue in the protein. From **Fig. 1**, it is seen that the average distance between $C_{\alpha 1} - N$ atoms is $d_1 = 1.47$ Å, the average distance between N–C atoms is $d_2 = 1.32$ Å and the angle α between $C_{\alpha 1} - N$ and N–C bonds is 123°. The distance between $C_{\alpha 1} - C$ atoms is therefore $d\left(C_{\alpha 1}, C\right) = \sqrt{d_1^2 + d_2^2 - 2d_1 d_2 \cos\alpha} = 2.453$. The distance between $C - C_{\alpha 2}$ atoms is $d\left(C, C_{\alpha 2}\right) = 1.53$, so the average distance between two C_α atoms is $d\left(C_{\alpha 1}, C_{\alpha 2}\right) \leq d\left(C_{\alpha 1}, C\right) + d\left(C, C_{\alpha 2}\right) = 2.453 + 1.57 = 4.023$. If the distance between two atoms is greater than 4.023, it is trimmed to 4.023. For a sliding window of size w, the lower bound of the distance between any two atoms is 0 and the upper bound is $4.023^* (w-1)$, so the distance between any pair of residues within a w length window is in the range $[0, 4.023^* (w-1)]$.

All the distances and angles are normalized and binned into an integer within the range $[0, b-1]$. We use the equation $d' = \left[\frac{d^* b}{4.023^*(w-1)} \right]$ to normalize and bin the distance and $\cos\theta' = \left[\frac{(\cos\theta+1)^* b}{2} \right]$ to normalize and bin the angle. **Table 1** shows three examples of normalized and binned feature vectors for $w = 3$ and $b = 10$. The size of each feature vector is $2^*(w-1) = 4$, and the normalized value is within $[0, 9]$.

After normalization and binning, each feature vector is defined as $p^s = \{p_0^s, p_1^s, \ldots, p_{2^*(w-1)-1}^s\}$, where p_i^s is an integer within the range $[0, b-1]$. Thus, the structure of each protein P is converted to a SF sequence $P^s = \{P_0^s, P_1^s \ldots P_{n-w+1}^s\}$, called the SF sequence, where P_i^s is the i-th normalized feature vector (p^s) along the backbone. Note that each symbol within an SF sequence is a vector of length $2(w-1)$, to which we assign a unique integer identifier as its label. Thus the SF sequences are over an alphabet of size $b^{2(w-1)}$.

2.3. Generalized Suffix Trees Construction

After obtaining the SF sequences for all proteins in the database, we use a generalized suffix tree (GST) as the indexing structure. GST is a compact representation of all the suffixes of multiple sequences and can be constructed in linear time *(27)*. A suffix can be located by following a unique path from the root to a leaf.

Table 1
Examples of normalized feature vectors for $w = 3$ and $b = 10$

	Feature vector			
	d	$\cos\theta$	d	$\cos\theta$
Original	3.55	0.29	5.4	−0.23
Normalized (a)	4	6	6	3
Original	4.04	0.11	5.75	−0.25
Normalized (b)	5	5	7	3
Original	3.60	0.45	5.29	0.21
Normalized (x)	4	7	6	6

The normalized features are represented by a set of new symbols such as a, b, and x.

To save the storage space for the suffix tree, we map each structure feature vector p^s to a unique key or symbol for the suffix tree construction and map it back to the normalized vector when we compute the distance between two feature vectors. For instance, the three feature vectors in **Table 1** could be mapped to the symbols a, b and x respectively.

Notation: Let GST be a generalized suffix tree, we use the following notation in the rest of the chapter. We use N for a node in the suffix tree, E for an edge, $C(E)$ for a child node of the edge E, $L(E)$ for the label on edge E, $L(E[i])$ for the i-th symbol of the edge label $L(E)$, $P(N)$ for the path label of the node N (formed by concatenating all the edge labels from the root node to N) and $P(E[i])$ for the path label of $L(E[i])$. Furthermore, each leaf node in GST contains a sequence-position pair (x, p), where x is a sequence identifier and p is the start position of the suffix within sequence x. For any node N, we use the notation sp-list(N) for the collection of the sequence-position pairs for all the leaves under N.

Example: **Figure 4** shows an example of GST for two SF sequences $S_1 = xabxa$ and $S_2 = babxba$, over the alphabet $\{a, b, x\}$, obtained by mapping each normalized feature vectors in **Table 1** to a unique letter symbol. For instance, the first normalized vector given in **Table 1**, namely [4, 6, 6, 3] may be mapped to a. Node 0 is the root node, nodes 1–7 are internal nodes, and the rest are leaves. "\$" is the unique termination character. The path label of

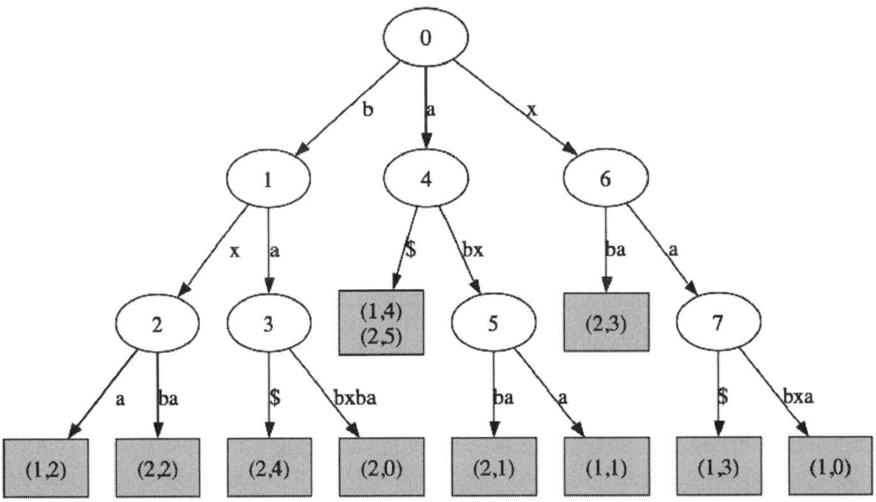

Fig. 4. Generalized suffix tree for sequences $S_1 = xabxa$ and $S_1 = babxba$.

node 7 is xa. The edge label $L(E)$ of the right edge out of node 7 is bxa, so its second character $L(E[2])$ is x, and the path label $P(E[2])$ is given as $xabx$. The sequence-position identifier $(1, 0)$for the right child of node 7 stands for $xabxa$, a suffix of sequence S_1 that starts at position 0. Thus sp - list $(7) = \{(1, 0), (1, 3)\}$ and the sp-list for node 6 is sp - list $(6) = \{(2, 3), (1, 3), (1, 0)\}$.

3. Querying

So far we have discussed how to build the suffix tree indexing based on the local structure features for each protein. In this section, we will present how to search for similar proteins.

Given a query (Q, ε), we first extract its feature vectors and convert it into a SF sequence Q^s as described above. Then three phases are performed: searching, ranking and post-processing. The searching phase retrieves all the matching segments/subsequences from the database within a distance threshold ε (on a per symbol basis), the ranking phase ranks all the proteins by chaining the matching segments and the post-processing step further uses Smith–Waterman *(28)* dynamic programming approach to find the best local alignment between the query and the selected proteins.

3.1. Searching

For a given query SF sequence $Q^s = \{Q_1^s Q_2^s \ldots Q_n^s\}$, maximum structure-feature distance threshold ε and a minimum match length threshold l, the search algorithm finds all maximal matching SF subsequences $P^s = \{P_1^s, P_2^s \ldots P_m^s\}$ that occur in both the query SF sequence and any database protein SF sequence. A maximal match has the following properties:

1. There exists a matching SF subsequence $Q_{i+1}^s \ldots Q_{i+m}^s$ of Q^s, such that dist $\left(Q_{i+j}^s, P_j^s\right) < \varepsilon$, where $j = 1, 2 \ldots m$, and Q_{i+j}^s and P_j^s are the normalized and discretized feature vectors of length $2^* (w - 1)$. Note that the distance threshold ε is applied on a pair of corresponding sequence features from Q^s and P^s. The distance function used in our algorithm is Euclidean distance.
2. The length of the match is at least as long as the length threshold, i.e., $m \geq l$.
3. If P^s is a SF subsequence of protein R^s, then neither $P^s v$ nor vP^s is a matching SF subsequence of Q^s and R^s, for any feature vector v (this ensures maximality).

For instance, abx is a maximal match between the SF sequences $xabxa$ and $babxba$ of **Fig. 4** Note that our approach differs from MUMmer genome alignment method presented in *(24)* which finds *exact* maximal*unique* matches between *two* genomes. Furthermore, in PSIST the distance between two symbols depends on the distance function dist and is not simply based on match/mismatch as in MUMmer.

To find all maximal matches within ε (on a per symbol basis) between the query Q^s and suffix tree GST_d built from the database proteins, one solution is to trace every SF subsequence of Q^s from the root of GST_d, but the common prefix of two subsequences will be searched twice and more comparisons will be performed. To reduce the number of comparisons, we build another suffix tree GST_q for Q^s and then traverse the two suffix trees simultaneously to retrieve all the maximal matches. In the discussion below, we use the subscript q for the query and d for the database. For instance, N_q stands for a query suffix tree node, while N_d stands for a database suffix tree node.

The matching algorithm starts with the MMS procedure as shown in **Fig. 5**, and its inputs are the root node (N_q) of the query suffix tree GST_q, the root node (N_d) of the database suffix tree GST_d, per sequence-feature distance tolerance ε and the minimum length of the maximal match l. For every edge out of the query node and database node, MMS calls the NodeSearch procedure (*see* **Fig. 6**) to match their labels and follow the path to find all the matching nodes.

In the NodeSearch procedure (*see* **Fig. 6**), for two edges from different suffix trees, the distance between the corresponding pair of label symbols ($L\left(E[i]_q\right)$ and $L(E[j]_d)$ is computed in step 2, **Fig. 6**. If the distance is larger than ε, which implies a mismatch, the procedure updates the MMSet (*see* **Fig. 7**) and proceeds to the next branch. If there is a match, the shorter edge will be the first to reach the end. If the child node of the short edge is a leaf, we need to update the MMSet. If the child node is an internal node, two different procedures are called recursively. (1) If the lengths of two edge labels are the same, then MMS procedure is called for two child nodes in step 3, **Fig. 6**. (2) If one of the edge has a shorter label, the algorithm NodeSearch will be called recursively with

Input	: query Node N_q, database Node N_d, distance ϵ, length threshold l
Output	: maximal matches set ($MMSet$)
Initialization	: $MMSet = \emptyset$

Procedure: MMS(N_q,N_d,ϵ,l)
foreach *edge* E_q *out of* N_q **do**
  foreach *edge* E_d *out of* N_d **do**
    NS($E_q, 0, E_d, 0, \epsilon, l$).

Fig. 5. MaximalMatches Search algorithm.

```
Input     : query Edge E_q, query Edge iterator i,
            database Edge E_d, database Edge iterator
            j, distance ε, length threshold l
Output    : maximal matches set (MMSet)

Procedure: NS(E_q,i,E_d,j,ε,l)
1 while i < L(E_q).len and j < L(E_d).len) do
    2 if dist(L(E[i]_q), L(E[j]_d)) > ε then
        updateMMS(C(E_q), C(E_d), P(E[i]_q).len − 1, l).
        return;
    else
        i=i+1, j=j+1
3 if i = L(E_q).len and j = L(E_d).len then
    if isleaf(C(E_q)) or isleaf(C(E_d)) then
        updateMMS(C(E_q), C(E_d), P(E[i]_q).len−1, l)).
    else
        MMS(C(E_q), C(E_d), ε, l).
4 if i = L(E_q).len and j < L(E_d).len then
    if isleaf(C(E_q)) then
        updateMMS(C(E_q), C(E_d), P(E[i]_q).len − 1, l).
    else
        foreach edge E_C out of C(E_q) do
            NS(E_C,0,E_d,j,ε,l).
5 if i < L(E_q).len and j = L(E_d).len then
    if isleaf(C(E_d)) then
        updateMMS(C(E_q), C(E_d), P(E[j]_d).len − 1, l).
    else
        foreach edge E_C out of C(E_d) do
            NS(E_q, i, E_C, 0, ε, l).
```

Fig. 6. NodeSearch algorithm.

the new input consisting of all the edges out of the child node of the short edge (*see* steps 4 and 5, **Fig. 6**).

Each matching SF subsequence s is defined by two triplets (x, p, m) and (y, q, m), where p and q are the start positions of s in the query sequence Q_x and the protein sequence P_y respectively, and m is the length of the match. If s is a maximal match, it will be added to the MMSet in the updateMMS procedure (*see* **Fig. 7**). To identify a maximal match, we need to compare whether any extension of the match will result in a mismatch. In

```
Input     : query Node N_q, database Node N_d, match
            length m, length threshold l
Output    : maximal matches set (MMSet)

Procedure: UpdateMMS(N_q, N_d, m, l)
if m >= l then
    foreach (x, a) ∈ sp-list(N_q) do
        foreach (y, b) ∈ sp-list(N_d) do
            if dist(Q_x[a − 1], P_y[b − 1]) > ε then
                add ((x, a, a + m − 1), (y, b, b + m − 1))
                to MMSet
```

Fig. 7. UpdateMaximalMatchesSet algorithm.

our algorithm, each common subsequence s is obtained either from characters mismatch or a leaf node, so we just need to compare the characters before the common subsequence ($Q_x[p-1]$) and $P_y[q-1]$ to identify the maximal match.

We can also process multiple query SF sequences at the same time by inserting them to the query suffix tree GST_q, so the nodes with the same path label are visited only once and the performance will be improved.

The complexity of matching GST_q and GST_d depends on the matching distance threshold ε. If $\varepsilon = 0$, the symbols match if and only if they are identical. Thus, searching the query tree against the database tree takes time $O(n|\Sigma|)$, where n is the length of the query and Σ is the alphabet size. If $\varepsilon > 0$, then for each edge of the query suffix tree we may have several matching edges in the database tree. In the worst case, each edge of the query tree matches each edge of the database tree, for a complexity of $O(nM)$, where n is query length and M is total length of all the SF sequences in the database. To verify maximality of a match takes additional time proportional to number of occurrences of the maximal matches in the query and database trees.

3.2. Ranking

The maximal matches are obtained for the query sequence and reference sequences in the database. Every maximal match is a diagonal run in the matrix formed by a query and reference sequence. We use the best diagonal runs described in the FASTA algorithm ***(29)*** as our ranking scheme. We calculate the alignment as a combination of the maximal matches with the maximal score. The score of the alignment is the sum of the scores of the maximal matches minus the gap penalty. We use the length of the maximal match as its score. Likewise, the length of the gap is used as the gap penalty. Two maximal matches can be chained together if there is no overlap between them. We use a fast greedy algorithm to find the chains of maximal alignments. At first, the maximal matches are sorted by their length. The longest maximal match is chosen first, and we remove all other overlapping matches. Then, we choose the second longest maximal match, remove its overlapping matches and repeat the above steps until no maximal matches are left. This way we find the longest chained maximal matches between the query and each retrieved database SF sequence. Finally, all the candidates with small alignment scores are screened out and only the top similar proteins are selected.

3.3. Post-processing

For each top protein SF sequence with a high score selected from the database, it is aligned with the query by running Smith–Waterman ***(28)*** dynamic programming method. The similarity score between two residues is set to 1 if the distance between their normalized feature vectors is smaller than ε, or else it is set to 0. Proteins are then ranked in decreasing order according to their new alignment scores, and the top proteins with the highest scores are reported to the user.

4. Experiments

The SCOP database ***(30)*** classifies proteins according to a four-level hierarchical classification, namely, family, super-family, fold and class. Because the SCOP database is curated by visual inspection, it is considered to be extremely accurate. For our tests, the target database we used has proteins from four classes of SCOP: all α, all β, $\alpha+\beta$ and α/β. Our data set D includes a total of 1810 proteins taken from 181 superfamilies that have at least 10 proteins, but only 10 proteins are chosen from each superfamily. One protein from each superfamily is chosen randomly as the query, so the size of the query set D_q is also 181. This is the same data set used in several previous indexing studies ***(16,18)***.

To evaluate our algorithm, we perform two different tests: The *retrieval* test finds the number of correct matching structures from the same superfamily as the query among the top k scoring proteins, and the *classification test* tries to classify the query at the superfamily and class levels. Our algorithm was implemented in C++ and all experiments reported below were done on a PC with 2.8 GHz CPU and 6 GB RAM, running Linux 2.6.6. To index the set of

1810 proteins (with average length 164) took a total time of 20.3 s. For the queries reported below, the query times ranged from 0.47 s to 4.41 s per query.

4.1. Retrieval Test

We compare our approach with one of the best previous indexing approach ProGreSS *(16)*, using the Java-based code provided by its authors. We also directly compare with a geometric hashing-based *(9)* indexing method, which we coded ourselves. For geometric hashing we take two consecutive C_α atoms along the backbone as the reference frame. Each remaining C_α atom and the reference frame form a triplet. The three pair-wise distances from a triplet are added to an R^*-tree if all of them are within 7 Å. For querying, we form query triplets in the same manner and find all matching triplets within ε range. Suppose there are n triplets with the same query reference frame and the matching protein has m triplets with the same reference frame, these two reference frames are considered to be a matching pair if the ratio between m and n is greater than a threshold, that is, if $m/n > 0.75$. The score of a protein is its number of matching reference frames with respect to the query, and the proteins are ranked based on their scores.

We ran the experiments using PSIST, ProGreSS and geometric hashing to obtain the number of proteins found from the same superfamily for each of the 181 queries. As each superfamily has 10 proteins, including the query, there can be at most 10 correct matching proteins from the same superfamily.

There are five parameters used in our approach. w is the size of the window used to index the local features, b is the range used to normalize the feature vectors, ε is the distance threshold based on the normalized feature vectors, l is the minimum length of the maximal matches and k is the number of top scoring proteins reported. We first show how PSIST performs for different values of w, ε, b, l and k.

Figure 8 shows the number of proteins found from the same superfamily for different top k cutoffs. Note that the number of correct matches is an average over all 181 SCOP superfamilies used in our test. The retrieval performance tapers off as k increases. We choose the largest cutoff as $k = 100$, as there is not much to be gained by using larger values.

We next study the effect of varying window size w, while keeping $b = 10$, $\varepsilon = 3$ and $l = 15$. **Figure 9** shows that a smaller window size of $w = 3$ yields the largest number of correct matches (on average 8 correct matches out of 10), and the retrieval rate drops as w increases. For a smaller window size, more matches are found in the database within the ε distance, and PSIST is able to find the best matches after finding the chain of maximal matches. For

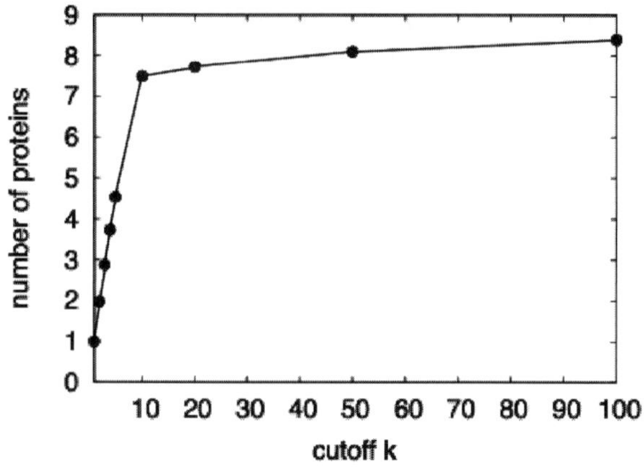

Fig. 8. Number of proteins found from same superfamily for different top k value ($w = 3$, $b = 10$, $\varepsilon = 3$ and $l = 10$).

larger windows, the number of matches drops and some of the correct proteins are missed. From this experiment, we conclude that $w = 3$ is the best for PSIST.

Figure 10 shows the effect of varying ε with $k = 100$. The larger the ε, the more the structures retrieved and then PSIST is able to find the correct ones by ranking the alignments. We find that $\varepsilon = 3$ works well for PSIST, and performance tapers off for larger values.

Figures 11 and **12** show that the varying normalization range b and the length of maximal match l have the similar effect on the number of proteins

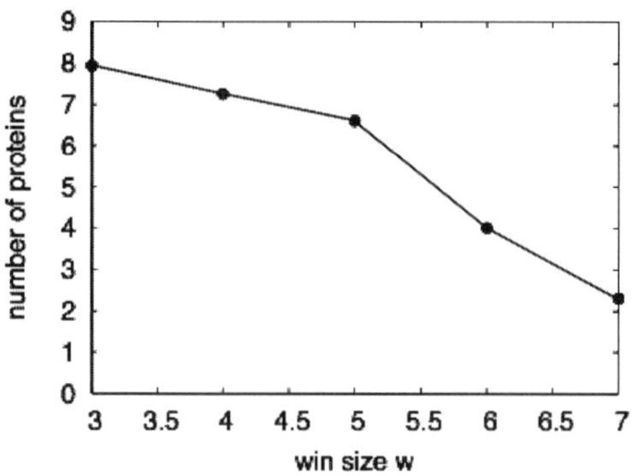

Fig. 9. Number of proteins found from the same superfamily for different window sizes when ($b = 10$, $\varepsilon = 3$ and $l = 15$).

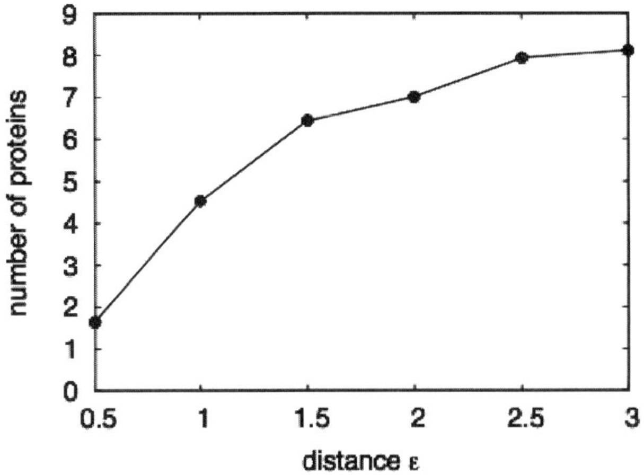

Fig. 10. Number of proteins found from the same superfamily for different ε ($w = 3$, $b = 10$, $\varepsilon = 3$ and $l = 15$).

found from the same superfamily. For smaller range b and maximal match length l, there can potentially be many incorrect proteins with similar match segments, but for larger b and l, fewer maximal matches, but correct proteins are found. PSIST obtains its best performance when the bin range is between 6 and 10, and the length between 9 and 12.

Table 2 shows the comparison of the number of proteins found from the same superfamily for different top k values. The table compares the performance of our approaches against geometric hashing and ProGreSS. Geometric hashing can find

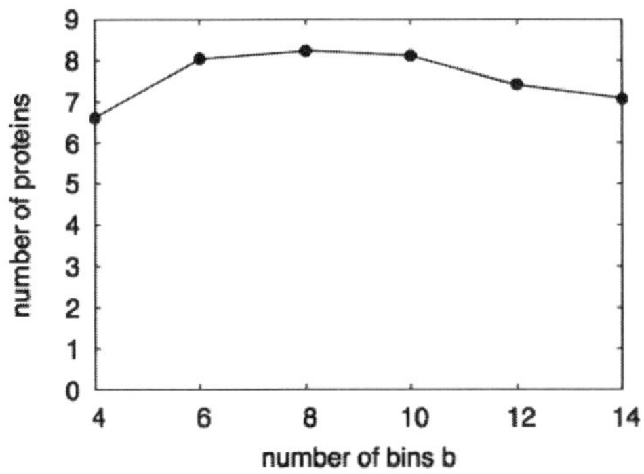

Fig. 11. Number of proteins found from the same superfamily for different b ($w = 3$, $\varepsilon = 2.5$ and $l = 15$).

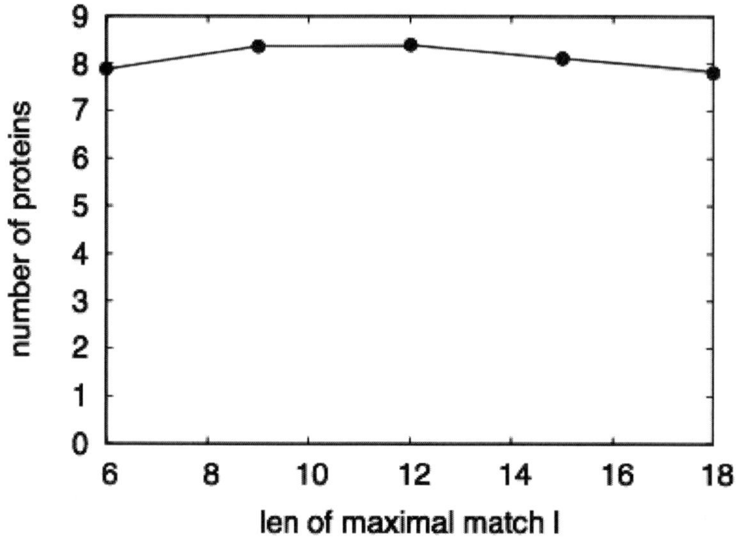

Fig. 12. Number of proteins found from the same superfamily for different length of maximal matches ($w = 3$, $\varepsilon = 2.5$ and $b = 10$)

only 2.43 correct proteins within the top 10 proteins (with $\varepsilon = 0.18$, which was the best value we determined empirically). It also has relatively poor performance for other values of k. Both ProGreSS and PSIST retrieve more than three correct proteins within the top four candidates. However, PSIST performs better than ProGreSS when the cutoff increases. For instance, PSIST could find 7.49 out of 10 proteins within the top 10 candidates. Note that based on the previous experiments, for the PSIST algorithm we set $w = 3$, $b = 10w = 3$, $\varepsilon = 3$ and $l = 9$. For fair comparison, we tuned the parameter settings for ProGreSS to report its best results (we use sequence distance threshold $\varepsilon_t = 0.05$, the structure distance threshold $\varepsilon_q = 0.01$ and window size $w = 3$).

Table 2
Overall Comparison of the Number of Proteins Found from the Same Superfamily Among the Top *k* Candidates

Algorithm	Top 4	Top 10	Top 50	Top 100
GeoHash	2.43	3.74	4.40	4.86
ProGreSS	3.53	6.17	6.69	7.09
PSIST	3.72	7.49	8.10	8.40

4.2. Classification Test

In the classification test, we assume we do not know the superfamily or the class to which a query protein belongs. For each query we then classify it into one of 181 SCOP superfamilies and one of the four SCOP classes (all α, all β, $\alpha+\beta$ and α/β) as follows. For each query, the top k similar proteins are selected from the database. The query itself is not counted in the top k matches. Each protein among the top k matches is assigned a score, a superfamily id, and a class id. The scores of the top k proteins from the same superfamily or class are accumulated. The query is assigned to the superfamily or class with the highest score. This classification approach can thus be thought of as k nearest neighbor classification. Below, we report results separately for the superfamily level and class level classification. For the performance, we report the percentage of correctly classified query proteins (out of the 181 queries). For the classification tests, we also compare with the numbers reported by PSI ***(18)*** and LFF ***(17)***, in addition to the results of ProGreSS and geometric hashing. For PSIST, ProGreSS and geometric hashing, we use the best parameter settings reported in **Section 4.1**.

Proteins are classified correctly if the proteins from the same superfamily have a better rank. Thus, the classification accuracy is proportional to the number of the correct proteins found in the top candidates. For instance, **Fig. 13** shows the percentage of query proteins correctly classified for different window sizes, when $\varepsilon = 3$ and using $k = 3$, at the superfamily and class levels. It has a

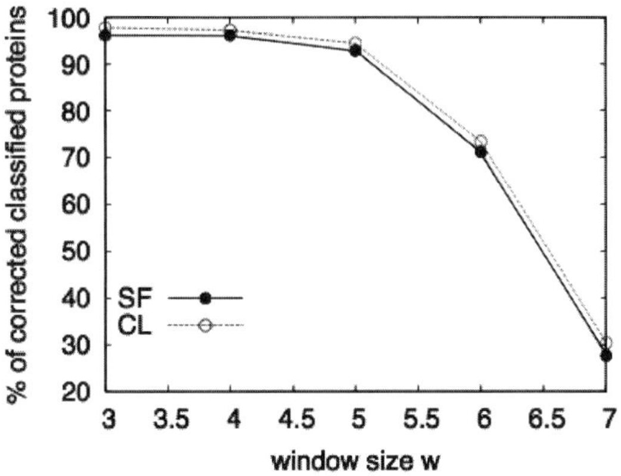

Fig. 13. Percentage of query proteins correctly classifed for different window sizes when $\varepsilon = 3$. (SF refers to the Superfamily level, and CL refers to the Class level).

similar shape as **Fig. 9**, the more the proteins found from the same superfamily the higher the accuracy obtained.

Table 3 shows the SCOP classification comparison with other algorithms at the superfamily and class levels respectively. Geometric hashing has the worst performance, it can only classify 60.2 and 72.9% proteins correctly at the superfamily and class levels. PSI *(18)* uses SSE-based features, and its accuracy for superfamily is 88%, but its class accuracy is unavailable. LFF profiles *(17)* only classify 68.5% of the superfamily correctly, but it agrees with SCOP classification at 93% for class level (note that LFF profiles use a different testing protein data set than ours). ProGreSS and PSIST could obtain more than three proteins within the top four candidates, so their accuracy is very close and much better than the others. ProGreSS uses both the structure and sequence features to classify the proteins, and its accuracy is 97.2 and 98.3% at the superfamily and class levels. Without considering the sequence features, PSIST has slightly better performance than ProGreSS, its accuracy is 97.8 and 99.4% at the superfamily and class levels.

4.3. Performance Test

We compare the running time of different approaches in this section. Suppose a protein has n residues, the window size is w, then the number of feature vectors is $n - w + 1$, so the complexity of our approach is $O(n - w - 1) = O(n)$ per protein. Assume the average number of neighbors of each reference frame is k, the complexity of our implementation of geometric hashing is $O(k^* n)$. Although they have the same complexity, geometric hashing is slower because of the coefficient k; it's running time is 1080.4 s per query for distance $\varepsilon = 0.18$.

Table 3
SCOP Classification Accuracy Comparison at the Superfamily and Class Level

Algorithm	Superfamily (%)	Class (%)
Geometric hashing	60.2	72.9
PSI	88	N/A
LFF	68.6	93.2
ProGreSS	97.2	98.3
PSIST	97.8	99.4

Table 4
Running Time Comparison

Algorithm	Superfamily (%)	Class (%)	Top 10	Time (s)
ProGreSS	97.2	98.3	6.17	1.67
PSIST-1	96.7	98.3	6.57	0.47
PSIST-2	97.2	99.4	7.19	4.41
PSIST-3	97.2	99.4	7.19	3.28

Both ProGreSS and PSIST provide a trade-off between the running time and the accuracy performance by adjusting the parameters such as window size and distance. For a fair algorithmic comparison, we compare the time performance of ProGreSS and PSIST based on their retrieval and classification test. **Table 4** shows the running time for ProGreSS and PSIST. For ProGreSS, we choose the best sequence and structure distance thresholds and set window size $w = 3$. We set $w = 3, b = 2, \varepsilon = 0$ and $l = 15$ for the first case of PSIST, and it is 3.5 times faster than ProGreSS with similar retrieval and classification perform ace. The last two cases have the same parameters: $w = 3, b = 6, \varepsilon = 2$ and $l = 15$, but the difference is that the third case builds a query suffix tree for every 20 queries and processes them together. They have the same retrieval and classification performance but the third case is faster. Although both cases are slower than ProGreSS, they retrieve on average more proteins (7.49 vs. 6.47) out of the top 10 matches and obtain slightly higher accuracy.

5. Conclusion

In this chapter, we present a new local feature representation of protein structures and convert the structure indexing to sequence indexing. We also propose a novel use of suffix trees to find the maximal matches between SF sequences and use the alignment between the query and database SF sequences to measure the structure similarity. Compared to ProGreSS, our approach either obtains higher accuracy or runs faster with similar classification accuracy.

Acknowledgment

We thank Tolga Can, Arnab Bhattacharya and Ambuj Singh for providing us the ProGreSS code and other assistance. We also thank Chris Bystroff and Nilanjana De for helpful suggestions. This work was supported in part by NSF CAREER Award IIS-0092978, DOE Career Award DE-FG02-02ER25538, NSF grant EIA-0103708 and NSF grant EMT-0432098.

References

1. B. Rost. Twilight zone of protein sequence alignments. *Protein Eng*, 12(2):85–94, 1999.
2. S. Altschul, T. Madden, J. Zhang, Z. Zhang, W. Miller, and D. Lipman. Gapped BLAST and PSI-BLAST: a new generation of protein database search programs. *Nucleic Acids Res*, 25(17):3389–3402, 1997.
3. I. Eidhammer, I. Jonassen, and W. Taylor. Structure comparison and structure patterns. *J Comp Biol*, 7(5):685–716, 2000.
4. L. Holm and C. Sander. Protein structure comparison by alignment of distance matrices. *J Mol Biol*, 233:123–138, 1993.
5. I. Shindyalov and P. Bourne. Protein structure alignment by incremental combinatorial extension (CE) of the optimal path. *Protein Eng*, 11(9):739–747, 1998.
6. T. Madej, J. Gibrat, and S. Bryant. Threading a database of protein cores.*Proteins*, 23:356–369, 1995.
7. K. Mizoguchi and N. Go. Comparison of spatial arrangements of secondary structural elements in proteins. *Protein Eng*, 8:353–362, 1995.
8. C. Orengo and W. Taylor. SSAP: sequential structure alignment program for protein structure comparisons. *Methods Enzymol*, 266:617–634, 1996.
9. Y. Lamdan and H. Wolfson. Geometric hashing: a general and efficient model-based recognition scheme. *International Conference on Computer Vision (ICCV)*, 238–249, 1988.
10. L. Holm and C. Sander. 3-d lookup: fast protein structure database searches at 90% reliability. *International Conference on Intelligent Systems for Molecular Biology (ISMB)*, 179–187, 1995.
11. R. Nussinov, N. Leibowit, and H. Wolfson. MUSTA: a general, efficient, automated method for multiple structure alignment and detection of common motifs: application to proteins. *J Comp Biol*, 8(2):93–121, 2001.
12. O. Dror, H. Benyamini, R. Nussinov, and H. Wolfson. MASS: multiple structural alignment by secondary structures. *Bioinformatics*, 19(12):95–104, 2003.
13. M. Shatsky, R. Nussinov, and H. Wolfson. Multiprot - a multiple protein structural alignment algorithm. *Proteins*, 56:143–156, 2004.
14. X. Yuan and C. Bystroff. Non-sequential structure-based alignments reveal topology-independent core packing arrangements in proteins. *Bioinformatics*, 21(7):1010–1019, 2005.
15. T. Can and Y. Wang. CTSS: a robust and efficient method for protein structure alignment based on local geometrical and biological features. *IEEE Computer Society Bioinformatics Conference (CSB)*, 169–179, 2003.
16. A. Bhattacharya, T. Can, T. Kahveci, A. Singh, and Y. Wang. Progress: simultaneous searching of protein databases by sequence and structure. *Pacific Symposium on Bioinformatics (PSB)*, 264–275, 2004.
17. I. Choi, J. Kwon, and S. Kim. Local feature frequency profile: a method to measure structural similarity in proteins. *Proc Natl Acad Sci*, 101(11):3797–3802, 2004.

18. O. Camoğlu, T. Kahveci, and A. Singh. Towards index-based similarity search for protein structure databases. *IEEE Computer Society Bioinformatics Conference (CSB)*, 148–158, 2003.
19. Z. Aung, W. Fu, and K. Tan. An efficient index-based protein structure database searching method.*International Conference on Database Systems for Advanced Applications (DASFAA)*, 311–318, 2003.
20. H. Täubig, A. Buchner, and J. Griebsch: A method for fast approximate searching of polypeptide structures in the PDB. *German Conference on Bioinformatics (GCB),* 2004.
21. D. Gusfield. *Algorithms on Strings, Trees, and Sequences: Computer Science and Computational Biology.* Cambridge University Press, New York, 1997.
22. E. Hunt, M. Atkinson, and R. Irving. Database indexing for large DNA and protein sequence collections. *International Conference on Very Large Data Bases (VLDB)*, 256–271, 2003.
23. C. Meek, J. Patel, and S. Kasetty. Oasis: an online and accurate technique for local-alignment searches on biological sequences. *International Conference on Very Large Data Bases (VLDB)*, 910–923, 2003.
24. A. Delcher, S. Kasif, R. Fleischmann, J. Peterson, O. White, and S. Salzberg. Alignment of whole genomes. *Nucleic Acids Res*, 27(11):2369–2376, 1999.
25. A. Delcher, A. Phillippy, J. Carlton, and S. Salzberg. Fast algorithms for large-scale genome alignment and comparison. *Nucleic Acids Res.*, 30(11): 2478–2483, 2002.
26. E. McCreight. A space-economic suffix tree construction algorithm. *J. of the ACM*, 23(2): 262–272, 1976.
27. E. Ukkonen. On-line construction of suffix trees. *Algorithmica*, 14(3):249–260, 1995.
28. F. Smith and M. Waterman. Identification of common molecular subsequences. *J Mol Biol*, (147):195–197, 1981.
29. W. R. Pearson and D. J. Lipman. Improved tools for biological sequence comparison.*Proc Natl Acad Sci*, 85:2444–2448, 1988.
30. A. Murzin, S. Brenner, T. Hubbard, and C. Chothia. SCOP: a structural classification of proteins database for the investigation of sequences and structures. *J Mol Biol*, 247:536–540, 1995.

IV

Protein Features Prediction

7

Hidden Markov Models for Prediction of Protein Features

Christopher Bystroff and Anders Krogh

Summary

Hidden Markov Models (HMMs) are an extremely versatile statistical representation that can be used to model any set of one-dimensional discrete symbol data. HMMs can model protein sequences in many ways, depending on what features of the protein are represented by the Markov states. For protein structure prediction, states have been chosen to represent either homologous sequence positions, local or secondary structure types, or transmembrane locality. The resulting models can be used to predict common ancestry, secondary or local structure, or membrane topology by applying one of the two standard algorithms for comparing a sequence to a model. In this chapter, we review those algorithms and discuss how HMMs have been constructed and refined for the purpose of protein structure prediction.

Key Words: Transmembrane; local; motif; Viterbi; Baum–Welch; profile; topology; folding.

1. Introduction

A hidden Markov Model (HMM) is a type of directed graph. The vertices of the graph are referred to as "states" or "Markov states," and the directed edges as "transitions." A state path through an HMM emits a symbol from each state, possibly depending on an input string of symbols. In this chapter, we will restrict the discussion to amino acids as the input symbols and to various structural features of proteins as the emitted, output symbols. The ways that the states are connected by transitions defines the topology of the model. Both

From: *Methods in Molecular Biology, vol. 413: Protein Structure Prediction, Second Edition*
Edited by: M. Zaki and C. Bystroff © Humana Press Inc., Totowa, NJ

the transitions and the emission symbols can be represented within the model as probability distributions, and those probabilities can be defined and refined based on protein sequence and structure data.

HMMs have been used extensively in bioinformatics. For proteins, one of the first applications was for sequence profile modeling, and that is probably still the most well-known application of HMMs to proteins *(1,2)*. In profile HMMs, each state emits either a single amino acid or a gap character, and the output emission string represents an alignment of the input string to the model. The model is generally constructed from a multiple sequence alignment, and as such, a profile HMM represents a family of homologous proteins. Profile HMMs have been discussed extensively in the literature, and so will be reviewed only briefly in this chapter. From a pedagogical point of view, profile HMMs are a bad place to start because they do not show the full modeling capabilities of HMMs. We will concentrate instead on the application of HMMs for the prediction of protein conformational and biochemical features. To show how simple HMMs actually are, we will start with an introduction to HMMs for non-specialists, using the prediction of membrane helices as an example. For people interested in implementing HMMs or going deeper into the theory, we recommend **ref.** ***3***, but this chapter will hopefully be a good introduction to the process of constructing, refining and using HMMs for protein sequences.

Algorithms for using HMMs will be discussed in the context of the transmembrane (TM) helix model. Predictions derived from HMMs consist of emission symbols, which may be calculated as either the most probable string, using the Viterbi algorithm, or as the probability of each emission symbol at each sequence position, using the forward–backward algorithm. In either case, the prediction consists of first defining a sequence of states, or pathway, then converting those states into emission symbols. This second step may be independent of the first. Early attempts to construct simple HMMs for the prediction of protein secondary structure will be reviewed next. In these models, each state emits a single position of either alpha helix, beta sheet or loop, where loop is defined as anything other than helix or sheet. Then an application to local structure prediction will be described, where the loop symbol is split into many, more specific symbols, representing different types of turns and loops in proteins. The concept of an HMM as a grammatical model is useful here, as short unbroken strings of states can be thought of as words in a sentence. HMMs have been used as grammatical models in other fields *(3)*.

2. Hidden Markov Models for Transmembrane Helices

In this section, HMMs will be explained from an example: the prediction of membrane helices in membrane proteins. Membrane helices are characterized by being more hydrophobic than average. Therefore, to locate TM helices in amino acid sequences, the traditional methods use some sort of hydrophobicity scale, which is typically averaged in windows and plotted along the protein. Instead, one could make a simple HMM such as the one shown in **Fig. 1**. The model consists of two states, one for TM helices, labeled by 'M,' and one for everything else, labeled by 'X' in the figure. State M has some associated emission probabilities that are simply the normalized frequencies of amino acids found in TM helices of known TM proteins. Similarly, state X has a set of probabilities estimated from all the rest of the amino acids in known TM proteins. The arrows are associated with the so-called transition probabilities. The M → X transition tell how likely it is to see a transition out of a TM helix, the M → M how likely it is to stay in the helix state, and so on. **Table 1** shows how the emission probabilities are calculated from a set of known TM proteins (which is the set of 160 TM proteins used in **ref. *4***. **Table 2** shows how the transition probabilities are calculated from the same set.

Why is it called a *hidden* Markov model? This is often a mystery to people, and from the way we introduced HMMs here, it is not obvious. Often, it is just the amino acid sequence that is known, so the states (in this case X or M) are unknown—the states are *hidden*. This is exactly the situation we are in when presented with a new protein and are asked to predict the membrane helices (if any). A lot of the theory about HMMs has to do with how to predict the hidden states, which in fact can give such predictions. In a (non-hidden) Markov model, there is a one-to-one correspondence between a state and an amino acid, or whatever type of symbols it models.

A labeling of a protein will denote an assignment of a class label to each amino acid—if the amino acid is in a TM helix, it is labeled M, otherwise it is labeled X. Using the HMM above, we can easily calculate the probability of a certain labeling of the sequence by just multiplying probabilities in the model.

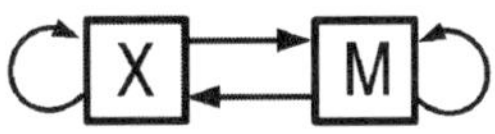

Fig. 1. A simple Hidden Markov Model for transmembrane helices. M represents transmembrane helix, X all other structures.

Table 1
The Counts and Frequencies for Amino Acids in Transmembrane Helices and Other Regions

Amino acids	TM helices		Other regions		Over-represented
	Count	Frequency	Count	Frequency	
I	1826	0.120	2187	0.046	2.61
F	1370	0.090	1854	0.039	2.31
L	2562	0.168	4156	0.087	1.93
V	1751	0.115	2935	0.061	1.89
M	616	0.040	1201	0.025	1.60
W	414	0.027	819	0.017	1.59
A	1657	0.109	3382	0.071	1.54
Y	615	0.040	1616	0.034	1.18
G	1243	0.082	3352	0.070	1.17
C	289	0.019	960	0.020	0.95
T	755	0.050	2852	0.060	0.83
S	806	0.053	3410	0.071	0.75
P	423	0.028	2640	0.055	0.51
H	121	0.008	1085	0.023	0.35
N	250	0.016	2279	0.048	0.33
Q	141	0.009	2054	0.043	0.21
D	104	0.007	2551	0.053	0.13
E	110	0.007	2983	0.062	0.11
K	78	0.005	2651	0.055	0.09
R	83	0.005	2933	0.061	0.08
Total	15214	1.000	47900	1.000	

The frequencies correspond to the emission probabilities in the two Hidden Markov Model states. The amino acids are sorted by over-representation in the TM helices (last column). Notice the most over-represented are hydrophobic amino acids and that the charged amino acids are in the bottom of the list.

More specifically, if s denotes the entire amino acid sequence and y denotes the associated sequence of class labels, the probability of the sequence is,

$$p(s|y) = a_{B,y_1} e_{y_1}(s_1) a_{y_1,y_2} e_{y_2}(s_2) \cdots e_{y_l}(s_l) a_{y_l,E} \tag{1}$$

where s_i is amino acid i, y_i is the corresponding class (H or X), and l is the sequence length. Here the e's are emission probabilities and the a's are the

Table 2
Calculation of Transition Frequencies for the Simple Model in Fig. 1

Sequences	160	
TM helices	696	
TM residues	15214	
Other residues	47900	
M → X	0.046	$=\frac{696}{15214}$
M → M	0.954	$=\frac{15214-696}{15214}$
X → X	0.982	$=\frac{47900-696-160}{47900-160}$
X → M	0.015	$=\frac{696}{47900-160}$

TM, transmembrane.

transition probabilities. For example, if $y_1 = \text{X}$ and $y_2 = \text{M}$, then $a(y_2, y_1) = 0.015$, corresponding to the X → M transition in **Table 2**. Now we can predict the most probable TM helices by finding the most probable labeling. For each possible labeling, we do the above calculation and choose the one with the highest probability. In principle it is straightforward, but there are 2^l possible labelings, so for long sequences it is tedious. Fortunately, there is a very simple way of doing this by dynamic programming, called the Viterbi algorithm. As with other dynamic programming algorithms, it reuses calculations as we walk along the sequence. It is outlined below.

2.1. The Viterbi Algorithm

Here the Viterbi algorithm is outlined for the simple two-state model. The task is to find the most probable labeling of a sequence, or the most probable "state path" as it is normally called. A (state) path means a sequence of states that the sequence can travel through. The path starts in the *begin* state. The probability that the best labeling is an M in position 1 is simply $v_{\text{M}}(1) = a_{B,\text{M}}\, e_{\text{M}}(s_1)$. The probability $v_{\text{X}}(1)$ of the best labeling being an *X* at position 1 is calculated analogously. The probability of the best labeling being an M in position 2 is obviously the best of extending the best labeling to position 1 ending in M and that ending in X. That is, $v_{\text{M}}(2)$ is the maximum of $v_{\text{M}}(1)a_{\text{M.M}}e_{\text{M}}(s_2)$ and $v_{\text{X}}(1)a_{\text{X.M}}e_{\text{M}}(s_2)$, and similarly for $v_{\text{X}}(2)$. This can be continued, so at amino acid *i* we would have

$$v_M(i) = \max\{v_M(i-1)\, a_{M,M} e_M(s_i),\, v_X(i-1)\, a_{X,M} e_M(s_i)\} \tag{2}$$

and

$$v_X(i) = \max\{v_M(i-1)\, a_{M,X} e_X(s_i),\, v_X(i-1)\, a_{X,X} e_X(s_i)\}. \quad (3)$$

To find the most probable labeling, one must do a trace-back. As we choose the best label at each position, we also record the previous label. Stringing these memories together starting at the end state produces the labeling. It is very similar to sequence alignment algorithms [e.g., the Needleman–Wunch algorithm *(5)*]. Note that the time it takes to do the complete calculation is proportional to l, as compared to 2^l for the more naive approach.

2.2. Posterior Probabilities

It turns out that the most probable labeling is usually not very close to the correct answer for a simple model like this. Alternatively, we might be interested in the probability that a given amino acid sits in a membrane helix or not. This probability is the total probability of getting an M at position i divided by the probability of having M or X at position i. These probabilities can be found by summing probabilities over all possible labelings, but again there are simple dynamic programming algorithms that do it more efficiently. The total probability is calculated by the so-called forward algorithm, which is like Viterbi, but sums instead of maximizing, so the forward variables f are calculated like

$$f_M(i) = f_M(i-1)\, a_{M,M} e_M(s_i) + f_X(i-1)\, a_{X,M} e_M(s_i) \quad (4)$$

and

$$f_X(i) = f_M(i-1)\, a_{M,X} e_X(s_i) + f_X(i-1)\, a_{X,X} e_X(s_i). \quad (5)$$

The final value obtained in the end state is the total probability of the sequence $P(S)$ given the model (the conditioning on the model is implicit in the equations). The value of f_M(i) is the probability of the sequence being in the M state at position i. A very similar algorithm called the "backward" algorithm starts at the other end of the protein and gives the probability of the sequence from i to the end if we are in state M at position i. The backward variable for membrane helix at position i, b_M is calculated as

$$b_M(i) = a_{M,M} e_M(s_{i+1})\, b_M(i+1) + a_{M,X} e_X(s_{i+1})\, b_X(i+1) \quad (6)$$

By multiplying the forward and backward variables and dividing by the total probability of the sequence as calculated by the forward algorithm, we obtain the probability of being in state M at position i. This so-called *posterior probability* tells us how likely it is that amino acid number i is in a membrane helix.

$$g_M(i) = \frac{f_M(i)\, b_M(i)}{P(S|\lambda)} \tag{7}$$

In **Fig. 2**, the result of the Viterbi algorithm and the posterior probability of being in the membrane are shown for a membrane protein (one of the 160 proteins in the data set used). In the example, the probability of being in a TM helix is much better correlated with the actual structure than the most probable structure predicted by Viterbi.

The posterior probability gives a result which is very similar to standard hydrophobicity plots except that it does not need some ad hoc window averaging, and the parameters of the model have been estimated from real TM proteins. More details on these calculations can be found in **ref. *6***.

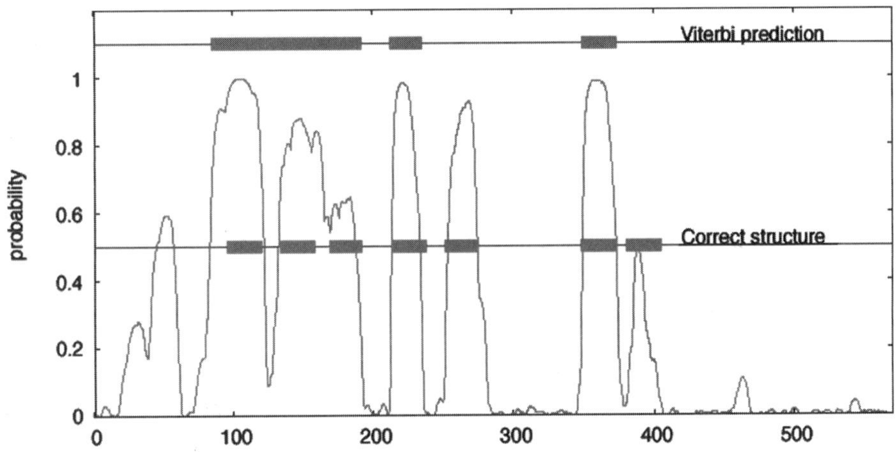

Fig. 2. A human adrenergic receptor (Alpha 1D-adrenoceptor, swiss-prot ID ADA1DHUMAN) was analyzed with the simple two-state Hidden Markov Model. The protein belongs to the G-protein-coupled receptor family and has seven transmembrane (TM) helices. The x-axis corresponds to amino acid positions in the protein, which has a length of 572. The thin line curve shows the posterior probability of being in a TM helix for each amino acid. The true TM helices are shown with the grey bars in the middle of the plot. The most probable labels as found by the Viterbi algorithm are shown in the top of the plot. Notice that the posterior probability actually corresponds quite well with the correct TM helices. Notice also that Viterbi makes some quite bad errors.

A very good introduction to the Viterbi, forward and backward algorithms, is found in **ref.** ***3***, which uses examples from signal processing and speech recognition.

2.3. More Complex Models

The model discussed above is the simplest possible. To model TM proteins, one could easily extend it by adding more states. For example, one could have states modeling the helix caps, different states for modeling cytoplasmic versus extracellular parts of the chain, and so on. Before going into the state of the art models, we will just discuss one simple extension of the simple model, which illustrates several important points.

In the simple model, there is an implicit and wrong assumption about the distribution of helix lengths. The M state has a transition to itself with probability $a_{M,M}$, which means the probability of staying in that state for l amino acids is proportional to $a_{M,M}{}^{l}$. This exponentially decaying function is very far from the real length distribution of TM helices, which would have a probability zero for a helix of length below around 15, a maximum probability around 22, and would decay to zero when the length gets to around 30. It is possible to model the length distribution more accurately by introducing more states. For instance, one could have an array of 35 states with each state having a transition to the next, and additionally, transitions from state 1 to all states after state 13. This makes it possible to model the observed length distribution exactly if we assume a minimum of 15 and a maximum of 35.

The state topology shown in **Fig. 3** implements an alternative length modeling, which relies on a bell-shaped probability distribution called a negative binomial ***(6)***. In this model, we have used four states with transitions to themselves to model the variable length. These states are tied, which means that they have the same emission and transition probabil-

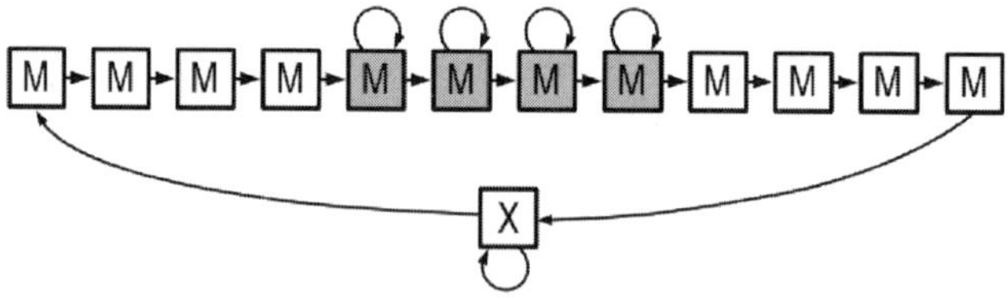

Fig. 3. A more sophisticated Hidden Markov Model, modeling the known length distribution in transmembrane helices. Only shaded M states have transitions to themselves.

ities. These states are flanked by four states to each side modeling the helix caps making the minimum length eight amino acids long, but the length modeling ensures that the distribution of lengths can fit the observed lengths.

After estimation of parameters as described below, we have used Viterbi and posterior decoding as for the simple model on the same protein, and the result is shown in **Fig. 4**. The Viterbi algorithm fails completely on this example. This is because the length modeling is dependent on summing over all the possible paths through the model, whereas Viterbi only finds one path. It is clear that the posterior probability correlated very well with the actual structure and that the more advanced model is better at discriminating the helices. The second and third helices, for instance, would have been predicted as one by the simple model (*see* **Fig. 2**), whereas the more complex model clearly separates them. The last model also has a more pronounced prediction of a wrong TM helix in the beginning, but has a clearer signal for the last.

2.4. Parameter Estimation

When making models such as the one shown in **Fig. 3**, a new problem arises: More than one state is associated with each label. Therefore, the estimation of parameters is not quite as straightforward. In the simple model, parameters

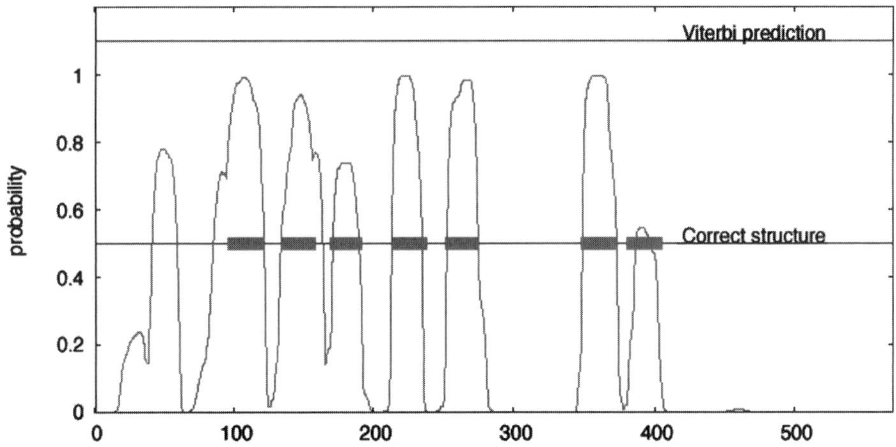

Fig. 4. Same plot as **Fig. 2**, but with the more sophisticated 15-state Hidden Markov Model in **Fig. 3**. Posterior probabilities are shown as thin lines. Transmembrane helices are now more well defined than with the simple model. But the Viterbi algorithm completely fails to predict helices.

could be estimated by counting, but that is not generally possible with more states per label (for the model in **Fig. 3** it could actually still be done, because of the tying of probabilities). The estimate of parameters from frequencies in the simple model can be shown to be the maximum likelihood estimate. In maximum likelihood, the parameters are chosen so as to maximize the total probability of the data (i.e., the set of membrane proteins used—sometimes called the training set). The same principle is used to estimate a more complex model.

The maximum likelihood parameters cannot be computed exactly in general, but an iterative estimation procedure for HMMs exists, which is called the Baum–Welch estimation procedure *(3,6)*, or expectation-maximization (EM). In this procedure, a set of start parameters are assigned (e.g., random numbers or uniform distributions). Based on this model, letter frequencies and transition frequencies are *estimated* for each state. These estimates are then used to obtain the new parameters. This procedure is then repeated until the parameters change very little in each update. It can be shown to give a local maximum of the likelihood, but not necessarily the global maximum.

First f, b, and g values (*see* **Eqs 4–7**) are calculated using the initial estimates for the parameters a and e. Then, a maximum likelihood re-estimate of the transition probability a'_{pq} would be calculated as

$$a'_{\mathrm{pq}} = \sum_t f_{\mathrm{p}}(t)\, a_{\mathrm{pq}} b_{\mathrm{q}}(t+1), \tag{8}$$

where t runs over the entire training set database. The maximum likelihood re-estimate of the emission values $e_{\mathrm{q}}(s)$ are calculated in a similar fashion.

$$e'_{\mathrm{q}}(m) = \frac{\sum_{t \in s_t = m} g_{\mathrm{q}}(t)}{\sum_t g_{\mathrm{q}}(t)} \tag{9}$$

The sum in the numerator runs over all positions t where the amino acid is m, while the sum in the denominator runs over all t. The method will converge to the parameters that maximize the probability of the data given the model.

In this section, we have illustrated some features of HMMs and hope that it has shown that HMMs are a very general and versatile sequence modeling framework. The models of TM proteins can be further refined by modeling the cytoplasmic and non-cytoplamic loops differently, by having two models of helices for the two directions through the membrane, and so on. The main limiting factor is the availability and quality of training data. This is important for all data-driven methods—it is not possible to reliably estimate arbitrary numbers of model parameters from a limited data set, so the complexity of the model is limited by the data.

2.5. Topology of Transmembrane Proteins

So far in this section HMMs were explained using the example of TM helix prediction, so it is a natural next step to discuss how to go from this example to actual systems for TM helix prediction. Computational prediction of TM helices started around 1981 with analysis of amino acid hydrophobicity profiles ***(7,8)***. The "positive inside rule" ***(9)***, which basically states that there is an abundance of positively charged amino acids (R and K) on the cytoplasmic side of the membrane, was used to predict the membrane protein topology ***(10)***. By the topology of a TM protein we mean the location of TM helices along the sequence and the location of the intervening regions as either cytoplasmic or non-cytoplasmic. Since then, a number of different non-HMM methods have been applied to the problem, e.g. **ref.** ***11–13***. Then, in 1998, the two first HMM-based methods appeared ***(14,15)***. It is probably fair to say that those methods building on HMMs are the most successful at present, and the purpose of this section is to review them and explain the main ideas in the HMMs used, in particular those features that distinguish them from each other. Most will deal with TM proteins of the helix bundle type, but in the end topology prediction with HMMs for beta barrel membrane proteins will also be reviewed, and we will also touch on signal peptide prediction.

There are several reasons why HMMs are very well suited for this problem. First, the HMM can nicely capture the compositional difference between the hydrophobic membrane helices and other regions as illustrated by the simple models above. Second, the HMM can capture the "grammar" of the problem. For instance, the length distribution can be modeled by an HMM in a natural way. This is not well modeled by sliding window approaches, such as neural networks (NNs). Similarly, the membrane helix separates inside (cytoplasmic) from outside, so one can build the model such that one can never predict, for example, inside-helix-inside, a cytoplasmic region followed by a membrane portion followed by another cytoplasmic region. This, again, is not handled elegantly by sliding window approaches. On the downside, HMMs are not as "non-linear" as NN, which means that they are not good at capturing non-linear correlations between amino acids, which may or may not be important.

The first two HMM approaches were developed independently and published in 1998. TMHMM ***(14,4)*** is a straightforward extension of the simple models described in the introduction with some sophistication in the estimation and decoding procedures. A total of 35 states were used for modeling a membrane helix with transitions allowing lengths from 15 to 35. Five states at each end of the TM region model the helix caps, with one set of emission probabilities for inside helix caps and another set for the outside caps (the states are tied). The states modeling the remaining helix (the core) are also tied and this is where the length can vary. There are two identical helix submodels, one for outgoing and one for incoming.

On the inside, 10 states model the last 10 amino acids before the start of a helix and 10 states model the first 10 amino acids after an ingoing helix. These 20 states have tied emissions and they have a transition pattern that allows for loop lengths shorter than 20. For loops longer than 20, the remaining amino acids are modeled by a single state denoted as the globular state. Loops on the outside of the membrane are modeled in exactly the same manner, except that loops longer than 100 amino acids were modeled in a separate branch. In total, there are seven different amino acid distributions:

1. globular part,
2. inside close to the membrane,
3. inside helix cap,
4. core TM helix,
5. outside helix cap,
6. outside close to the membrane for long loops, and
7. outside close to the membrane for short loops.

TMHMM was trained and cross-validated on a set of 160 membrane proteins with known topologies. The exact boundary between the TM helix and loop region is not well defined because the experimental data are not very accurate in that respect, and even for TM proteins with known 3D structures the exact boundaries are hard to define. Therefore, a procedure was adopted where a region of ±3 amino acids were "unlabelled"' during the initial training to let the model itself find the optimal boundary. After the initial training, the model was used to re-label the data allowing the boundaries to move by ±5 amino acids. That is, new boundaries were defined by the model, which were more accurate or at least more optimal for the model (*see* **Fig. 5**). This new labeling was used to train the model again with fixed boundaries. In a final stage of training, the conditional maximum likelihood was used. For prediction of the topology, the N-best algorithm was used ***(16)***. TMHMM was shown to have a better performance than other methods at the time.

Interestingly, the HMMTOP model was developed at the same time and used a completely different approach ***(15)***. One of the main differences is that the model was developed for prediction on a family of similar proteins, so the

```
            MCYGKCARCIGHSLVGLALLCIAANILLYFPNGETKYASENHLSRFVWFFS
Correct     iiiiiiiiiMMMMMMMMMMMMMMMMMMMMMoooooooooooooooMMMMMM
Unlabeled   iiii..........MMMMMMMMMMM..........ooooo..........M
Relabeled   iiiiiiiiiiMMMMMMMMMMMMMMMMMMMMMMMooooooooooooooMMMM

            GIVGGGLLMLLPAFVFIGLEQDDCCGCCGHENCGKRCAMLSSVLAALIGIA
Correct     MMMMMMMMMMMMMMMMMMMiiiiiiiiiiiiiiiiiiMMMMMMMMMMMMMM
Unlabeled   MMMMMMMMMMMMMM..........iiiiiiii..........MMMMMMMMM
Relabeled   MMMMMMMMMMMMMMMMMMMiiiiiiiiiiiiiiiiiiiiMMMMMMMMMMMM

            GSGYCVIVAALGLAEGPLCLDSLGQWNYTFASTEGQYLLDTSTWSECTEPK
Correct     MMMMMMMMMMMMMMooooooooooooooooooooooooooooooooooooo
Unlabeled   MMMMMMMMM..........oooooooooooooooooooooooooooooooo
Relabeled   MMMMMMMMMMMoooooooooooooooooooooooooooooooooooooooo

            HIVEWNVSLFSILLALGGIEFILCLIQVINGVLGGICGFCCSHQQQYDC
Correct     ooooooooMMMMMMMMMMMMMMMMMMMMMMMMMMMMMMMiiiiiiiiii
Unlabeled   ooo..........MMMMMMMMMMMMMMMMMMMMM..........iiiii
Relabeled   oooooooooooMMMMMMMMMMMMMMMMMMMMMMMiiiiiiiiiiiiiii
```

Fig. 5. The protein "transmembrane 4 L6 family member 1," T4S1 HUMAN from Swiss-Prot is shown with the membrane labeling from Swiss-Prot ("Correct") below. Here the line called "Unlabeled" shows the diluted labeling, which is used to constrain the prediction using the initial model. The last line ("Relabeled") is the labeling predicted according to the constraints and used for the remaining rounds of training.

first step is to identify sequences that are similar to the query sequence using BLAST. The model architecture resembles that of TMHMM, but instead of estimating and decoding in the usual way (as described in **Subheading 1.**), HMMTOP uses retraining when predicting a family of TM proteins. When estimating models from a small set of data, it is customary to use pseudocounts, which are numbers that are added to the estimated counts in the Baum–Welch procedure. The pseudo counts ensure that the model parameters do not depart too much from values set a priori. In HMMTOP, a data set of 83 known structures was used to set pseudo counts for all the model parameters, and these guided the estimation of the model from a set of homologous membrane proteins. Because the TM topology is not known for the set, the training is unsupervised, but because of the pseudo counts, it is ensured that the model does not develop into something meaningless. After training the model, the Viterbi algorithm was used to decode the original query protein. The advantage of the approach is that the model is adapted to the particular type of protein in question and that it can use evolutionary information without having to align sequences, but there is a danger that features having nothing to do with membrane helices take over parts of the model. The performance of the method is very similar to that of TMHMM (see below).

Since HMMTOP and TMHMM, several methods have emerged, which improves and extends on the HMM framework, and most of these shows gains in performance. Kahsay et al. developed *(17)* a TMHMM-like model, and an improved pseudo count scheme is used to achieve a better prediction performance. Some methods deal with inclusion of evolutionary information through a pre-calculated multiple alignment. In the work by Viklund and Elofsson *(18)*, the TMHMM architecture is used with multiple alignment columns used as observations instead of amino acids. The probability of an alignment is essentially the geometric mean of the probabilities of the individual sequences. In the work by Käll et al. *(19)*, predictions on individual sequences are averaged and "maximum accuracy decoding" is used to obtain a consensus prediction from the sequences.

2.6. Prediction of Signal Peptides and Other Features

Signal peptides are another important feature of proteins that are well suited for modeling by HMMs. Signal peptides and their cleavage sites have been predicted by many different methods. The most used is probably SignalP, which is based on NN, and in later versions combined with an HMM. The latest version of SignalP is described in **ref. *20***, which also reviews the earlier literature.

A signal peptide, which is cleaved off the protein during the transport over the membrane, is the n-terminal part of the protein and contains a region that is typically positively charged (the n-region) followed by a hydrophobic region (the h-region) and then the c-region, where there is some conservation of amino acids around the cleavage site. This structure is very well suited for HMM modeling, because it is easy to make submodels corresponding to these regions. In the work by Nielsen and Krogh *(21)*, a fairly simple model was made containing a few states for each of the regions. The model can be used to predict whether a protein contains a signal peptide based on the total probability of the sequence given the model, or it can be used to predict the cleavage site by calculating the posterior probability of the state modeling the last amino acid before the cleavage site. For the first task, it was better than the original SignalP NN, whereas the NN was better at detecting the exact cleavage site. In later versions of SignalP, the NN and HMMs are combined. A similar HMM was later used to discriminate lipoprotein signal sequences from signal peptides in Gram negative bacteria *(22)*.

Signal sequences and N-terminal membrane helices are difficult to discriminate from each other. Many signal peptides are predicted as TM helices and vice versa. This fact is unfortunately often over-looked in the literature, which results in quite different performance evaluations. For instance, in the work by Klee and Ellis ***(23)***, membrane proteins were excluded when evaluating the performance of signal peptide prediction, which gives a fairly unrealistic result, and similarly, some comparisons of TM helix prediction methods is done on proteins without signal peptides. To try to deal with this difficulty, Phobius was developed, which is an HMM that models both signal peptides and TM helices ***(19)***. Phobius was shown to better discriminate the two features than any combination of individual predictors.

Membrane proteins that form a beta barrel rather than a helix bundle have more recently also been modeled with models similar to the TMHMM structure ***(24,25)***. The main problem is that very few of these protein structures have been determined, which limits the amount of data for training and testing. Finally, it should be mentioned that also coiled-coil domains have been successfully modeled with HMMs ***(26)***.

3. Secondary Structure

Students of protein structure are very familiar with the three types of secondary structure: alpha helix (H), extended beta strand (E), and loop or coil (L). The programs DSSP ***(27)*** and STRIDE ***(28)*** assign protein secondary structure to each position in a known structure based on their characteristic hydrogen bonding patterns, successive $i \rightarrow i+4$ H-bonds for H and either parallel hydrogen bonds ($i \rightarrow j, i+1 \rightarrow j+1$, etc.) or antiparallel ones ($i \rightarrow j, i-1 \rightarrow j+1$, etc) for E. The L assignments are simply all positions that are not H or E. The accuracy in three-state prediction is measured using "Q3," which is simply the fraction correct of all discrete three-state predictions.

The most successful methods for secondary structure prediction have been NN ***(29,30)*** and, more recently, support vector machines ***(31)***, both which detect correlations between secondary structure and patterns in the surrounding sequence. But HMMs provide more insight into the underlying constraints of the system. To illustrate the differences between NNs and HMMs, consider the case of alpha helices. Helices cannot occur as single-position predictions, as by definition, a helix must contain at least four consecutive amino acids. The NN approach would attempt to learn this property by training the network on known structure, where presumably no single character helix occurs. Q3 would be higher, and the model would be rewarded, for a singleton H prediction relative to no H's when the true structure is H. But this prediction is locally wrong. An HMM model for helix such as the one in **Fig. 6** guarantees that no single-H predictions are ever made, as the first H position can transition only to at least two more H's. HMMs can impose constraints such as these that enforce and reproduce known structural features such as length distributions, similar to the case for TM helix length previously illustrated in **Fig. 3**.

Rationally designed HMMs succeeded in producing predictions that satisfied protein-like length distributions and secondary structure content *(32)*. When measured using Q3, HMM-based secondary structure did not compare well to NN-based methods, but rather than call this a failure, we become conscious of the insensitivity of the Q3 metric to certain types of accuracy, especially length distribution accuracy. The entrenchment of one metric for accuracy can have a stifling effect on innovation. Fortunately, this has been recognized by participants in CASP, and now, a large number of new metrics for accuracy are being used and actively discussed *(33)*.

Models such as the one in **Fig. 6** partially hard-wire the known periodicity of the sequence in amphipathic alpha helices. The inward facing side of the helix is dominated by non-polar side chains while the outward facing side chains are polar, giving a polar/non-polar periodicity of three to four residues. This is modeled as a cycle (states 1, 2, and 3 in **Fig. 6**). When iteratively trained against sequences with true secondary structure assignments, one or more of these five states will adopt a non-polar character and one or two will adopt a polar character. Similarly, beta strands have a polar/non-polar periodicity of 2, which can be enforced by the design of the HMM. The main shortcoming of HMMs for secondary structure prediction seemed to be the need for a rational

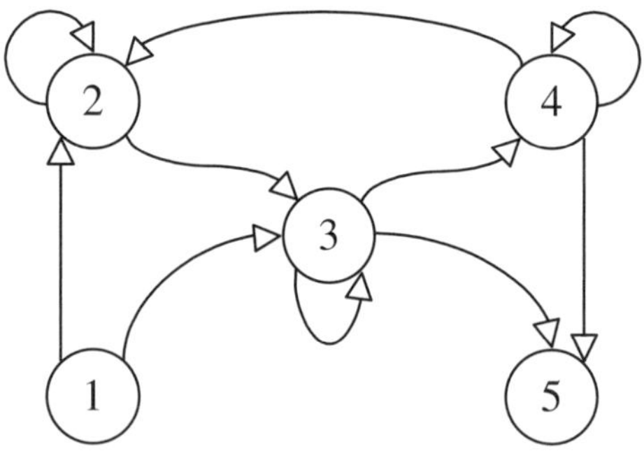

Fig. 6. The topology of the helix unit used by Asai et al. *(32)* to predict secondary structure. The periodicity of amphipathic helices is approximately modeled by the cycle of states. States 1 and 5 represent the start and end of the helix, respectively.

design of the topology of the model. NNs, on the contrary, require no such rational input as training could be initiated using a generic network.

4. Local Structure

In **Subheading 3.**, we described protein structure using three states, H, E, and L. Because of the L state, a secondary structure prediction cannot be mapped to a unique local conformation, where "local" is used to mean proximal in the sequence. The L state is a catch-all state that includes helix cap and reverse turn backbone angles as well as angles normally found in helices and strands. Unlike for secondary structure, there remains to date no widely accepted metric for accuracy in local structure prediction. However, several attempts have been made to go beyond three states in one-dimensional protein structure prediction using HMMs.

Temple Smith and his colleagues ***(34,35)*** experimented extensively with state-space models, close cousins of HMMs, which distinguished between tight turns and long loops. Helices were modeled as amphipathic patterns of exposed and buried amino acids, and the allowed lengths of secondary structural elements were enforced. Models were designed to recognize the class of the protein and were successful in that goal. State-space models preceded the popularity of HMMs in protein science and, despite their similarity to HMMs, were never formalized as such and consequently were not refined by EM.

Local structure is unambiguous if the backbone angles at each position are uniquely defined, even if the definition is not precise. In several studies, it was found that a fairly small number of ways exist to fold a short peptide of four to six residues ***(35–37)*** and that conservation of backbone angle types was a sufficient condition for the conservation of the hydrogen bonds and side chain interactions that stabilized the structure ***(38)***. In clusters of similar peptide conformations, a surprisingly broad range of backbone angles were often permitted, ranging upwards of 90° ***(38,39)*** for some positions.

HMMs have been constructed for local structure prediction by defining states to represent short-peptide (four amino acids) conformations as defined by their inter-alpha carbon distances ***(40)***. These structure-based clusters were shown to have sequence preferences. But as the mapping of sequence patterns to structure is many-to-one, sequence specificity was lost when the structures were clustered together. For example, the cluster belonging to alpha helix did not show positional specificity for buried and exposed residues because the backbone distances are not different between the buried and exposed sides. The buried side and the exposed side were clustered together. The same can be said for beta strands, which also have a buried/exposed sequence signature.

Another weakness of the model is that it does not formally define a probability distribution over structures, because consecutive distance vectors are dependent and may be in conflict. It means that a "structure" emitted by the model is often non-physical. However, this work demonstrated that proteins could be modeled as sequences of structural building blocks, akin to a grammar.

To overcome some of the above limitations, recently a model was developed, which emits angles of C-alpha traces, which always yields a valid structure. In this model, the angles are also sequence dependent, so it gives a very good basis for sampling of protein conformations ***(41,42)***.

4.1. HMMSTR, a Model for Local Structure

Here, we discuss in detail the process of constructing an HMM for local structure called HMMSTR. In this model, protein sequences are treated as sentences composed of words. In language, the order of letters within each word is highly invariant, but words may be arranged in different orders to form sentences. In proteins, words are recurrent local structure types such as a type-1 beta hairpin. Every type-1 beta hairpin has a canonical sequence of backbone angles, and although type-1 beta hairpins do not all have identical amino acid sequences, they conserve a common pattern of amino acid types. In addition to type-1 beta hairpins, there are dozens of recurrent short sequences of amino acid types that may be thought of as the words that make up protein sentences. Using this principle, it was straightforward to identify short sequence patterns that correlated strongly with angle patterns by co-clustering. The resulting sequence/structure mappings were called I-sites ***(38)***, and the HMM based on these mappings was dubbed HMMSTR (an HMM for STRucture) ***(43)***.

These I-site motifs were treated as words in sentences, and an HMM was constructed to model all of the ways these words could be strung together, much like the grammatical rules of language. Unlike written language, however, motifs in proteins are not delimited by spaces or any type of punctuation. In fact, they overlap to a great extent. To create an HMM based on motifs, the locations of all of the occurrences of all 182 I-site motifs were found in all proteins of known structure. Each Markov state represented a motif position, and transitions were drawn between motif positions that were adjacent at any position in the database. Many of the I-site motifs share segments in common. For example, an alpha helix motif and a helix capping motif share the helix segments; therefore, the locations of occurrences of these two motifs in the database would often overlap. If positions in any two motifs frequently overlapped in proteins, then the corresponding states were merged to one state.

The analogy to this process would be the process of building an HMM for language. The database would be a library of English literature, and the states would be all of the words in the dictionary. Words that were found adjacent to each other anywhere in the literature would be connected by a transition. When the words were more frequently adjacent, the probability of the transition would be correspondingly higher. In this example, there would be a one-to-one correspondence between states and emissions (words), making this model a Markov Chain with nothing "hidden" about it. But if we consider that the words are made of letters and that they may be mis-spelled or the punctuation misplaced, then the letter sequence does not uniquely define the state sequence and the model is an HMM, not a Markov chain.

4.1.1. Training and Topological Modification of HMMSTR

The process of training the HMM involves changing the emission and transition probabilities to maximize the probability of the model (λ) given the data (S), or $P(\lambda|S)$. Using the Baum–Welch algorithm, or EM, the parameters are iteratively re-estimated from the posteriori state probabilities as explained earlier for the TM helix model (*see* **Eqs 8** and **9**). However, EM is incapable of resetting any value that is zero. This means that new transitions cannot be added to the model, they can only be modified and removed.

In refining the HMMSTR model, a technique was used to identify "missing" transitions by finding the maximum value of z,

$$z_{\mathrm{pq}} = \sum_{t} f_{\mathrm{p}}(t)\, b_{\mathrm{q}}(t+1) \qquad (10)$$

over all states p and q that are unconnected in the current model. A new transition, a_{pq} was set to a small non-zero value, after which subsequent interactions of EM would converge on an optimal value for a_{pq}. This proved to be a very useful technique for automatically modifying the topology of the HMM.

Another trick that was used was to create a flexible HMM topology was the introduction of a non-emitting state that connected all "sink" states with all "source" states. Sinks are states that have no outgoing forward transitions, and sources are states with no incoming transitions. The HMMSTR model after merging I-site motifs contained several sink and source states because some motifs occurred at the beginning or end of sequences or adjacent to unstructured loops. A transition to the non-emitting "naught" state is equivalent to a set of transitions from all sink states to all source states. By including the naught state, we guaranteed that the model had no dead ends. No protein sequence could be assigned a zero probability. The naught state also doubled as the beginning and ending state.

4.1.2. Application of HMMSTR to Understanding Misfolding

Local structure predictions made by HMMSTR have been used for structure prediction at the global level using Rosetta ***(44)*** or for prediction of contact maps using HMMSTR-CM ***(45)***. Another application of local structure predictions is to understand the folding pathways of proteins. When proteins fold, some parts of the protein fold earlier and some parts later. The order of events is called the "pathway" of folding. Errors in folding can lead to pathological misfolded states that can aggregate in the cell. Human prion protein is an all alpha-helical protein in its globular form, but a misfolded state of prion protein forms beta sheet-rich amyloid fibers under certain conditions and when certain point mutations are present. How the globular, helix-rich structure might convert to beta strands is of great interest.

The structure of the soluble form of human prion was solved first by NMR ***(46)***, which showed it to be a monomeric 3-helix bundle, then later by X-ray crystallography ***(47)***, which showed it to be a domain-swapped dimer. Posterior probabilities were calculated for each state in the HMMSTR model according to **Eq. 7**. For the purposes of analysis, all states that correspond to alpha helix and all states that correspond to beta strand were summed separately, and the resulting propensities for secondary structure were compared to the true secondary structure (*see* **Fig. 7**), looking for discrepancies. The HMMSTR predictions showed that one of the three helices, helix 2, has a low statistical propensity for helix. This is also a part of the sequence having point mutations that are associated with familial amyloid encephalopathy ***(48)***.

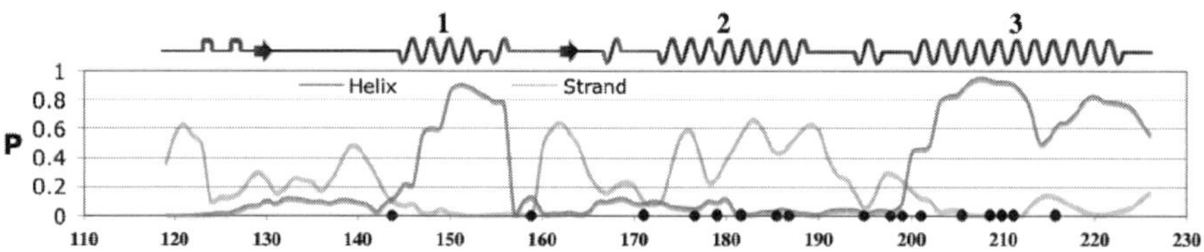

Fig. 7. Posterior probabilities for HMMSTR states condensed to secondary structure probabilities for human prion protein (PDB code 1I4M): helix (darker line) and strand (lighter line). Figure above the graph represents the true secondary structure for the protein. The second helix is mis-predicted, showing a strong tendency for beta strand. The locations of known pathogenic point mutations are marked as black dots ***(24)***.

The implication is that helix 2 (residues 173–188) is energetically unstable as an alpha helix and is most likely the site of amyloid initiation, as amyloid is composed of stacked beta strands. The sequence patterns in the region 173–188 do not match the canonical patterns associated with helix, and as these patterns are canonical for reasons of energetic natural selection, we conclude that the structure of helix 2 is unstable. As is true for the case in point, mutations that lead to structural instability can lead to disease. In this case, the mutation data preceded the structure, but a conclusion that the second helix would be sensitive to mutation would have been possible without knowing the location of the mutations. Pathogenic mutations on helix 3 are explained by close interactions with helix 2 in the structure.

5. Profile HMMs

Profile HMMs have been reviewed extensively elsewhere ***(1,2)***, but a short discussion will serve as a reminder. Profile HMMs greatly improved remote homology detection over the previous state-of-the-art, mostly due to their ability to capture position-specific gap probabilities and amino acid probabilities.

All protein structure prediction algorithms have their roots in one of two basic principles: energy and evolution. Energy guides the folding process; evolution produced the diversity of proteins that exist today. Energy underlies the prediction of local structure using HMMSTR or TM helices using TMHMM. These models contain sequence patterns for either local motifs or TM helices. The sequence patterns within these models predict structural properties because sequences that match these patterns have a low free energy when they fold into the corresponding structures. These are not evolutionary models because sequences that match the same pattern are not assumed to have a common ancestor. Global sequence similarity was not used to build these models.

Profile HMMs, on the contrary, are rooted in the principle of common ancestry. Sequences that have greater than about 25% sequence identity are very likely to have diverged from the same ancestral sequence. In fact, nowhere in evolutionary history have sequences with common ancestors been found to adopt different folded structures. It is a rule without exceptions. Common ancestry directly implies a common structure. But, homology is often difficult to detect. There are many sequences with even less than 25% identity are known to have structural homology and therefore probably have a common ancestor sequence. But sequence identity is a poor metric in this "twilight zone" of low sequence similarity. The vast majority of matches below this cutoff are non-homologous. To increase the sensitivity of sequence alignment as a structure prediction tool, a family of sequences can be modeled as an amino acid probability distribution, or more simply, a "profile." Comparing a sequence to a profile is much more sensitive than comparing two single sequences.

Comparing any two sequences, or a sequence and a profile, requires finding the highest scoring alignment. In traditional dynamic programming algorithms, a position-independent gap penalty is used as part of the scoring function, but upon inspection of multiple sequence alignments for large families, it is abundantly clear that gaps are far more likely in some positions and less likely in others. Profile HMMs attempt to capture this information by allowing each position in the profile to have a different probability of initiating a gap or insertion.

The topology of a profile HMM is fixed, having three types of states, Match, Gap, and Insertion, one of each per position in the profile. Match states emit amino acids from a profile and have transitions forward to the next Match state, a Gap state and an Insertion state. Gap states do not emit but only connect Match states that are not adjacent. Insertion states emit amino acids and have transitions to themselves, to Gap states, and to Match states. The transition probabilities are initialized based on a set of aligned sequences, possibly using phylogenetic sequence weighting to correct for redundancy among the sequences. **Figure 8** shows a small segment of a multiple sequence alignment and a segment of a profile HMM that models it. States that have zero probability are dimmed in this image.

Libraries of profile HMMs, such as Pfam ***(49)***, increase sensitivity in remote homology detection, as shown in CASP experiments ***(50,51)***. Profile HMMs

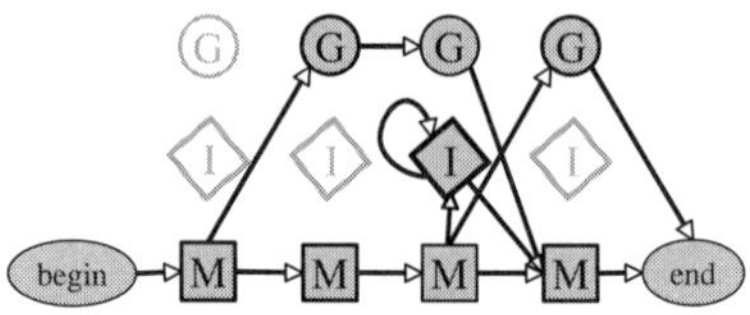

Sequence alignment	State paths
ADG••K	MMMM
A••••K	MGGM
ADG•••	MMMG
ADGHIK	MMMIIM

Fig. 8. A profile Hidden Markov Model for a short fragment of a multiple sequence alignment. M states represent matches to the parent sequence (1st sequence), G states are non-emitting gaps states, and I states are insertions, relative the parent. Begin and end states are obligatory non-emitting states that normally would occur before and after the full-length alignment. Each sequence can be represented as a state pathway.

are considered by many to be the state-of-the-art in modeling protein sequence families for the purpose of remote homology detection. This high status is a result of the flexible and yet constraining nature of HMMs in general. This section only serves as a reminder. For more details, *see* **refs *1*,*6*,** and ***52***.

6. Conclusions

HMMs may be used to predict protein structure by modeling their state topology and training their parameters against a set of sequences on known structure. We have described models where the states take on different structural meaning—structure and location relative to a membrane (TMHMM), or backbone angles (HMMSTR), or common ancestry in a sequence family (Pfam). Algorithms were introduced that find the best state path through an HMM (Viterbi), find the posterior probability of any state at any position in the sequence (forward/backward), and re-estimate the parameters of the model from data (Baum–Welch). All HMMs use these three algorithms. The topology of an HMM may be predefined using expert knowledge, or may be at least partially defined by the data itself.

Acknowledgments

This material is based upon work supported by the National Science Foundation under Grant DBI-0448072 to C.B.

References

1. Eddy, S. Profile hidden Markov models. Bioinformatics, 14:755–763.
2. Madera, M. et al. (2004). The SUPERFAMILY database in 2004: additions and improvements. Nucleic Acids Res, 32(90001):235–239.
3. Rabiner, L. (1989). A tutorial on hidden Markov models and selected applications in speech recognition. Proceedings of the IEEE, 77(2):257–286.
4. Krogh, A., Larsson, B., von Heijne, G., and Sonnhammer, E. (2001). Predicting transmembrane protein topology with a hidden Markov model: application to complete genomes. J Mol Biol, 305(3):567–580.
5. Needleman, S. and Wunsch, C. (1970). A general method applicable to the search for similarities in the amino acid sequence of two proteins. J Mol Biol, 48(3):443–453.
6. Durbin, R., Eddy, S., Krogh, A., and Mitchison, G. (1998). Biological Sequence Analysis: Probabilistic Models of Proteins and Nucleic Acids. Cambridge University Press.
7. Kyte, J. and Doolittle, R. (1982). A simple method for displaying the hydropathic character of a protein. J Mol Biol, 157(1):105–132.

8. Argos, P., Rao, J., and Hargrave, P. (1982). Structural prediction of membrane-bound proteins. Eur J Biochem, 128:565–575.
9. von Heijne, G. (1990). The signal peptide. J Membr Biol, 115(3):195–201.
10. von Heijne, G. (1992). Membrane protein structure prediction. Hydrophobicity analysis and the positive-inside rule. J Mol Biol, 225(2):487–494.
11. Jones, D., Taylor, W., and Thornton, J. (1994). A model recognition approach to the prediction of all-helical membrane protein structure and topology. Biochemistry, 33(10):3038–3049.
12. Rost, B., Casadio, R., and Fariselli, P. (1996). Refining neural network predictions for helical transmembrane proteins by dynamic programming. Proc Int Conf Intell Syst Mol Biol, 4:192–200.
13. Yuan, Z., Mattick, J., and Teasdale, R. (2004). SVMtm: support vector machines to predict transmembrane segments. J Comput Chem, 25(5):632–636.
14. Sonnhammer, E., von Heijne, G., Krogh, A., et al. (1998). A hidden Markov model for predicting transmembrane helices in protein sequences. Proc Int Conf Intell Syst Mol Biol, 6:175–182.
15. Tusnady, G. and Simon, I. (1998). Principles governing amino acid composition of integral membrane proteins: application to topology prediction. J Mol Biol, 283(2):489–506.
16. Chow, Y. and Schwartz, R. (1989). The N-Best algorithm: an efficient procedure for finding top N sentence hypotheses. Proceedings of the DARPA Speech and Natural Language Workshop, 199–202.
17. Kahsay, R., Gao, G., Liao, L., and Journals, O. (2005). An improved hidden Markov model for transmembrane protein detection and topology prediction and its applications to complete genomes. Bioinformatics, 21(9):1853–1858.
18. Viklund, H. and Elofsson, A. (2004). Best α-helical transmembrane protein topology predictions are achieved using hidden Markov models and evolutionary information. Protein Sci, 13:1908–1917.
19. Käll, L., Krogh, A., and Sonnhammer, E. (2005). An HMM posterior decoder for sequence feature prediction that includes homology in formation. Bioinformatics, 21(1):i251–i257.
20. Bendtsen, J., Nielsen, H., von Heijne, G., and Brunak, S. (2004). Improved prediction of signal peptides: SignalP 3.0. J Mol Biol, 340(4):783–795.
21. Nielsen, H. and Krogh, A. (1998). Prediction of signal peptides and signal anchors by a hidden Markov model. Proc Int Conf Intell Syst Mol Biol, 6:122–130.
22. Juncker, A., Willenbrock, H., von Heijne, G., Brunak, S., Nielsen, H., and Krogh, A. (2003). Prediction of lipoprotein signal peptides in Gram-negative bacteria. Protein Sci, 12:1652–1662.
23. Klee, E. and Ellis, L. (2005). Evaluating eukaryotic secreted protein prediction. BMC Bioinformatics, 6(1):256.

24. Martelli, P.L., Fariselli P., and Casadio, R. (2003). An ENSEMBLE machine learning approach for the prediction of all-alpha membrane proteins. Bioinformatics, 19(Suppl 1):I205–I211.
25. Fariselli, P., Finelli, M., Marchignoli, D., Martelli, P.L., Rossi, I., and Casadio, R. (2003). MaxSubSeq: an algorithm for segment-length optimization. The case study of the transmembrane spanning segments. Bioinformatics, 19:500–505.
26. Delorenzi, M. and Speed, T. (2002). An HMM model for coiled-coil domains and a comparison with PSSM-based predictions. Bioinformatics, 18(4):617–625.
27. Kabsch, W. and Sander, C. (1983). How good are predictions of protein secondary structure? Biopolymers, 22:2577–2637.
28. Heinig, M. and Frishman, D. (2004). STRIDE: a web server for secondary structure assignment from known atomic coordinates of proteins. Nucleic Acids Res, 32:500–502.
29. Rost, B. and Sander, C. (1993). Prediction of protein secondary structure at better than 70% accuracy. J Mol Biol, 232(2):584–599.
30. Jones, D. (1999). Protein secondary structure prediction based on position-specific scoring matrices. J Mol Biol, 292(2):195–202.
31. Ward, J., McGuffin, L., Buxton, B., and Jones, D. (2003). Secondary structure prediction with support vector machines. Bioinformatics, 19(13):1650–1655.
32. Asai, K., Hayamizu, S., and Handa, K. (1993). Prediction of protein secondary structure by the hidden Markov model. Bioinformatics, 9:141–146.
33. Zemla, A., Venclovas, C., Moult, J., and Fidelis, K. (2001). Processing and evaluation of predictions in CASP 4. Proteins, 45(Suppl 5):13–21.
34. Stultz, C. (1993). Structural analysis based on state-space modeling. Protein Sci, 2(3):305–314.
35. Bienkowska, J., He, H., and Smith, T. (2001). Automatic pattern embedding in protein structure models. Intelligent Systems, IEEE [see also IEEE Expert], 16(6):21–25.
36. Rooman, M.J., Kocher, J.P., and Wodak, S.J. (1991). Prediction of protein backbone conformation based on seven structure assignments. Influence of local interactions. J Mol Biol, 221(3):961–979.
37. de Brevern, A.G., Valadie, H., Hazout, S., and Etchebest, C. (2002). Extension of a local backbone description using a structural alphabet: a new approach to the sequence-structure relationship. Protein Sci, 11:2871–2886.
38. Bystroff, C. and Baker, D. (1998). Prediction of local structure in proteins using a library of sequence-structure motifs. J Mol Biol, 281(3):565–577.
39. Unger, R., Harel, D., Wherland, S., and Sussman, J. (1989). A 3D building blocks approach to analyzing and predicting structure of proteins. Proteins, 5:355–373.
40. Camproux, A., Tuffery, P., Chevrolat, J., Boisvieux, J., and Hazout, S. (1999). Hidden Markov model approach for identifying the modular framework of the protein backbone. Protein Eng, 12(12):1063–1073.

41. Kent, J. T. and Hamelryck, T. (2005). Using the Fisher-Bingham distribution in stochastic models for protein structure. In Barber, S., Baxter, P. D., V.Mardia, K., and Walls, R. E., editors, Proceedings of the 24th LASR Workshop, 57–60. Leeds University Press.
42. Hamelryck, T., Kent, JT, Krogh, A. (2006) Sampling realistic protein conformations using local structural bias. PLoS J Comput Biol., 2(9):e131.
43. Bystroff, C., Thorsson, V., and Baker, D. (2000). HMMSTR: a hidden Markov model for local sequence-structure correlations in proteins. J Mol Biol, 301(1):173–190.
44. Bystroff, C. and Shao, Y. (2002). Fully automated ab initio protein structure prediction using I-SITES, HMMSTR and ROSETTA. Bioinformatics, 18(1):54–61.
45. Shao, Y. and Bystroff, C. (2003). Predicting interresidue contacts using templates and pathways. Proteins, 53(Supple 6):497–502.
46. Zahn, R., Liu, A., Luhrs, T., Riek, R., von Schroetter, C., Garcia, F., Billeter, M., Calzolai, L., Wider, G., and Wuthrich, K. (2000). NMR solution structure of the human prion protein. Proc Natl Acad Sci USA, 97(1):145–150.
47. Knaus, K., Morillas, M., Swietnicki, W., Malone, M., Surewicz, W., and Yee, V. (2001). Crystal structure of the human prion protein reveals a mechanism for oligomerization. Nat Struct Biol, 8:770–774.
48. Kovacs, G., Trabattoni, G., Hainfellner, J., Ironside, J., Knight, R., and Budka, H. (2002). Mutations of the prion protein gene. J Neurol, 249(11):1567–1582.
49. Bateman, A., Coin, L., Durbin, R., Finn, R.D., Hollich, V., Griffiths-Jones, S., Khanna, A., Marshall, M., Moxon, S., Sonnhammer, E.L., Studholme, D.J., Yeats, C., and Eddy, S.R. (2004). The Pfam protein families database. Nucleic Acids Res 32: D138–D141.
50. Karplus, K., Sjoelander, K., Barrett, C., Cline, M., Haussler, D., Hughey, R., Holm, L., and Sander, C. (1997). Predicting protein structure using hidden Markov models. Proteins, 29(Suppl 1):134–139.
51. Tsigelny, I., Sharikov, Y., and Ten Eyck, L. (2002). Hidden Markov Models-based system (HMMSPECTR) for detecting structural homologies on the basis of sequential information. Protein Eng, 15(5):347–352.
52. Krogh, A., Brown, M., Mian, I. S., Sjölander, K., and Haussler, D. (1994). Hidden Markov Models in computational biology: applications to protein modeling. J Mol Biol., 235:1501–1531.

8

The Pros and Cons of Predicting Protein Contact Maps

Lisa Bartoli, Emidio Capriotti, Piero Fariselli, Pier Luigi Martelli, and Rita Casadio

Summary

Is there any reason why we should predict contact maps (CMs)? The question is one of the several 'NP-hard' questions that arise when striving for feasible solutions of the protein folding problem. At some point, theoreticians started thinking that a possible alternative to an unsolvable problem was to predict a simplified version of the protein structure: a CM. In this chapter, we will clarify that whenever problems are difficult they remain at least as difficult in the process of finding approximate solutions or heuristic approaches. However, humans rarely give up, as it is stimulating to find solutions in the face of difficulties. CMs of proteins are an interesting and useful representation of protein structures. These two-dimensional representations capture all the important features of a protein fold. We will review the general characteristics of CMs and the methods developed to study and predict them, and we will highlight some new ideas on how to improve CM predictions.

Key Words: Protein structure prediction; Protein contacts; Small world; Structure reconstruction; Machine learning; Contact map; Protein folding.

1. From Protein Structures to Contact Maps

Proteins structures are described by the coordinates (CO-representation) of the atoms that constitute the macromolecule. For a protein with n atoms we need $3n$ numbers (x, y and z coordinates for each atom) to specify its three-dimensional (3D) structure. An alternative view is to consider the distance matrix (DM), a symmetric matrix that contains the Euclidean distance between each pair of atoms. If the number of atoms is n we need n^2 elements; because

From: *Methods in Molecular Biology, vol. 413: Protein Structure Prediction, Second Edition*
Edited by: M. Zaki and C. Bystroff © Humana Press Inc., Totowa, NJ

the matrix is symmetric (the distance between atoms i and j is the same of that between j and i), the real number of elements is only $n(n - 1)/2$. Both representations, namely the coordinates and the DM, are equivalent, that is, we can convert each representation into the other. DM can be computed from the CO-representation simply by evaluating the Euclidean distance between each pair of atoms: values stored in the appropriate DM cell uniquely identify the pair i and j. Conversely, to go from DM to CO is not so trivial. There exists a Lagrange theorem ***(1)*** that states that once that the Gram matrix derived from DM is diagonalized, the three eigenvectors that correspond to the three highest eigenvalues are the atom coordinates in a 3D cartesian reference. Actually, there are two solutions, but the chirality of the molecule routinely can help in selecting the correct one (***1*** and references therein).

DM representation has far more elements than the coordinate-based representation, so why adopt it? The main advantage of DM representation arises when only a part of the data is known (i.e., in low-resolution NMR experiments). Still solutions can be found, thanks to DM properties ***(1)***. Another advantage of DM is that the protein is represented in a framework that automatically incorporates translational and rotational invariance and this in principle is more suitable for learning approaches.

Quite often in order to simplify the protein representation not all protein atoms are taken into account and residues are considered as unique entities. In this case, the DM has a number of rows (and columns) equal to the residue numbers. Each DM entry is then the distance between residue i and j. The distance between two residues can be defined in different ways, such as the following:

- the distance between a specific pair of atoms (i.e., CA–CA or CB–CB),
- the shortest distance among the atoms belonging to i residue and those belonging to residue j, and
- the distance between the centres of mass of the two residues.

Even though these choices are quite different and structurally minimal, they provide enough information to build the protein backbone, or at least the CA trace ***(1,2)***.

Starting from the protein DM and selecting an arbitrary distance cut-off, a further simplified representation can be obtained: the protein contact map (CM). CMs are binary symmetric matrices, whose non-zero elements represent the contacts between residues (*see* **Fig. 1**). In more details, given a DM and a defined threshold T the corresponding CM can be computed as:

$$\mathrm{CM}[i, j] = 1 \text{ if DM } [i, j] < T$$
$$\mathrm{CM}[i, j] = 0 \text{ if DM } [i, j] \geq T$$

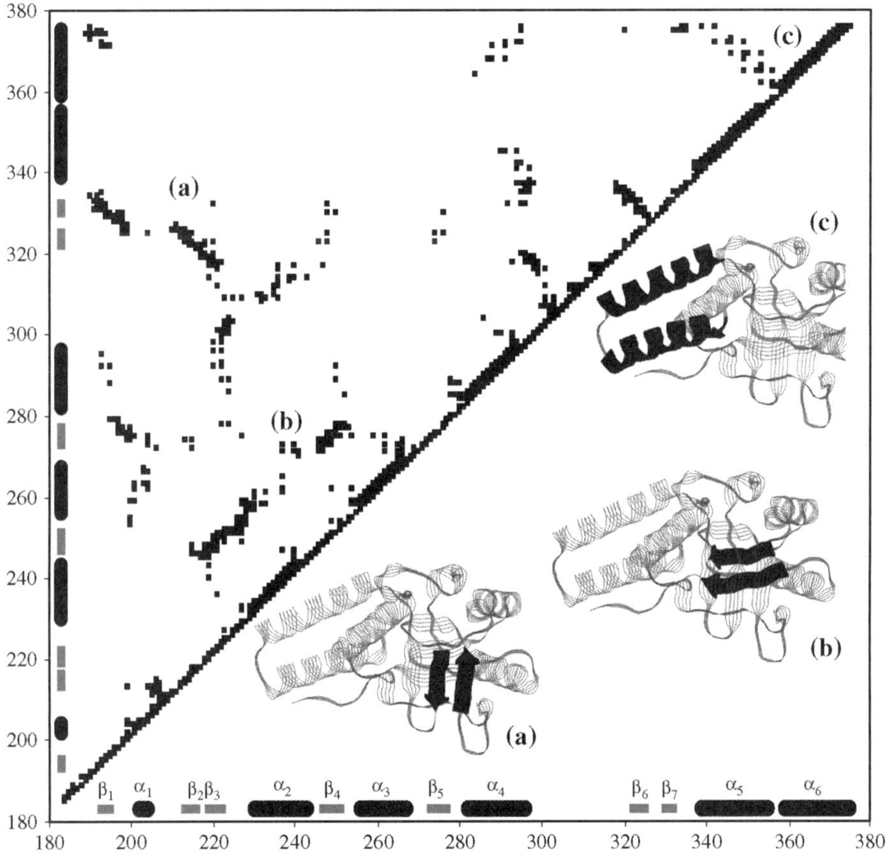

Fig. 1. Contact map of HSP-60 protein fragment (PDB code: 1KID). The secondary structure elements are highlighted along the x axis and y axis. α-helices and β-strands are represented in black and grey, respectively. On the left side of the plot the black dots indicate the contact regions (cut-off radius 8.0 Å centered at CB atoms). On the right side, the structural protein features are shown: (a) Anti-parallel sheet contacts; (b) parallel sheet contacts; (c) contacts between helical regions.

While the problem of reconstructing the protein coordinates from the DM has a well known solution, there are no analogous theorems for CM. However, some empirical applications have been built to address this issue. The results indicate that (at least for the tested proteins) it is possible to reconstruct the CO-representations from CMs *(2–5)*.

Protein CM representation has some pros and cons.

Pros:

- Unlike other protein representations such as secondary structure, CM conveys strong information about the protein 3D structure.
- The CM representation is translation and rotation invariant and more compact than the DM representation.
- CM is more suited than DM for learning problems. The binary CM nature can be regarded as a classical problem of a two-state classification and this has been thoroughly studied. There are several machine learning methods available to address the problem of the prediction of CM from the protein residue sequence *(6)*.
- It has been shown that the empirical reconstruction algorithms are quite insensitive to high levels of random noise in CMs, so that for reconstructing the 3D structure of the protein it is not necessary to correctly predict all contacts *(2,4)*.

Cons:

- There is no theory on CM that can help to define the limits and the strength of this representation. For instance, the effect of the contact threshold on the information content is not theoretically assessable. For this reason, different researchers adopt different protein representations and contact thresholds.
- The problem of CM comparison is very hard, as it is that of a sub-graph isomorphism, which is NP-hard *(7)*.
- CMs of real proteins are a tiny subset of the possible binary symmetric matrices *(2)*; however, no simple and fast algorithm has been found to sort out the protein-feasible CM from the others.
- CM prediction is an intrinsically non-local problem. Also, this is a very difficult problem to deal with, as a contact between two residues poses constraints on the feasibilities of all other contacts.
- Although the reconstruction programs are very insensitive to random noise, they are not as robust when the prediction errors are correlated, as is the case with current prediction algorithms.

CMs can be regarded both as symmetric matrices and as graphs. Actually, the CM representation is an adjacency matrix, where the contacts are the edges and the residues are the nodes. It is useful to distinguish between short-range and long-range contacts. The distinction between short-range (sometimes called 'local') and long-range ('non-local') contacts is not due to the type of interaction, nor the spatial distance, but it is due to the relative sequence separation. Contacts between residues that are separated less than a given number of residues S ($|i-j| \leq S$) are said to be short-range. Conversely, if the sequence separation is greater than S, they are said to be long-range. The choice of S is arbitrary, but it is commonly accepted that $|i-j| \leq 7-10$ represents short-range contacts, while $|i-j| > 7-10$ represents long-range ones.

2. Properties of Protein Contact Maps

When CMs are analyzed, one of the first features is that the number of contacts increases almost linearly with the protein length, independently of the adopted distance measures (CA–CA, CB–CB, etc.) and of the threshold cut-off used *(8)*. More formally, if L is the protein length and nc is the number of contacts, the real number of contacts can be quite accurately estimated using the linear equation

$$nc = A_T \times L$$

where A_T is a constant that only depends on the contact threshold (T). In practice, a change in the contact threshold T (in a reasonable range) has the only effect of modifying the slope of the line. This finding, together with the fact that the number of possible contacts ***NCM***, which is the number of independent CM elements ($NCM = L(L-1)/2$), increases with the square of the protein length, implies that the contact densities in the map (nc/NCM) decrease as the inverse of the protein length. In other words, long proteins have a lower contact density than short ones *(8)*.

Protein CMs have also more contacts in the short-sequence separations than those obtained using random graphs with the same number of contacts *(8)*. This is an indication that protein structures have a high tendency to form contacts with sequence neighbours.

Studying the properties of the CM eigenvectors, it has been found that there is a high correlation between the eigenvector corresponding to the highest eigenvalue (first eigenvector) and the residue coordination numbers *(5,9)*. The residue coordination number (or contact vector) is the number of contacts of each given residue with all the others in the protein space *(10)*. This figure can be easily computed from the contact matrix by summing up the rows (or the columns) of CM.

Galaktionov and Marshall *(5)* reported that from the knowledge of the real residue coordination numbers, it is possible to reconstruct to some extent (about 4 Å of Root Mean Square Deviation (RMSD)) the 3D structure of the protein.

A further surprising property of the first eigenvector of CM is the fact that a CM can be reconstructed using only the information contained in this vector coupled along with the information derived from the protein backbone constraints *(9)*. However, this is not a general property of all binary symmetric matrices, only of the subset comprising single-domain proteins *(9)*.

3. Reconstructing Protein Structures From Contact Maps

As outlined above, a CM contains a simplified representation of the protein conformation and it is unambiguously computed from the structure by a binary simplification of the DM. It is well known that a protein structure can be reconstructed from its DM by means of the Lagrange theorem *(1)*. This procedure is unambiguous, except for the ambiguity due to chiral symmetry. The questions are these: is it possible to recover the structure starting from its real CM as well? And from a predicted CM?

Bohr et al. *(3)* implemented a method based on the definition of a continuous function that measures the distance of a protein structure from a given CM. By adding some terms for assuring the connectivity and the compactness of the protein structure, a target function was obtained and then minimized using a simple steepest descent algorithm. The optimal computed structure satisfies as many contacts as possible.

At an 8 Å threshold for the distance between two CA atoms, the algorithm recovers the structure starting from the real CM with a RMSD less than 3 Å. It is worth noticing that the threshold value for the contact definition can be chosen within a wide range without greatly affecting the deviation of the recovered structure with respect to the real one. The optimal threshold for the minimization depends on the protein size.

The algorithm is efficient when a real CM is adopted; however, it fails when predicted CMs are considered for defining the target function. When the rate of error on the predicted map is only about 5%, it leads to structures with a RMSD $\geq$ 5 Å. This is due not only to the low quality of the prediction but also to the fact that a physical CM needs to satisfy complex constraints in order to represent a real structure.

When predicting contacts between each pair of residues in a sequence, the computation is independent of the other assigned contacts and then the resulting map is likely to be non-physical. In these cases, the recovering algorithm has to deal with the noise introduced by the inconsistency of the predicted contacts. This issue was thoroughly discussed by Vendruscolo and Domany *(2)* who implemented a stochastic algorithm for building a structure satisfying the protein CM. The algorithm builds a structure adding residues one at a time, trying different random conformations and then randomly adapting the preceding portion of the chain. In each step, the number of fulfilled contact constraints is the objective function for selecting the best conformations. By this, starting from the real map with a threshold distance value equal to 9 Å, the protein structure is reconstructed with a RMSD between 1 and 2 Å. The authors introduce noise in the physical map by flipping randomly chosen positions in the map and their algorithm results more robust than that of Bohr et al. *(3)*. Indeed even when about 20% of the map is randomly inverted, the algorithm reconstructs structures with a 4 Å RMSD to the real protein. However, this kind of non-physical CMs are likely to contain much more information than the predicted ones, as the randomness of the flipping conserves most of the original protein structure representation. Unfortunately, in a predicted map, errors are often more correlated and then recovering of the 3D structure is far more difficult.

In short, the implemented algorithms to reconstruct protein structure starting from CM prove that for a wide range of distance cut-offs, the CM is a good representation of the protein backbone conformation. It is possible to reconstruct the structure in the best cases with a deviation of less than 3 Å. Nevertheless, it should be considered that presently it is still impossible to deal with predicted maps, as in this case the level of noise is too high.

4. The Prediction of Protein Contact Maps

In these years, several researchers have been predicting CMs starting from protein sequence information. This interest grew after it was shown that it is possible to reconstruct protein structures from their CMs (*see* Section 3). Among the first attempts to predict residue contacts in proteins, there are methods based on correlated mutations *(11,12)*. In this case, the basic idea is that the maintenance of protein functions constrains the evolution of residue sequences. This fact can be exploited to interpret correlated mutations, observed in a sequence family, as an indication of a probable physical contact in 3D. On this basis, if a given residue mutates in a position, it is likely that a residue in contact with it will mutate too, in order to compensate the previous change. Also, strong hydrophobic conserved residues have a high probability of being in contact *(11)*.

An alternative approach is to learn the correlation between sequence and CM using machine learning tools. In this respect, several methods have been introduced: neural networks that exploit multiple sequence alignments ***(4,8, 13,14)***, hidden Markov models ***(15)***, support vector machines ***(16)***, genetic programming ***(17)*** and recurrent neural networks ***(18)***. Neural network-based methods incorporate several sequence features related to the local environment of two residues for their prediction of being or not in contact, including in some cases correlated mutations and residue conservation ***(4,8)***. More recently, Punta and Rost have improved the neural network prediction accuracy by adding information relative to the segment that connects the two residues undergoing prediction. This is done by coding also the sequence environment of the residue that falls exactly in the middle between the two residues considered. More precisely, if the contact propensity for the pair i, j ($j > i$) is predicted, they also code the environment for position $k = (j+i)/2$. This information seems to improve the neural network prediction accuracy up to 32% when sequence separation is six residues long ***(14)***, and this is the highest score reported so far. Similar to other predictors, this accuracy is obtained using a number of predicted contacts equal to half of the protein length ***(14)***.

Another method codes the protein underlying grammar for hidden Markov models to find residue contact patterns among different pairs of segments by adopting an approach that can be regarded as an extension of threading methods ***(15)***.

Recently, machine learning methods have tried to incorporate information relative to the geometric properties of CMs. It seems that the introduction of the information relative to the prediction of the first eigenvector density components helps the prediction of the final CM ***(19,20)***.

During the last Critical Assessment of Technique for Protein Structure Prediction (CASP6), some methods and servers were mainly evaluated on long-range contact predictions for a set of about 10 proteins belonging to the new fold targets ***(21)***. The assessors found that three approaches, including PROFcon ***(14)***, with similar levels of accuracy and coverage performed a little better than others ***(14,17,21)***. Comparisons of the predictions of the three best methods with those of CASP5/CAFASP3 suggested some improvement, although there were not enough targets in the comparison set to make this statistically significant. Irrespective of the CM prediction accuracy, they are still better than constraints from the best de novo 3D prediction methods ***(20)***.

How a predicted CM looks like? As an example, in **Fig. 2**, we show the prediction of an all-alpha protein. For this specific protein, accuracy is 44%, a quite satisfactory value when it is considered that this protein structural type is the most difficult to be predicted. Prediction in this case was computed with an updated version of our CORNET method ***(8)***.

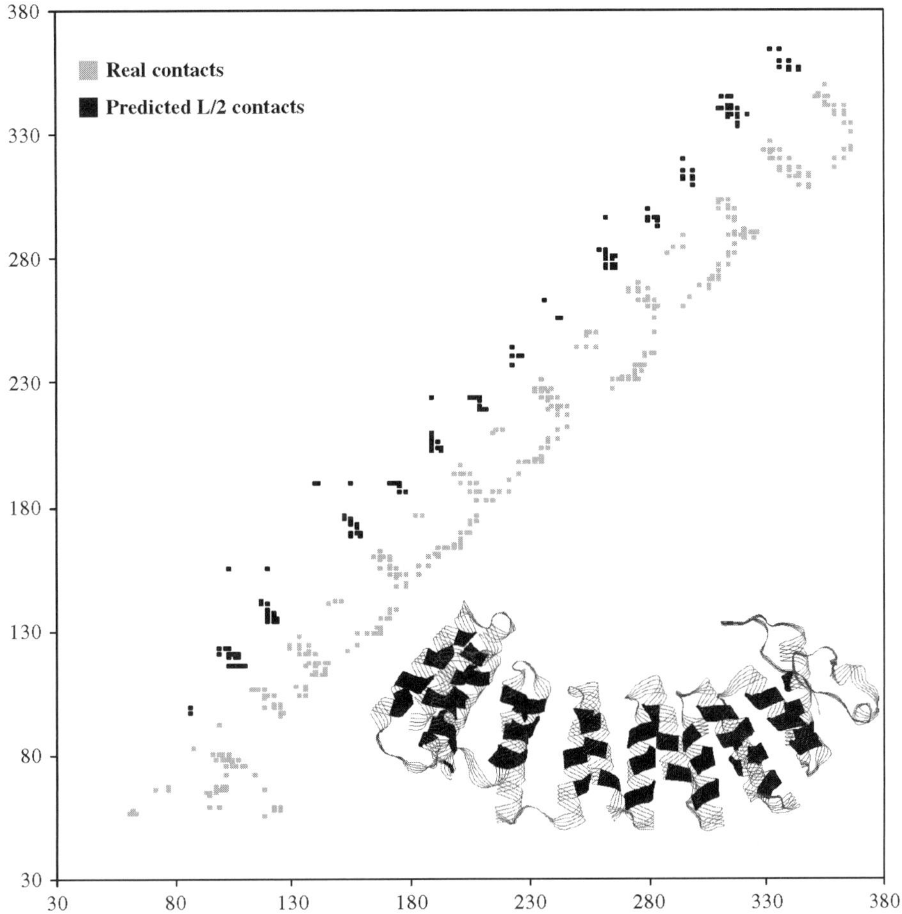

Fig. 2. Real versus predicted contact map of the α-subunit of the human Farnesyltransferase (PDB code: 1LD8 chain A). On the left side of the plot the black dots indicate the predicted L/2 contact residues. On the right side, in grey, the real residue contacts are shown (cut-off radius is 8.0 Å centered at the CB atoms and sequence separation ≥ 6). In the corner on the right, the protein structure is shown, highlighted in black, the correctly predicted contacts. On this protein, our neural network-based predictor reaches an accuracy equal to 44%.

5. Small World and Contact Maps

Unfortunately we use CMs, we predict them, but we are still unhappy. How do we improve our methods and our prediction? The solution is still to be found. In the meantime, we suggest another perspective in the following sections.

5.1. Small World

To overview some recent literature on proteins, we should introduce a few concepts explaining what 'small world' is and how it has been used to highlight protein folding properties.

In the mid-1990s, Duncan Watts, while studying for his PhD in Applied Mathematics, was invited to study a very particular problem: how crickets synchronize their singing *(22)*. He was convinced that, to deeply understand this problem, he had to observe the way the crickets pay attention to each other. This is the starting point of the study of networks under a different perspective than that of random networks that were previously introduced by Erdós and Rényi (*(22)* and references therein). Watts started his study on social networks trying to answer to a simple question: how many probabilities are there that two persons, both my friends, know each other? With his Professor Steven Strogatz, he found that social networks were clustered and not randomly distributed and that the same paradigm could model dynamical relations in many different systems *(22)*.

To explain the omnipresence of clustering in real world networks, Watts and Strogatz *(23)* proposed a new connection topology called a 'small world' network, showing that it can be interpolated between regular and random networks with a random rewiring procedure. According to this model, small world systems can be highly clustered, like regular graphs, and at the same time they are endowed with a small average path length, as it is for random networks.

Watts and Strogatz *(23)* introduced two numbers to describe the characteristics of small world networks: the characteristic path length L and the clustering coefficient C. L is given by the number of edges in the shortest path between two vertices, averaged over all pairs of vertices:

$$L = \frac{2}{N(N-1)} \sum_{i=1}^{N-1} \sum_{j=i+1}^{N} L_{ij},$$

where L_{ij} is the shortest path length between vertices i and j.

Supposing that a vertex k has N_k neighbours, then at the most $N_k(N_k - 1)/2$ edges can exist between them. If n_k is the actual number of edges among the neighbours, then C is defined as:

$$C = \frac{1}{N} \sum_{k=1}^{N} \frac{n_k}{N_k (N_k - 1)/2}.$$

L measures the typical separation between two vertices in the graph (a global property) and C is a measure of local clustering or cliquishness of a typical neighbourhood (a local property) *(23)*.

5.2. Small World and Protein Structures

The extension of the small world view to proteins was straightforward. Vendruscolo et al. *(24)* showed that protein structures have small world topology. The small world behaviour of protein structures is reflected by the presence in their graph of a relatively small number of vertices with many connections *(24)*. Two residues are considered as connected if the distance between their CA atoms is less than a threshold distance fixed at 8.5 Å. By analysing a data set of 978 representative proteins, it was found that the average value of L is 4.1 ± 0.9 and that of C is 0.58 ± 0.04. These values were compared with those obtained for random and regular graphs. By assuming that K is the average number of links in the graph (the average number of contacts in a protein) and N is the number of vertices (protein residues), then $L_{\text{random}} \sim \ln N / \ln K$ and $C_{\text{random}} \sim K/N$; $L_{\text{regular}} \sim N(N+K-2)/2K(N-1)$ and $C_{\text{regular}} \sim 3(K-2)/4(K-1)$ *(25)*. Values of 2.4 ± 0.3 and 0.08 ± 0.06 were reported for L_{random} and C_{random} respectively; L_{regular} and C_{regular} were 10.4 ± 7.0 and 0.67 ± 0.04, respectively *(24)*.

In this chapter for sake of clarity and with the specific aim of relating the small world representation to CMs (see below), we perform the same type of analysis on a new and a more selected data set of non-redundant mono-domain proteins (497 proteins) (*see* **Fig. 3**). We reached similar conclusions as before *(24)*, obtaining L and C equal to 3.9 ± 0.9 and 0.57 ± 0.03, respectively. For our data set, L_{random} is 2.1 ± 0.2, C_{random} is 0.08 ± 0.04, L_{regular} is 8.7 ± 4.2 and C_{regular} is 0.67 ± 0.01, confirming again that $L_{\text{random}} < L < L_{\text{regular}}$ and that $C_{\text{random}} < C < C_{\text{regular}}$, a key conclusion for resorting small world behaviour.

Small world view was adopted also for homopolymers obtained with a CM dynamics *(26)* and for atomic clusters obtained with Lennard–Jones interactions with a Monte Carlo method *(27)*. In both cases, the values of C and L were found similar to those of proteins, indicating a small world topology also for these systems. It was therefore concluded that protein chain connectivity plays a minor role in the small world behaviour and that for a globular protein the small world character would mainly arise from the overall geometry (surface to volume ratio) *(24)*.

What we did in house was substantially to add to these concepts by analysing other properties of our non-redundant protein set that have been related to small world behaviour. Another tendency that shows this property is that L increases linearly with log N (as a measure of the protein length) and that the slope is higher than the random reference case (*see* **Fig. 4**). This type of plot is frequent in the pertinent literature *(28,29)*. In our case, we add to the conclusion by analysing a non-redundant set of mono-domain proteins.

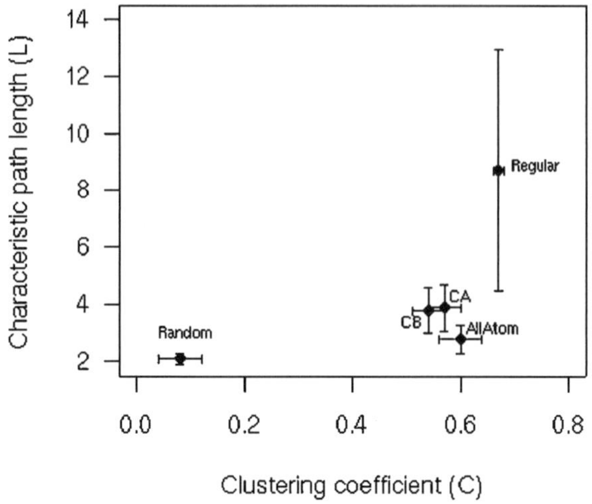

Fig. 3. Plot of the average path length versus the clustering coefficient computed on our data set of non-redundant set of mono-domain proteins (comprising 497 protein chains with sequence identity < 25%). Average values are reported with the associated standard deviation. Proteins are represented by CA, CB and all-atom (cut-off radius is 8.5 Å). Random: corresponding random graphs; Regular: corresponding regular graphs. See text for details.

As observed in the work of Atilgan et al. *(28)*, the average value of *C* remains nearly constant with increasing protein size. We found the same trend on our data set (*see* **Fig. 5**). It should be however noticed that for each protein the tendency is that *C* decreases at increasing protein size. This fact is viewed as indicative of the modular nature of the small world networks. When globular and fibrous proteins are compared, no relevant difference arises, and a general belief is that 'small worldness' persists irrespectively of structural differences *(28–30)*.

Atilgan et al. *(28)* studied 595 proteins with sequence homology <25%, a set described before *(13)*. The protein core local organization (residues residing at depths greater than 4 Å) is the same even if the size of the protein is different. Beyond a depth of approximately 4 Å from the protein surface, the clustering coefficient approaches a fixed value of approximately 0.35, irrespective of the size of the protein at hand. The same small world organization seems therefore to live throughout the protein, despite the heterogeneous density distribution that it may be found in different folds pertaining to different proteins.

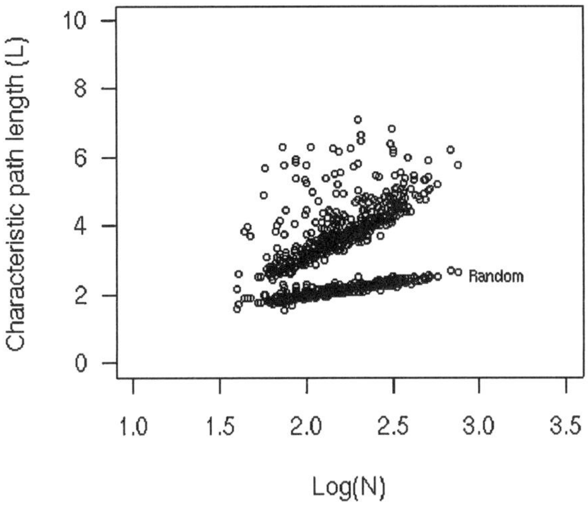

Fig. 4. Characteristic path length as a function of logarithm of the protein length [Log (N)]. L is shown for each protein of our data set. Real protein values cluster above those of corresponding random networks.

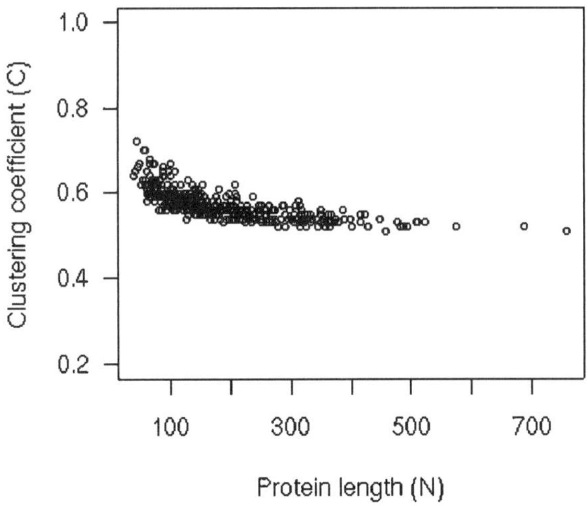

Fig. 5. Clustering coefficients of the different proteins as a function of the protein length (see text for details).

5.3. Local Versus Global Contacts

Greene and Higman *(30)* adopted an all-atom representation of the proteins instead of the less informative CA simplified representation. A contact was allowed between two residues when at least one pair of their atoms is within 5 Å from each other. By this, multiple links between residues are allowed. The small world property was analysed on a set of 65 non-redundant proteins divided into nine highly populated fold types representing the four SCOP protein classes: all-α, all-β, α/β, α + β (http://scop.mrc-lmb.cam.ac.uk/scop/). Interestingly Greene and Higman *(30)* found a difference of the behaviour between what they called networks of short-range and long-range contacts. Interactions are considered short range or long range if they occur between residues that are ≤ 10 and more than 10 residues apart in the protein sequence, respectively. A long-range interaction graph does not differ from a random graph; however, when also short-range contacts are taken into consideration the small world behaviour emerges. By following the short-range and long-range contact distinction, we compute C and L values for our protein set. The results are shown in **Fig. 6**, confirming that long-range contacts

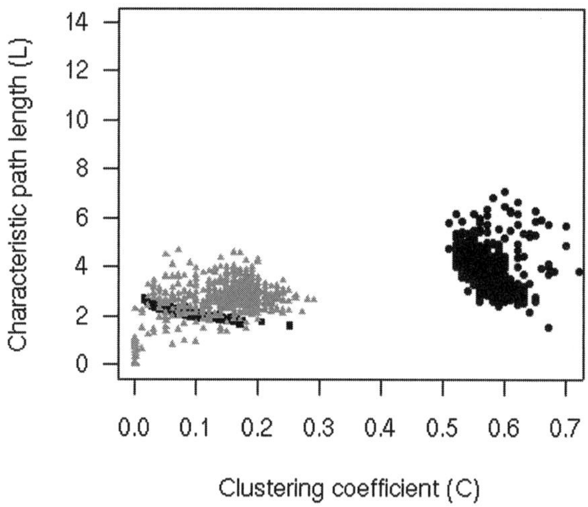

Fig. 6. The characteristic path length versus the clustering coefficient for each protein in the data set considering long-range contacts and complete contact maps. Black circles: complete protein contact maps. Grey triangles: long-range contacts. Black squares: random networks. Apparently, long-range contacts overlap with corresponding random networks.

can be modelled by a random graph and that small world properties emerge only when the whole CM is considered.

5.4. All-α Versus All-β Contacts

Several authors inspected how small world behaviour is dependent on the protein structural type, routinely following the SCOP classification *(28–30)*. A thorough investigation study reveals a marginal but consistent difference in the C index value of all-α and all-β proteins. We show our results in **Fig. 7**. When considering the average C values, we find that they are 0.597 for all-α and 0.551 for all-β proteins, respectively. These values confirm the difference previously reported *(29)*. This difference may be due to the larger geometrical compactness of α-helices as compared to β-sheets. Our data set contains 113 all-α proteins and 110 all-β proteins.

5.5. Scale-Free Networks and Contact Maps

Scale-free networks are small world; however, small world networks are not necessarily scale-free *(31)*. In the protein world, CMs are not scale-free networks. A scale-free connectivity follows a power law $p(k) \sim k^{-\gamma}$ (where k is the number of links of a node and p is the probability of a node to have

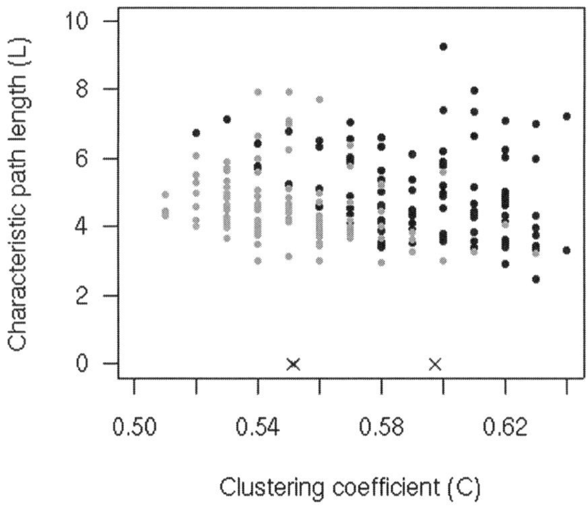

Fig. 7. The characteristic path length versus the clustering coefficient for 113 all-α (black dots) and 110 all-β proteins (grey dots). The two crosses indicate the average C values for the two groups: 0.597 and 0.551 for all-α and all-β proteins, respectively (see text for details).

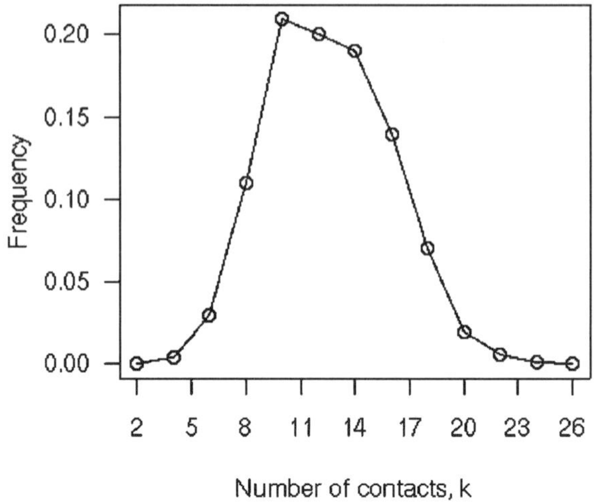

Fig. 8. Small world networks are not scale-free: frequency of residues (vertices) as a function of the number of contact per residue (k) in our protein data set.

k links). In a typical scale-free network $2 \leq \gamma \leq 3$. The distribution of both long-range and short-range contacts reveals a tendency to a bell-shaped Poisson curve which is typical of random networks and not of scale-free ones *(30)*. The plot shown in **Fig. 8** is the result of a study on our data set of complete CMs, confirming the non-scale-free behaviour of contact distribution in our protein set.

6. Exploiting Small World Properties of Contact Maps

In Section 5, we showed that protein CMs are peculiar graphs that exhibit small world properties. The question arises whether predicted CMs behave similarly. Thus, we predicted some 100 mono-domain proteins using PROFcon *(14)* that has been demonstrated to be one of the best performing available methods *(21)*. However, PROFcon assigns predictions only to pair of residues that are more than five residues apart, and therefore, in order to compare the predicted CMs with the observed ones, we also added the trivial connectivity to the predictions (which consists of the CM diagonals i, i + 1 and i, i + 2). The trivial contacts are due to the backbone connectivity when a CB threshold is set to 8 Å (as was in this case). The results are reported in **Fig. 9**, where it is evident that also the predicted CMs generate graphs with small world behavior. Nevertheless, the predicted CMs have lower values of both characteristic path length (L) and clustering coefficient (C) with respect to real proteins. Prediction

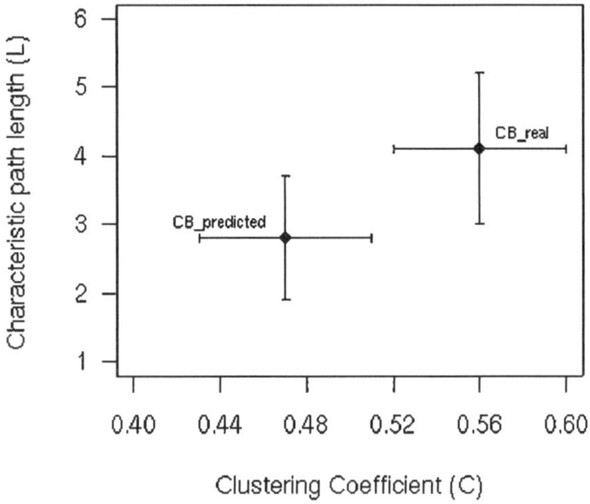

Fig. 9. Plot of the average characteristic path length versus the average clustering coefficient computed on 113 contact maps of all-α proteins predicted with the PROFcon prediction method ***(14)*** (CB predicted) compared to physical ones (CB real). Predicted contact maps are non-random but still different from real contact maps.

therefore generates CMs that are different from random but still far from the real proteins. Eventually, this perspective may help in filtering out spurious assignments.

7. Conclusions

Writing a review article is always an effort, especially when piled up results in a field are still promising results. In this chapter, we hope to have addressed the old and present problems in CM predictions and highlighted why we are still willing to devote our effort to this field. Also, we have suggested that possibly by merging small world view of proteins and CMs, new optimization algorithms may be developed to reduce signal-to-noise ratio. This will eventually help us also in finally reconstructing the 3D protein structure from predicted CMs.

Acknowledgments

We thank MIUR for the following grants: PNR-2003 grant delivered to PF, a PNR 2001–2003 (FIRB art.8) and PNR 2003 projects (FIRB art.8) on Bioinformatics for Genomics and Proteomics and LIBI-Laboratorio Internazionale di BioInformatica, both delivered to RC. This work was also supported by the

Biosapiens Network of Excellence project (a grant of the European Unions VI Framework Programme).

References

1. Havel, T. F. (1998). Distance geometry: theory, algorithms, and chemical applications. *Encyclopedia of Computational Chemistry*. John Wiley & Sons, New York.
2. Vendruscolo, M. and Domany, E. (1999). Protein folding using contact maps. *arXiv* cond-mat, 9901215.
3. Bohr, J., Bohr, H., Brunak, S., Cotterill, R. M., Fredholm, H., Lautrup, B. and Petersen, S. B. (1993). Protein structures from distance inequalities. *Journal of Molecular Biology* **231**, 861–869.
4. Fariselli, P., Olmea, O., Valencia, A. and Casadio, R. (2001). Progress in predicting inter-residue contacts of proteins with neural networks and correlated mutations. *Proteins* **Suppl 5**, 157–162.
5. Galaktionov, S. G. and Marshall, G. R. (1994). *27th Annual Hawaii International Conference on System Sciences (HICSS-27)*, Maui, Hawaii.
6. Baldi, P. and Brunak S. (2001). *Bioinformatics: The Machine Learning Approach, A Bradford Book,* Second edition. MIT Press, Cambridge.
7. Goldman, D., Istrail, S. and Papadimitriou, C. (1999). Algorithmic aspects of protein structure similarity. *Proceedings of the 40th IEEE Symposium on Foundations of Computer Science*, New York, (USA), 512–522.
8. Fariselli, P., Olmea, O., Valencia, A. and Casadio, R. (2001). Prediction of contact maps with neural networks and correlated mutations. *Protein Engineering* **14**, 835–843.
9. Porto, M., Bastolla, U., Roman, H. E. and Vendruscolo, M. (2004). Reconstruction of protein structures from a vectorial representation. *Physical Review Letters* **92**, 218101.
10. Pollastri, G., Baldi, P., Fariselli, P. and Casadio, R. (2002). Prediction of coordination number and relative solvent accessibility in proteins. *Proteins* **47(2)**, 142–153.
11. Goebel, U., Sander, C., Schneider, R. and Valencia, A. (1994). Correlated mutations and residue contacts in proteins. *Proteins* **18**, 309–317.
12. Olmea, O. and Valencia, A. (1997). Improving contact predictions by the combination of correlated mutations and other sources of sequence information. *Folding & Design* **2**, S25–S32.
13. Fariselli, P. and Casadio, R. (1999). A neural network based predictor of residue contacts in proteins. *Protein Engineering* **12**, 15–21.
14. Punta, M. and Rost, B. (2005). PROFcon: novel prediction of long-range contacts. *Bioinformatics* **21**, 2960–2968.

15. Bystroff, C. and Shao, Y. (2002). Fully automated ab initio protein structure prediction using I-SITES, HMMSTR and ROSETTA. *Bioinformatics* **18 Suppl 1**, S54–S61.
16. Zhao, Y. and Karypis, G. (2003). *3rd IEEE International Conference on Bioinformatics and Bioengineering (BIBE).*
17. MacCallum, R. M. (2004). Striped sheets and protein contact prediction. *Bioinformatics* **20 Suppl 1**, I224–I231.
18. Pollastri, G. and Baldi, P. (2002). Prediction of contact maps by GIOHMMs and recurrent neural networks using lateral propagation from all four cardinal corners. *Bioinformatics* **18 Suppl 1**, S62–S70.
19. Vullo, A., Walsh, I. and Pollastri, G. (2006). A two-stage approach for improved prediction of residue contact maps. *BMC Bioinformatics* **7**, 180.
20. Eyrich, V. A., Przybylski, D., Koh, I. Y., Grana, O., Pazos, F., Valencia, A. and Rost, B. (2003). CAFASP3 in the spotlight of EVA. *Proteins* **53 Suppl 6**, 548–560.
21. Grana, O., Baker, D., MacCallum, R. M., Meiler, J., Punta, M., Rost, B., Tress, M. L. and Valencia, A. (2005). CASP6 assessment of contact prediction. *Proteins* **61 Suppl 7**, 214–224.
22. Barabasi, A. L. (2003). *Linked: The New Science of Networks*, Perseus Publishing, Cambridge, Massachusetts.
23. Watts, D. J. and Strogatz, S. H. (1998). Collective dynamics of 'small-world' networks. *Nature* **393**, 440–442.
24. Vendruscolo, M., Dokholyan, N. V., Paci, E. and Karplus, M. (2002). Small-world view of the amino acids that play a key role in protein folding. *Physical Review. E, Statistical, Nonlinear, and Soft Matter Physics* **65**, 061910.
25. Watts, D. J. (1999). *Small Worlds. The Dynamics of Networks Between Order and Randomness*, Princeton University Press, Princeton, New Jersey.
26. Vendruscolo, M. and Domany, E. (1998). Efficient dynamics in the space of contact maps. *Folding & Design* **3**, 329–336.
27. Andricioaei, I., Voter, A. F. and Straub, J. E. (2001). Smart Darting Monte Carlo. *The Journal of Chemical Physics* **114**, 6994–7000.
28. Atilgan, A. R., Akan, P. and Baysal, C. (2004). Small-world communication of residues and significance for protein dynamics. *Biophysical Journal* **86**, 85–91.
29. Bagler, G. and Sinha, S. (2005). Network properties of protein structures. *Physica A* **346**, 27–33.
30. Greene, L. H. and Higman, V. A. (2003). Uncovering network systems within protein structures. *Journal of Molecular Biology* **334**, 781–791.
31. Barabasi, A. L. and Albert, R. (1999). Emergence of scaling in random networks. *Science* **286**, 509–512.

9

Roadmap Methods for Protein Folding

Mark Moll, David Schwarz, and Lydia E. Kavraki

Summary

Protein folding refers to the process whereby a protein assumes its intricate three-dimensional shape. This chapter reviews a class of methods for studying the folding process called roadmap methods. The goal of these methods is not to predict the folded structure of a protein, but rather to analyze the folding kinetics. It is assumed that the folded state is known. Roadmap methods maintain a graph representation of sampled conformations. By analyzing this graph one can predict structure formation order, the probability of folding, and get a coarse view of the energy landscape.

Key Words: protein folding, folding kinetics, roadmap methods, conformation sampling techniques, energy landscape

1. Introduction

Protein folding refers to the process whereby a protein assumes its intricate three-dimensional shape. Different aspects of this problem have attracted much attention in the last decade. Both experimental and computational methods have been used to study protein folding, and there has been considerable progress *(1–7)*.

This chapter reviews a class of methods for studying protein folding called roadmap methods *(8–19)*. These methods are relatively new and are still under active development. Roadmap methods are computational methods that have been developed to understand the process or the mechanism by which a protein folds or unfolds. It is typically assumed that the folded state is already known.

From: *Methods in Molecular Biology, vol. 413: Protein Structure Prediction, Second Edition*
Edited by: M. Zaki and C. Bystroff © Humana Press Inc., Totowa, NJ

Note that this is not a comprehensive survey of all existing computational protein-folding methods. In particular, it does not cover molecular dynamics (MD) methods *(20)*, Monte Carlo (MC) methods *(21)*, the use of coarse grain models in simulations, and many others.

Many papers *(20–22)* have discussed the advantages and disadvantages of traditional computational methods for studying protein folding. Some of the drawbacks include the facts that classical MD/MC simulations typically compute only one trajectory, that escaping local minima can be very difficult, and that the process has no memory to recognize whether conformations have been visited in the past or not. These issues led some researchers to develop enhanced versions of MC and MD methods, which take advantage of laboratory data, non-uniform or accelerated timescales, modified energy functions, parallelism, biases away from previously generated conformations, and other modifications *(23–26)*. Other researchers, inspired by advancements in robot modeling and by the need for alternative protein-modeling methods, began to build so-called roadmaps to explore the conformational space of proteins. A roadmap is a representation of many conformations and the transitions between them as a graph data structure. Roadmap-based methods were originally developed in robotics *(27)* where the configuration (conformation) space of a robot is explored to find a collision-free path that will take the robot from an initial position to a final position. By taking advantage of the analogy between robots and molecules, in which the main molecular chain of a protein corresponds to an articulated robot, roadmap methods were adapted to study how a protein can attain a known final shape. Roadmap methods were significantly modified and enhanced to address the folding problem. Their application to the folding problem is still relatively new and not as well-understood as MD/MC simulations. They seem to offer vast computational improvements and potentially increased coverage of the conformational space compared with traditional methods. This could mean that "interesting" areas of the conformational space can quickly be discovered, and, if necessary, further explored with traditional methods. Yet, it is not clear how much (if anything) is lost by the use of coarse approximations. This chapter surveys some of the most promising road map methods for protein folding *(9–19)*.

2. Background

2.1. Protein Representation

The simplest representation of a protein is a vector that contains the Cartesian coordinates of all atoms in a conformation. This is the representation used in

MD/MC simulations; molecular potential energy functions are almost always parameterized by atomic coordinates in Cartesian space *(28)*.

The drastic changes in the conformation of a protein occur, however, with rotations about certain bonds. Often, a vector of bond rotations is used as a more compact representation of a protein. The amount of rotation about a single bond relative to some reference state is called the dihedral angle. This representation ignores the stretching of bond lengths and bond angles, but these effects are often negligible compared with the bond rotations. Efficient ways to calculate the Cartesian coordinates of all atoms given the dihedral angles of a protein are given in **ref.** ***29***.

Another way to represent a protein is to model flexibility at the level of secondary structure. A molecule is divided into α-helices, β-sheets, and connecting loops. The sequence of secondary structure elements is represented by a sequence of vectors. Rotational degrees of freedom are assigned at the junctions where the vectors meet. The α-helices and β-sheets can twist about their axis, and the loop regions are allowed to extend in the direction of their vector. In this representation, traditional energy functions cannot be used, but it is possible to approximate molecular energy using a simple potential function *(30)*.

In roadmap methods for protein folding, all the above representations have been used, but the most popular one is the representation of conformations by dihedral angles. As will be explained in the next section, roadmaps sample the conformation space of a protein. The dihedral angle representation of a protein readily allows the generation of samples that have properties suitable for road-map-based methods.

2.2. RoadMap Algorithms for Robot Motion Planning

The idea of using a roadmap to represent properties of a complex space originated in robotic motion planning *(27,31)*. In motion planning, a collision-free path between a start and goal configuration of a robot is computed. Consider a long articulated robot for the moment. The degrees of freedom of such a robot correspond to moving its joints. The set of all configurations of a robot is called its configuration space. Each point in this space corresponds to a robot configuration. A simple, two-dimensional robotic configuration space is illustrated in **Fig. 1A**. The subset of configurations where the robot does not collide with any obstacles (including the robot itself) is called the free space and is shown in white in **Fig. 1A**. The set of configurations in which the robot collides with itself or a workspace obstacle is called the occupied space and is shown in black in **Fig. 1A**. Motion planning can thus be phrased as

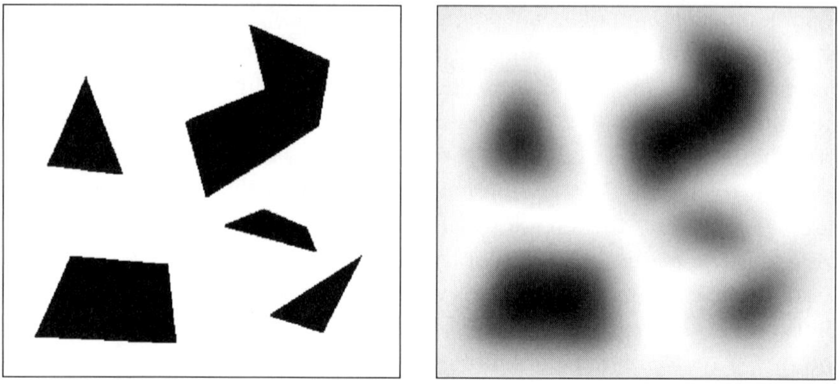

(A) A two-dimensional robotic configuration space. Black shapes represent sets of configurations that place the robot in collision with obstacles.

(B) A two-dimensional molecular conformation space, which could correspond to a molecule with two rotatable bonds. White regions are low-energy, black high-energy, and gray intermediate-energy. The higher the energy of a conformation, the less likely a molecule is to assume that conformation.

Fig. 1. Robotic configuration space versus molecular conformation space. **(A)** A two-dimensional robotic configuration space. Black shapes represent sets of configurations that place the robot in collision with obstacles. **(B)** A two-dimensional molecular conformation space, which could correspond to a molecule with two rotatable bonds. White regions represent low energy, black high energy, and gray intermediate energy. The higher the energy of a conformation, the less likely a molecule is to assume that conformation.

the problem of finding a curve (a path) that lies completely in the free part of the configuration space.

Computing the free space *exactly* is a very hard problem. The size of the configuration space and the complexity of the motion-planning problem grow exponentially with the number of degrees of freedom *(32)*. Sampling-based techniques called Probabilistic Roadmap Methods (PRMs) *(27)* build a roadmap: a graph representation of the free space, where nodes correspond to configurations and edges to paths between them. This roadmap is computed as follows. First, a large number of collision-free configurations are sampled. Next, for each configuration, an attempt is made to find a path to some of its nearest neighbors. These local paths can simply be straight lines in the configuration space. If the path between two configurations lies entirely in the free space, it is added to the roadmap. The motion planning problem is now easily solved. The start and goal configurations are connected to their nearest neighbors in the road map. The path is then found by performing a simple graph search to connect the start to the goal. Note that the roadmap has to be computed only once for a given robot and that many motion- planning queries can be solved with the same roadmap. PRMs are able to solve motion-planning problems in very high-dimensional configuration spaces, but they do not guarantee completeness, i.e., they do not always find a path if one exists. Instead, they have been shown to be probabilistically complete, i.e., if a path exists, then with high probability the PRM algorithm will find it. This probability goes to one as the number of sampled configurations increases. Many variations of the basic PRM algorithm have been proposed to increase the sampling of configurations in difficult areas (such as narrow passages). A discussion of the PRM algorithm and its variations can be found in **ref. *31***.

For certain applications, it is known a priori that only one motion-planning query will need to be solved, so sampling the entire configuration space may be unnecessary. This observation leads to a different class of sampling-based path-planning algorithms in which a tree of configurations is grown from the start to the goal configuration and/or vice versa. The three main variations within this class are called rapidly exploring random trees (RRTs) ***(33)***, expansive spaces trees (ESTs) ***(34)***, and path-directed subdivision trees (PDSTs) ***(35)***. RRTs grow a tree of configurations as follows. First, a random configuration, which may be in collision, is sampled. Next, the nearest configuration in the existing tree to the one just sampled is found. Initially, the tree consists of just the start configuration. From the nearest configuration, a new configuration is found at some distance in the direction of the randomly sampled configuration. This process is repeated until the tree is close to the goal configuration. This algorithm tends to "pull" the tree growth in the direction of unexplored parts of the configuration space. ESTs, on the contrary, can be thought of as "pushing" the tree growth in promising areas. During each iteration of the EST algorithm, a previously sampled configuration is selected at random and a new configuration is sampled in a neighborhood of it. The key in the algorithm is the probability distribution function used to sample the previous configurations. The EST assigns a probability to each configuration that is proportional to the distance to the k nearest neighbors and inversely proportional to the number of times the configuration has been selected before. Sampling using this distribution expands the trees toward unexplored areas of the configuration space. PDSTs represent the trees somewhat differently from other tree-based and roadmap methods. Rather than maintaining a set of nodes and edges, a PDST consists of a set of edges, representing paths, joined at branches. It also maintains a cell decomposition of the configuration space and assigns paths to cells. At each step of the PDST exploration, an edge is selected based on an estimate of how well the area around each edge has already been explored (measured using the cell decomposition), and a new edge is created starting from a random point along the selected edge. In this way, the tree expands outward from its origin and the updating of the cell decomposition leads the expansion of the tree to less well- sampled areas.

Both roadmap and tree-based path-planning and exploration algorithms have been used to study the dynamic properties of proteins, including their folding behavior but also their interactions with other molecules ***(36–40)***. In order to apply these robotics-based methods to complex molecular systems, however, some adaptations of the algorithms are necessary, as will be presented in the following sections.

3. Roadmaps for Protein Folding

Conceptually, there is an analogy between high-energy areas in the conformation space of a molecular system and obstacles, and between low-energy areas and free space (*see* **Fig. 1B**). There may not be a single cutoff energy threshold, however, to separate the conformation space in black and white regions. Molecular conformation spaces therefore have a fuzzier notion of collision and free space than robotic configuration spaces, as is shown in **Fig. 1B**, and there are other important differences between exploring the free space of a robot and the free space of a biomolecule. In a biochemical context, low-energy paths are of specific interest, rather than paths in general. In folding, in particular, if it is assumed that the folded state of a protein is known, then researchers would like to find how the protein unfolds and refolds and determine some aggregate properties of these pathways, such as the overall folding rate and probability of any given structure to proceed to a folded state. It is important to note that the goal is not to predict the folded state from a sequence of amino acids. The interest is in folding kinetics: the aim is to get a better understanding of the process or mechanism by which a protein folds and unfolds. It is assumed that the folded state has already been determined.

The essential ingredients of any roadmap method are the choice of degrees of freedom, the conformation sampling technique, and the way to connect conformations to form a roadmap. Another important ingredient for roadmaps of molecular systems is the energy model. So far, simplified energy models have been used. It remains to be seen how accurate these models are for complex problems. This section will review how road-map-based methods can provide new insights into folding kinetics.

Before getting into the details of specific methods, it is worth mentioning that the idea of using roadmap methods to study problems in molecular biology originated with Singh et al. ***(41)***, who adapted the PRM algorithm to study the docking of a ligand to a protein. Nodes in the roadmap represented conformations and poses of the ligand and were sampled at random around the protein and kept or rejected based on their energy. Neighboring nodes were connected with an edge if a set of conformations sampled on a straight line in configuration space between them were all below an energy cutoff, and edges were labeled with transition probabilities depending on the energy difference between the nodes at either end. This work permitted the identification of active sites in proteins.

Several research groups extended and adapted this work, refocusing it on protein-folding mechanisms ***(9–19)***. The general trends of this ongoing research include tweaking the energy function, edge weights, and/or node sampling schemes. The goal of such work is ultimately to develop methods in which the final energy distribution of the set of nodes and paths in the roadmap corresponds to the energy distribution predicted by statistical mechanics (Boltzmann-like). Given a high-quality roadmap, it should be possible to determine properties of the protein's motion and folding behavior from all-path analyses.

In general, the folding kinetics can be analyzed by looking at many paths in the roadmap. There are two fundamentally different ways to construct and interpret the roadmap. In the first method (described in **Subheading 3.1.**), the object is to compute the most energetically favorable paths between the folded state and denatured states and to consider those the folding pathways. This is the approach taken by Amato et al. ***(9–13)***. In the second method, the weights of edges in a roadmap are interpreted as probabilities and the roadmap gives rise to a Markov chain. The folding pathways are analyzed by performing random walks on the roadmap or by computing the limit distribution from the matrix of state transition probabilities. This is the approach taken by Apaydin et al. ***(16–19)***. This work is described in **Subheading 3.2.** Finally, in **Subheading 3.3.**, we describe the third method, proposed by Singhal et al. ***(14,15)***, which combines roadmap methods with MD/MC methods.

3.1. PRMs for Protein-Folding Pathways

In the work of Amato et al. ***(9–13)***, the backbone ϕ and ψ dihedral angles are taken to be the degrees of freedom. The side chains are assumed to be rigidly attached to the backbone. For a protein consisting of n residues, there are $2(n-1)$ degrees of freedom (the first and last rotational angles do not contribute). Conformations can be sampled by randomly picking angles from the allowable range. The sampling can be based on Ramachandran plots ***(42)***, but this approach has a very small probability of producing conformations without steric clashes. In early work, Amato et al. ***(13)*** used Gaussian sampling around the folded state with various standard deviations to create new conformations. This works well for proteins with approximately 60 residues, but it still does not scale up to larger proteins with over 100 residues. A more successful strategy is the following: instead of sampling only around the native state, conformations are sampled around all previously sampled conformations. This is done in a way that creates a "wavefront" of conformations growing outward from the native state. The conformations are partitioned into bins based on the number of native contacts. A native contact is defined as a pair of C_α atoms that are within 7 Å of each other in the native state. The bins are equal-sized and the number of bins is proportional to the number of native contacts in the native state. A conformation q is accepted based on its energy $E(q)$. When a structure is generated, it is checked for collision of side chains and rejected if any are found. If it passes that test, the energy consists of a term favoring documented secondary structure through known backbone hydrogen and disulfide bonds, and a term for hydrophobic interactions.

The probability of accepting a conformation q is

$$P\,(\text{accept } q) = \begin{cases} 1 & \text{if } E(q) < E_{\min} \\ \frac{E_{\max}-E(q)}{E_{\max}-E_{\min}} & \text{if } E_{\min} \leq E(q) \leq E_{\max} \\ 0 & \text{if} E(q) > E_{\max} \end{cases}$$

Thus, all low-energy conformations are kept, as well as some of the medium-energy conformations, in order to connect the low-energy areas. The energy thresholds $E_{\min}$ and $E_{\max}$ are set at 50,000 and 89,000 kJ, respectively. The accepted conformations are put in the appropriate bin. The sampling process iteratively tries to fill all bins, starting with the bin with 100% native contacts. Once a neighboring bin has at least n conformations, sampling is performed around conformations in that bin, in order to fill the succeeding bins. Although this sampling method does not seem to correspond to a Boltzmann distribution of states, it still may capture some of the essential folding properties such as contact formation order ***(11)***.

The second phase in the roadmap construction is the connection of the sampled conformations. For each conformation, the method attempts to connect each node to its k nearest neighbors. The ϕ and ψ angles are linearly interpolated, and energy is checked along the line in conformation space connecting a conformation q_0 and one of its neighbors q_1. If the energy does not exceed some threshold, the edge connecting q_0 and q_1 is added to the roadmap. The edge is given a weight that depends on the energy along the line connecting q_0 and q_1. Suppose the energy of the sequence of conformations $q_0 = c_0, c_1, c_2, \ldots, c_{n-1}, c_n = q_1$ along the line connecting q_0 and q_1 has been computed. The probability of moving from c_i to c_{i+1} is

$$P_i = \begin{cases} e^{-\frac{\Delta E_i}{kT}} & \text{if } \Delta E_i > 0, \\ 1 & \text{if } \Delta E_i \leq 0. \end{cases}$$

Here, $\Delta E_i = E(c_{i+1}) - E(c_i)$. The weight of the edge between q_0 and q_1 is then defined as

$$w(q_0, q_1) = \sum_{i=0}^{n-1} - \log P_i.$$

The edge weight is intended to encode the likelihood of going from one conformation to another given the energy profile of the path.

After the roadmap is constructed, the folding pathways can be extracted. Starting from the native structure, the shortest path to every other conformation can be found using Dijkstra's algorithm ***(43)***.

This roadmap construction method was tested on 14 proteins with 56–110 residues, including protein G and protein A ***(11)***. Roadmaps were constructed in 2–15 hours. From this, many folding pathways can be extracted and their properties analyzed. Of particular interest is the order of secondary structure formation along each path between the stable unfolded states and the folded state. This order provides a rough overview of the folding mechanism of the protein and can often be determined by laboratory experiment, thereby providing a criterion that is used to validate the roadmap method.

Using a constructed roadmap, we determined the order of secondary structure formation for a single path from an unfolded to folded state by, for each native contact in a secondary structure element, finding the first conformation along the path that contains that contact. Along a single path, the appearance time for a secondary structure element is computed as the mean of the appearance times for all of its contacts. Overall, the predicted secondary structure formation order is the order with the greatest frequency over all paths. For the experimental set of 14 proteins, this analysis of the roadmap correctly predicted the formation order of secondary structure in all cases where laboratory data were available for comparison.

In later work ***(10)***, the same group that developed the original methods did a more detailed study of proteins L and G. These proteins both consist of an α-helix and a four-stranded β-sheet. Despite this structural similarity, the secondary structures are experimentally documented to form in different orders. PRM analysis correctly predicted these differences in secondary structure formation order.

In their latest work *(9)*, Thomas et al. noticed that even the bin-based construction method described above often requires 10,000 or more samples to construct a complete roadmap for relatively small (60–100 residue) proteins. For more typical protein sizes, this poor scaling rapidly becomes prohibitive. As a result, Thomas et al. *(9)* developed a new sampling method based on rigidity analysis of each sampled conformation. Using information about constraints on motion such as disulfide bridges and hydrogen bonds, this analysis classifies each bond as independently flexible, dependently flexible, or rigid. Independently flexible bonds may be rotated without any effect on other degrees of freedom. Dependently rotatable bonds may rotate but necessarily cause other related bonds to rotate also. Rigid bonds, as the name suggests, generally cannot rotate because they are part of a fully constrained cluster of atoms. Dependently flexible bonds form sets with fewer than the expected number of degrees of freedom.

Under rigidity-based sampling, new samples are generated by perturbing the dihedral angles of existing conformations in a non-uniform way. Specifically, independently flexible bonds are rotated with a high probability, P_{flex}. Rigid bonds are rotated with a low but non-zero probability P_{rigid}. For sets of dependently flexible bonds with k internal degrees of freedom, k are selected at random and rotated with probability P_{flex}, and the remaining bonds are rotated with probability P_{rigid}. Thomas et al. found that allowing rigid bonds to rotate helps the method attain better coverage of the conformation space, while biasing rotations to occur most often for flexible bonds focuses the sampling on regions of the conformation space most likely to be accessible to a real protein.

When tested on a set of 26 proteins, it is reported *(9)* that rigidity-based sampling yielded roadmaps with substantially better connectivity (measured as edges per node) than earlier sampling methods. In many cases, this could often be accomplished using a quarter to half as many nodes as were necessary to produce the roadmap under Gaussian sampling. In addition to correctly predicting the secondary structure formation order of proteins G and L, analysis of roadmaps created using rigidity sampling also correctly predicted the order of secondary structure formation of NuG1 and NuG2. PRM analysis by Thomas et al. without rigidity sampling had previously failed to predict the order of structure formation in these proteins.

3.2. Stochastic roadmap Simulation

Stochastic RoadMap Simulation (SRS), developed by Apaydin et al. ***(16–19)*** is a general technique to study molecular motion. The method derived its early inspiration from the work of Singh et al. ***(41)***, who were attempting to find a way to predict active sites in proteins using roadmap methods.

The roadmap construction in SRS is straightforward. First, a number of conformations are sampled independently at random from the conformation space. Each conformation is connected to its k nearest neighbors. The transition probability P_{ij} of an edge connecting nodes v_i and v_j is defined as

$$P_{ij} = \begin{cases} \frac{1}{d_j} e^{-\frac{\Delta E_{ij}}{k_B T}} & \text{if } \frac{\varepsilon_j / d_j}{\varepsilon_i / d_i} < 1, \\ \frac{1}{d_i} & \text{otherwise,} \end{cases}$$

where ε_i and ε_j are the Boltzmann factors for conformations c_i and c_j, and d_i and d_j are the number of neighbors for v_i and v_j. The Boltzmann factor of a conformation c is defined as $\varepsilon = \exp(-E(c)/k_b T)$. A self-transition is added with probability $P_{ii} = 1 - \sum_{i \neq j} P_{ij}$, so that all transition probabilities of a node add up to 1. The energy $E(c)$ is a hydrophobic–polar (H–P) energy function ***(30)***, in which each amino acid residue is classified as hydrophobic or polar, and favorable energy is computed for hydrophobic residues in contact with (within a cutoff distance of) each other. Conformations are also checked for steric clashes (overlapping atoms) and rejected if necessary.

A random walk on this roadmap is defined as follows. Starting at node v_i, a neighbor v_j is chosen uniformly at random. A move from v_i to v_j is accepted with probability

$$A_{ij} = \begin{cases} \frac{d_i}{d_j} e^{-\frac{\Delta E_{ij}}{k_B T}} & \text{if } \frac{\varepsilon_j / d_j}{\varepsilon_i / d_i} < 1, \\ 1 & \text{otherwise.} \end{cases}$$

Each neighbor of v_i has a probability of $1/d_i$ of being chosen. So the probability of a transition from v_i to v_j is $\frac{1}{d_i} A_{ij} = P_{ij}$.

If a random walk is made on this roadmap, then each state i has a probability π_i of being visited. As a random walk continues for an infinitely long time, assuming the Markov chain is ergodic, the probabilities π_i converge to fixed values that are the same for *any* random walk. Moreover, if the conformation space is sampled more and more finely, it can be shown that the limit distribution of the roadmap is the same as the limit distribution of an MC simulation ***(18)***. In other words, the resulting distribution is theoretically consistent with the Boltzmann distribution of energies predicted by statistical mechanics, and, equivalently, with the results of a large number of MC simulations.

Once constructed, the roadmap can be interpreted as a Markov chain and therefore be analyzed using techniques from Markov-chain theory. This can be used to calculate a quantity for each node called P_{fold}, the probability that the structure at that node will become completely folded before it becomes completely unfolded. This quantity can be used to estimate which structures constitute the transition state of the folding process, as well as to estimate the folding time for the protein.

Let F denote the set of nodes that correspond to conformations that are considered folded. Now suppose there is another stable state called the unfolded state. Let U denote the set of nodes corresponding to conformations close to the unfolded state. The probability of folding, P_{fold}, also called the transmission coefficient *(44)*, for a given node v_i can be written as

$$P_{\text{fold}}^{(i)} = \sum_{v_j \in \mathtt{F}} P_{ij} \cdot 1 + \sum_{v_j \in \mathtt{U}} P_{ij} \cdot 0 + \sum_{v_j \notin (\mathtt{F} \cup \mathtt{U})} P_{ij} \cdot P_{\text{fold}}^{(j)}.$$

The probability of folding is conditional on the first transition. If a node in F is reached, then F has been reached before U with probability 1. Similarly, if a node in U is reached, then F has been reached before U with probability 0. Otherwise, $P_{\text{fold}}^{(i)}$ depends on the probability of $P_{\text{fold}}^{(j)}$. Fast iterative solvers for linear systems can be used to compute P_{fold} for all nodes. For their initial work, Apaydin et al. used the Jacobi method as their linear system solver but noted that other approaches might provide faster performance.

SRS has been applied to the ColE1 repressor of primer and the homodomain of Engrailed, a developmental protein, which are stored in the Protein Data Bank *(45)* as 1rop and 1hdd, respectively *(19)*. The vector model described in **Subheading 2.1** was used to represent the degrees of freedom. With this model, 1rop has six degrees of freedom and 1hdd has 12 degrees of freedom. Energy was computed by the H–P energy model *(30)* mentioned previously. P_{fold} was computed for about 45 randomly selected conformations using SRS and using MC simulations. The correlation between the P_{fold} values of the two methods quickly converged to 1 as the number of nodes was increased, but SRS was roughly four orders of magnitude faster than the MC simulations. With SRS, the roadmap captures a substantial sampling of all folding and unfolding pathways simultaneously, and P_{fold} was computed for all nodes, not just the 45 that were randomly selected. Thus, SRS appears to be a promising alternative to running many independent MC simulations for examining protein-folding behavior.

In recent work, it has been demonstrated that SRS can be used to estimate the transition state ensemble (TSE), folding rate, and Φ-values of proteins ***(16)***. All of these values are of interest because they are quantities that can be measured by laboratory experiment and thus can be used to verify how well a simulation method such as SRS models the true behavior of a protein. Additionally, the TSE, if accurately determined, can provide insight into the overall folding mechanism of the protein.

The TSE is the set of conformations that represent the peak of the energy barrier that must be crossed by the protein in transitioning between the unfolded and native states. Alternatively, they are the states whose true P_{fold} is 0.5; the structures that have an equal probability of proceeding either to the folded or unfolded state. To account for modeling error, the TSE is taken to be the set of all conformations with P_{fold} between 0.45 and 0.55.

Apaydin et al. tested the method's ability to calculate the folding rate on a test set of 16 proteins and compared the results with the dynamic programming algorithm of Garbuzinskiy et al. ***(46)***. Intuitively, the folding rate is the fraction of unfolded molecules in some bulk set that transition to the folded state per unit of time. SRS-based estimates of the folding rate were found to correlate well with experimentally determined values and were consistently lower than those found by the other method. This indicates a consistent and significant difference between the TSEs found by the two methods and therefore their predicted folding rates. The difference appeared to be because of a less restrictive definition of the TSE by the dynamic programming method. Eighty percent of the structures identified as members of the TSE by the dynamic programming method were not considered part of the TSE by SRS. The more restricted set found by SRS led to more accurate estimation of measurable folding properties.

Φ-values are per-residue numbers between 0 and 1 indicating the degree to which the corresponding residue has reached its native conformation in the transition state of the protein ***(47)***. They are measured in the laboratory by mutating specific residues of the protein and determining the effect of each mutation on its folding rate and therefore, indirectly, the free energies of intermediate structures in the folding process. A Φ-value of 1 indicates that the mutation affects the folded state and transition state by the same amount and that the transition state of that residue therefore is essentially the same as the folded state. A Φ-value of 0 means the residue is unfolded in the transition state.

The developers of SRS found Φ-values for each residue of their 16-protein test set ***(16)***. The results were mixed but promising. For some proteins, such as CheY and the RNA-binding domain of U1A, their results correlated well with experiment, but their average error for Φ-values of the whole set of proteins was 0.21, which is quite large given the 0–1 range of Φ-values. Some of this error may be accounted for by the difference between the true free-energy variation of folding, as measurable in a laboratory, versus the approximation of free energy used in simulations.

3.3. Markovian State Models

A different way to construct a roadmap is by sampling small MD/MC trajectories rather than individual conformations, generating a Markovian State Model (MSM) ***(15)***. The use of MD/MC simulations for sampling suggests, among other things, that it is reasonable to expect that the resulting samples will have a realistic distribution of energies consistent with the predictions of statistical mechanics.

Suppose an initial MD or MC simulation trajectory starts in the folded state and ends in the unfolded state. Let $\{c_0, c_1, \ldots, c_n\}$ be a sequence of conformations along this trajectory separated by some fixed time step. A conformation c_i is selected uniformly at random from this sequence and a new MD/MC simulation is started from here. If the simulation does not reach the folded or unfolded state within some time limit, the trajectory is rejected. Otherwise, the trajectory is kept and a new current trajectory is created. Let the generated trajectory be denoted by $\{c'_0, c'_1, \ldots, c'_m\}$. If c'_m is in the folded state, the current trajectory becomes $\{c'_m, c'_{m-1}, \ldots, c'_0, c_i, \ldots, c_n\}$. If c'_m is in the unfolded state, the current trajectory becomes $\{c_1, c_2, \ldots, c_i, c'_0, c'_1 \ldots, c'_m\}$. Again, a conformation is selected uniformly at random from the current trajectory, and this procedure of generating new trajectories is repeated a set number of times.

Each conformation and each transition in each sampled trajectory is represented by a node and an edge, respectively, in the roadmap. Each edge has associated with it a simulation time t_{ij} required to make the corresponding transition. The trajectories are simulated such that this time-step between adjacent conformations in the trajectory is constant. How this is done depends on the type of simulation being run. Each edge also has a probability P_{ij} that is initialized to 1. The next step is to merge nodes that are within some cutoff distance of each other, because they represent the same conformation. This step amounts to clustering of the nodes into conformational substates. To merge two nodes, one of the nodes is removed from the roadmap and all of its edges are added to the node it is merged with. If this results in multiple edges between a pair of nodes, the edges need to be merged as well. The probability and time of the merged edge are defined as

$$P_{ij}^{\text{new}} = P_{ij}^1 + P_{ij}^2, \qquad t_{ij}^{\text{new}} = \frac{P_{ij}^1 t_{ij}^1 + P_{ij}^2 t_{ij}^2}{P_{ij}^1 + P_{ij}^2}.$$

After all nodes are merged that are within the cutoff distance of each other, the probabilities are renormalized so that the sum of the probabilities of all outgoing edges at a node is equal to 1. Singhal et al. *(15)* show that it is possible to derive a roadmap for a different temperature simply by reweighting the edges.

As with SRS, one can apply standard Markov-chain techniques to compute P_{fold} from the roadmap. One can also compute the average time it takes to reach the folded state. The validity of this roadmap construction method was tested on a two-dimensional artificial model system and on a small protein, the 12-residue tryptophan zipper β-hairpin, TZ2. TZ2 has previously been simulated on Folding@home *(48)*. Some of this data was used to build a stochastic roadmap. The predicted P_{fold} values and the average times to reach the folded state were in agreement with experimental data.

One problem with both the SRS and the MSM is that, because a road map of a conformation space is a discretization of a continuous space, the transition probabilities between nodes are only an approximation of reality. In a finite set of simulations, some states and transitions that occur with relatively low probability may never be sampled. Because the transition probabilities out of each node are forced to sum to 1, the transitions that are found are overrepresented because of the absence of others. This can lead to error in the computation of ensemble properties, including the predicted folding rate.

The developers of the MSM method proposed a method to estimate the error in the set of transition probabilities found by their sampling and therefore the error (or uncertainty) in their calculated folding rates *(14)*. Furthermore, by isolating which states contribute the most to this uncertainty, it becomes possible to adaptively select which states to generate sample simulations from at each step in building the roadmap so as to minimize the final uncertainty of the folding rate.

In analysis of MSMs, the folding rate is measured by estimating the mean first passage time (MFPT) from the unfolded state, x_1, to the folded state. This requires estimation of the MFPT, x_i, for all nodes in the roadmap, as follows:

$$x_i = \begin{cases} \Delta t + \sum_{j=1}^{K} x_j p_{ij} & i \neq K, \\ 0 & i = K, \end{cases}$$

where K is the index of the folded state, Δt is the size of the time interval between successive structures in the simulations used to construct the MSM, and p_{ij} is the probability of transitioning from state i to state j in time Δt. The MFPT from the first state, x_1, can be used to estimate the folding rate of the protein under the simulated conditions.

The problem is that it is not possible to determine the exact values of p_{ij} and therefore not possible to calculate exact values of MFPT. The maximum likelihood estimate, given the roadmap built through series of simulations, is $\hat{p}_{ij} = \frac{z_{ij}}{n_i}$, where z_{ij} is the observed number of transitions from state i to state j, and n_i is the total number of transitions out of state i. The observations z_{ij} follow a multinomial distribution that depends on the true transition probabilities. Ideally, the method would be able to estimate not just the most likely transition probabilities for a state but also the distribution of all possible sets of transition probabilities, and therefore our uncertainty of these estimates. Singhal et al. ***(14)*** show that this uncertainty follows a Dirichlet distribution, and based on that observation, provide a number of algorithms for finding the distribution of x_1, and therefore estimating the error of the calculated MFPT.

The basic idea of all of the algorithms is to sample a set of transition probabilities from a Dirichlet or approximation of a Dirichlet distribution whose parameters are based on the observed transition counts. Distributions for x_i and for MFPT, specifically x_1, are then inferred from the distributions of these samples. For details of the algorithms, please see the original paper ***(14)***.

The resulting uncertainty distribution for x_1 is a multivariate normal distribution, with calculable mean and variance. This distribution expresses how much confidence may be placed in the estimate of MFPT, but it also has implications for the construction of MSMs. It is possible to break the variance down into contributions from each state in the roadmap and furthermore to estimate the amount by which the variance due to any given state will decrease given some number of new MD/MC simulations starting from that state. The selection of which state to use for the next simulation need no longer be uniform at random, as described initially, but can instead be based on which choice of state is most likely to reduce the overall uncertainty of the MFPT by the greatest amount. This greatly increases the confidence of folding rate estimates and other properties calculated from an MSM generated by a set number of MD/MC simulations, versus undirected sampling.

Singhal et al. validated their error analysis method by again testing it on a set of simulations of TZ2, with a total of 87 distinct states. Using this example, they verified that all error estimation methods give comparable results for the mean and variance of the MFPT and that using the error estimates for adaptively focusing their sampling gave them a 20-fold improvement in certainty of their estimate of the MFPT for a given number of samples.

4. Discussion

Roadmap methods have been developed in recent years to study how a protein folds into its final known configuration. These roadmaps are generated by sampling conformations of a protein and connecting the sampled configurations in a number of ways. Various methods for generating and connecting roadmap nodes can only be expected to increase as time goes on. The same kind of growth was observed when roadmap methods became popular in robotics for solving the robot motion-planning problem as researchers began to understand how to better target their methods to the characteristics of the problems being address *(31)*. All existing approaches struggle to understand how to use energy estimates in the construction of the roadmap and the interpretation of the results. A number of questions are raised about how to compute the free energy for proteins of interest, which is a serious issue and a topic in need of further study.

Although the performance of roadmap methods is often compared with that of MD/MC methods, for now roadmaps are not necessarily meant to be a substitute for MD/MC simulations. Rather, the hope is that with a simplified energy model and clever sampling techniques, roadmap methods could quickly provide a coarse view of the energy landscape. Of course, much depends on the energy function used. The areas of interest identified in this landscape can provide a starting point for traditional MD/MC simulations.

Roadmap methods have also been applied to the study of other biological problems, including docking. In docking, the goal is to find low-energy conformations of a receptor–ligand complex. Recent examples of this work include *(36,37)*. Structure prediction is another area where roadmap methods have been applied *(38,39)*. By a combination of cleverly sampling and pruning conformations, Brunette and Brock *(38,39)* build up a compact model of the molecular energy landscape for a given protein. Finally, a road-map-based method for the generation of loop conformations was developed in **ref.** ***40***. Clearly, there are attractive features in road-map-based approaches for exploring high-dimensional spaces arising from geometric problems, which have prompted researchers to use them in various biological problems. Although roadmap-based methods are well understood in robotic problems, it is the authors' belief that a number of issues that mainly relate to the interplay of energy and geometry are still poorly understood for biological problems. Nevertheless, promising results are emerging that will no doubt fuel further advancements.

References

1. M. Gruebele. Protein folding: the free energy surface. *Current Opinion in Structural Biology*, 12:161–168, 2002.
2. T. Head-Gordon and S. Brown. Minimalist models for protein folding and design. *Current Opinion in Structural Biology*, 13:160–167, 2003.
3. X. Zhuang and M. Rief. Single-molecule folding. *Current Opinion in Structural Biology*, 13:88–97, 2003.
4. M. Vendruscolo and E. Paci. Protein folding: bringing theory and experiment closer together. *Current Opinion in Structural Biology*, 13:82–87, 2003.
5. C. M. Dobson. Protein folding and misfolding. *Nature*, 426:884–890,2003.
6. J. N. Onuchic and P. G. Wolynes. Theory of protein folding. *Current Opinion in Structural Biology*, 14:70–75, 2004.
7. C. M. Dobson. Principles of protein folding, misfolding and aggregation. *Seminars in Cell & Developmental Biology*, 15:3–16, 2004.
8. M. S. Apaydin. Stochastic roadmap simulation: an efficient representation and algorithm for analyzing molecular motion. PhD thesis, Stanford University, Stanford, CA 94305 USA, Aug 2004.
9. S. L. Thomas, X. Tang, L. Tapia, and N. M. Amato. Simulating protein motions with rigidity analysis. In *Proceedings of the ACM International Conference on Research in Computational Molecular Biology (RECOMB)*, pages 394–409, 2006.
10. S. Thomas, G. Song, and N. M. Amato. Protein folding by motion planning. *Physical Biology*, 2:S148–S155, 2005.
11. N. M. Amato, K. A. Dill, and G. Song. Using motion planning to map protein folding landscapes and analyze folding kinetics of known native structures. *Journal of Computational Biology: A Journal of Computational Molecular Cell Biology*, 10(3–4):239–255, 2003.
12. G. Song. A motion planning approach to protein folding. PhD thesis, Dept. of Computer Science, Texas A&M University, December 2003.
13. N. M. Amato and G. Song. Using motion planning to study protein folding pathways. *Journal of Computational Biology: A Journal of Computational Molecular Cell Biology*, 9(2):149–168, 2002.
14. N. Singhal and V. S. Pande. Error analysis and efficient sampling in Markovian state models for molecular dynamics. *The Journal of Chemical Physics*, 123(20):204909,2005.
15. N. Singhal, C. D. Snow, and V. S. Pande. Using path sampling to build better Markovian state models: predicting the folding rate and mechanism of a tryptophan zipper beta hairpin. *The Journal of Chemical Physics*, 121(1):415–425,2004.
16. T.-H. Chiang, M. S. Apaydin, D. L. Brutlag, D. Hsu, and J.-C. Latombe. Predicting experimental quantities in protein folding kinetics using stochastic roadmap simulation. In *Proceedings of the ACM International Conference on Research in Computational Molecular Biology (RECOMB)*, pages 410–424, 2006.

17. M. S. Apaydin, D. L. Brutlag, C. Guestrin, D. Hsu, and J.-C. Latombe. Stochastic conformational roadmaps for computing ensemble properties of molecular motion. In J. D. Boissonnat, J. Burdick, K. Goldberg, and S. Hutchinson, editors, *Algorithmic Foundations of Robotics V*, pages 131–147. Springer,2004.
18. M. S. Apaydin, D. L. Brutlag, C. Guestrin, D. Hsu, J.-C. Latombe, and C. Varma. Stochastic roadmap simulation: an efficient representation and algorithm for analyzing molecular motion. *Journal of Computational Biology: A Journal of Computational Molecular Cell Biology*, 10(3–4):257–281, 2003.
19. M. S. Apaydin, C. E. Guestrin, C. Varma, D. L. Brutlag, and J.-C. Latombe. Stochastic roadmap simulation for the study of ligand- protein interactions. *Bioinformatics*, 18 Suppl 2:18–26, 2002.
20. M. Karplus and J. Kuriyan. Molecular dynamics and protein function. *Proceedings of the National Academy of Sciences of the United States of America*, 102:6679–6685, 2005.
21. D. R. Ripoll, J. A. Vila, and H. A. Scheraga. Folding of the villin headpiece subdomain from random structures. Analysis of the charge distribution as a function of pH. *Journal of Molecular Biology*, 339(4):915–925, 2004.
22. W. F. van Gunsteren and H. J. C. Berendsen. Computer simulation of molecular dynamics: methodology, applications and perspectives in chemistry. *Angewandte Chemie International Edition in English*, 29(9):992–1023, 1990.
23. T. Huber, A. E. Torda, and W. F. van Gunsteren. Local elevation: a method for improving the searching properties of molecular dynamics simulation. *Journal of Computer-Aided Molecular Design*, 8(6):695–708, 1994.
24. B. G. Schulze, H. Grubmueller, and J. D. Evanseck. Functional significance of hierarchical tiers in carbonmonoxy myoglobin: conformational substates and transitions studied by conformational flooding simulations. *Journal of the American Chemical Society*, 122(36):8700–8711, 2000.
25. Y. Zhang, D. Kihara, and J. Skolnick. Local energy landscape flattening: parallel hyperbolic Monte Carlo sampling of protein folding. *Proteins*, 48(2): 192–201, 2002.
26. K. Lindorff-Larsen, R. B. Best, M. A. DePristo, C. M. Dobson, and M. Vendruscolo. Simultaneous determination of protein structure and dynamics. *Nature*, 433(7022):128–132, 2005.
27. L. E. Kavraki, P. Švestka, J.-C. Latombe, and M. H. Overmars. Probabilistic roadmaps for path planning in high-dimensional configuration spaces. *IEEE Transactions on Robotics and Automation: A Publication of the IEEE Robotics and Automation Society*, 12(4):566–580, 1996.
28. A. D. MacKerell, Jr. Empirical force fields for biological macromolecules: overview and issues. *Journal of Computational Chemistry*, 25(13):1584–1604, 2004.
29. M. Zhang and L. E. Kavraki. A new method for fast and accurate computation of molecular conformations. *Journal of Chemical Information and Computer Sciences*, 42:64–70, 2002.

30. S. Sun, P. D. Thomas, and K. A. Dill. A simple protein folding algorithm using a binary code and secondary structure constraints. *Protein Engineering*, 8(8):769–778, 1995.
31. H. Choset, K. M. Lynch, S. Hutchinson, G. Kantor, W. Burgard, L. E. Kavraki, and S. Thrun. *Principles of Robot Motion: Theory, Algorithms, and Implementations*. MIT Press, 2005.
32. J.-C. Latombe. *Robot Motion Planning*, chapter 7, pages 295–353. Kluwer, Dordrecht; Boston, 1991.
33. S. M. LaValle and J. J. Kuffner. Randomized kinodynamic planning. *The International Journal of Robotics Research*, 20(5):378–400, 2001.
34. D. Hsu, J.-C. Latombe, and R. Motwani. Path planning in expansive configuration spaces. *International Journal of Computational Geometry and Applications*, 9(4–5):495–512, 1999.
35. A. Ladd and L. E. Kavraki. Fast exploration for robots with dynamics. In *Workshop on the Algorithmic Foundations of Robotics*, 2004.
36. M. Moll, M. D. Schwarz, A. Heath, and L. E. Kavraki. On flexible docking using expansive search. Technical Report 04-443, Rice University, Houston, TX, 2004.
37. J. Cortés, T. Siméon, V. R. de Angulo, D. Guieysse, M. Remauld-Siméon, and V. Tran. A path planning approach for computing large-amplitude motions of flexible molecules. *Bioinformatics*, 21 Suppl. 1:i116–i125,2005.
38. T. J. Brunette and O. Brock. Improving protein structure prediction with model-based search. *Bioinformatics*, 21 Suppl. 1:i66–i74, 2005.
39. T. J. Brunette and O. Brock. Model-based search to determine minima in molecular energy landscapes. Technical Report 04-48, Dept. of Computer Science, University of Massachusetts, Amherst, MA, 2005.
40. J. Cortés, T. Siméon, M. Remaud-Siméon, and V. Tran. Geometric algorithms for the conformational analysis of long protein loops. *Journal of Computational Chemistry*, 25(7):956–967, 2004.
41. A. P. Singh, J.-C. Latombe, and D. L. Brutlag. A motion planning approach to flexible ligand binding. In *Proceedings of Seventh International Conference on Intelligent Systems for Molecular Biology (ISMB)*, pages 252–261, 1999.
42. G. N. Ramachandran and V. Sasisekharan. Conformation of polypeptides and proteins. *Advances in Protein Chemistry*, 23:283–438, 1968.
43. T. H. Cormen, C. E. Leiserson, R. R. Rivest, and C. Stein. *Introduction to Algorithms*. McGraw-Hill, second edition, 1990.
44. R. Du, V. Pande, A. Y. Grosberg, T. Tanaka, and E. Shakhnovich. On the transition coordinate for protein folding. *The Journal of Chemical Physics*, 108:334–350, 1998.
45. H. M. Berman, J. Westbrook, Z. Feng, G. Gilliland, T. N. Bhat, H. Weissig, I. N. Shindyalov, and P. E. Bourne. The protein data bank. *Nucleic Acids Research*, 28:235–242, 2000.

46. S. O. Garbuzynskiy, A. V. Finkelstein, and O. V. Galzitskaya. Outlining folding nuclei in globular proteins. *Journal of Molecular Biology*, 336:509–525, 2004.
47. A. Fersht. *Structure and Mechanism in Protein Science: A Guide to Enzyme Catalysis and Protein Folding*. W.H. Freeman & Company, 1999.
48. M. Shirts and V. S. Pande. Screen savers of the world unite *Science*, 290:1903–1904, 2000.

V

Methods for de novo Structure Prediction

10

Scoring Functions for De Novo Protein Structure Prediction Revisited

Shing-Chung Ngan, Ling-Hong Hung, Tianyun Liu, and Ram Samudrala

Summary

De novo protein structure prediction methods attempt to predict tertiary structures from sequences based on general principles that govern protein folding energetics and/or statistical tendencies of conformational features that native structures acquire, without the use of explicit templates. A general paradigm for de novo prediction involves sampling the conformational space, guided by scoring functions and other sequence-dependent biases, such that a large set of candidate ("decoy") structures are generated, and then selecting native-like conformations from those decoys using scoring functions as well as conformer clustering. High-resolution refinement is sometimes used as a final step to fine-tune native-like structures. There are two major classes of scoring functions. Physics-based functions are based on mathematical models describing aspects of the known physics of molecular interaction. Knowledge-based functions are formed with statistical models capturing aspects of the properties of native protein conformations. We discuss the implementation and use of some of the scoring functions from these two classes for de novo structure prediction in this chapter.

Key Words: De novo; physics-based; knowledge-based; potential; protein folding.

1. Introduction

The success of large-scale genome sequencing efforts has spurred structural genomic initiatives, with the goal of determining as many protein folds as possible *(1–4)*. At present, structural determination by crystallography and nuclear magnetic resonance (NMR) techniques are still slow and expensive in terms of manpower and resources, despite attempts to automate the

From: *Methods in Molecular Biology, vol. 413: Protein Structure Prediction, Second Edition*
Edited by: M. Zaki and C. Bystroff © Humana Press Inc., Totowa, NJ

processes. Computational structure prediction algorithms, while not providing the accuracy of the traditional techniques, are extremely quick and inexpensive and can provide useful low-resolution data for structure comparisons ***(5)***. Given the immense number of structures that the structural genomic projects are attempting to solve, there would be a considerable gain even if the computational structural prediction approach were applicable only to a subset of proteins.

Most current research in protein structure prediction is based on Anfinsen's thermodynamic hypothesis that the native structure of a protein can be determined entirely from its amino acid sequence ***(6)***. The two main categories of methods for predicting protein structure from sequence are comparative and de novo modeling. In the comparative modeling category, the methodologies rely on the presence of one or more evolutionarily related template protein structures that are used to construct a model. Traditionally, the evolutionary relationship can be deduced from sequence similarity ***(7–9)*** or by "threading" a sequence against a library of structures and selecting the best match ***(10,11)***. However, because of the improved sensitivity of the sequence similarity based methods, the threading approach has essentially been supplanted ***(12,13)***. In the de novo category, structure prediction methods attempt to predict tertiary structures from sequences based on general principles that govern protein-folding energetics and/or statistical tendencies of conformational features that native structures acquire, without the use of explicit templates ***(14–16)***. A general paradigm for de novo structure prediction involves sampling the conformational space, guided with scoring functions and other sequence-dependent biases, such that a large set of candidate ("decoy") structures are generated, and then selecting native-like conformations from those decoys using scoring functions and conformer clustering as filters ***(17,18)***. As a final step, detailed energy potentials are sometimes employed to perform high-resolution refinement on these native-like structures. Although the first papers on protein structure prediction appeared some thirty years ago, de novo structure prediction remains a difficult challenge today ***(12,13,19–21)***.

Scoring functions are employed in all stages of de novo structure prediction. For the conformational search stage, a selected combination of scoring functions approximates the energy landscape of the protein conformational space. Search methodologies such as Monte Carlo simulated annealing (MCSA) and molecular dynamics (MD) then generate trajectories leading to the minima of the landscape. As the conformational search process needs to evaluate new conformations encountered at every step, it is computationally intensive, and the scoring functions used in this stage need to be computationally efficient. Because none of the existing scoring functions can faithfully reproduce the true energy landscape of the conformational space, the search process often leads to many false minima. Thus, one usually repeats the search process many times with many different starting conditions and random seeds and obtains a collection of candidate ("decoy") structures. Then, a second set of (possibly different) scoring functions are used in the decoy selection stage as filter to eliminate non-native structures and retain the native-like ones. Conformer clustering is often used as an additional step to further refine the collection of the native-like conformations, followed by high-resolution refinement of the few remaining candidate structures. Compared to the functions used in the conformational search stage, the functions employed in the decoy selection stage can be algorithmically more complex and more detailed, because the number of candidate conformations to evaluate is much less than the number of conformations encountered during the search process. Scoring functions used in the high-resolution refinement stage are usually computational expensive functions formulated from detailed mathematical models of short-range interactions among atoms, allowing small local perturbations to fine-tune native-like structures.

There are two broad classes of scoring functions. The first class of functions are largely based on some aspects of the known physics of molecular interaction, such as the Van der Waals force, electrostatics, and the bending and torsional forces, to determine the energy of a particular conformation ***(22–27)***. The second class of functions is knowledge-based. Each of these knowledge-based functions tries to capture some aspects of the properties of protein native conformations, for example, the tendencies of certain residues to form contact with one another or with the solvent. These knowledge-based functions are usually compiled based on the statistics of a database of experimentally determined protein structures ***(28–34)***. In essence, the physics-based functions aim at predicting the native structure of a given sequence by mimicking the energetics of protein folding, whereas the knowledge-based functions bypass this intermediate step by directly making statistical inferences on what are observed in the database. Thus, the accuracy of the physics-based functions is determined by how realistic the underlying physical models are, whereas the accuracy of the knowledge-based functions is determined by the quality of the database as well as the validity of the statistical assumptions.

In an earlier edition, we introduced scoring functions for de novo structure prediction ***(35)***. In this chapter, we revisit physics-based and knowledge-based scoring functions in the context of their roles in the current state of the art structure prediction efforts. For the physics-based approach, the often-called Class I force field, which is a common foundation among the widely used molecular modeling force fields such as AMBER, CHARMM, OPLS, and ENCAD, is discussed. Extensions to this force field and the role of modeling solvent effects are also described. For the knowledge-based approach, we study the Bayesian (conditional) probability formalism, using it to derive the all-atom distance-dependent conditional probability discriminatory function (RAPDF) *(34)*. As an additional illustration, we delineate how one can combine the Bayesian probability formalism with the neural network methodology to construct neural network-based scoring functions. Then, a few other novel knowledge-based scoring functions from the recent literature are highlighted. Although it is not strictly a physics- or knowledge-based methodology, we briefly discuss the use of conformer clustering to further enhance decoy selection, as this technique has been shown to be useful in de novo structure prediction. Finally, a sophisticated combined physics- and knowledge-based potential used for high-resolution refinement is described.

2. Theoretical Background and Methods

2.1. An Overview of Physics-Based Energy Functions

Using quantum mechanical techniques, highly accurate energies can be calculated for small organic and inorganic molecules *(36,37)*. However, because of their sizes and flexibility as well as the presence of solvent molecules, proteins are much more difficult systems to model. The polar aqueous environment vastly complicates the calculation of the electrostatic energies. For instance, although there is no dispute that the largest driving force for protein folding is the hydrophobic effect *(38,39)*, which is associated with the decrease of water entropy upon the solvation of non-polar groups, the exact structural configuration of water molecules hydrating the solute remains unknown.

Although a full quantum mechanical treatment for a complete protein is not feasible, approximations and simplifications can be made to derive empirical physics-based energies. For example, hydrogen bond geometries that are applicable to those found in proteins can be determined from quantum mechanical calculations of simple systems *(40)*. Electrostatics calculations can be approximated using classical point charges and modifying the dielectric constant to approximate the polarizability of the protein and the solvent. Van der Waals interactions are often approximated by Lennard–Jones potentials. The first use of these approximate functions was in MD simulations, where fast and easily calculated energies were required to determine the force fields. Some prototypes for these types of energies are AMBER *(41)*, CHARMM *(42)*, OPLS *(24)*, and ENCAD *(43)*. Parameters for these energies have been obtained by fitting equations and results of computer simulations to data from experiments and from quantum mechanical calculations. These physics-based energies perform adequately for perturbations around a known native conformation ***(44,45)***, because the electrostatic and solvent-dependent information is implicit in the initial conformation itself. In combination with experimental NMR constraints ***(46,47)***, these force fields enable the determination of accurate structures, so long as there are enough constraints to define the fold. Unfortunately, in isolation, the solvent and electrostatic modeling is insufficient for full and reliable simulation of protein folding. As a result, producing accurate protein folding simulations from physics-based energies alone is still a very challenging and active area of research.

2.1.1. Class I Physics-Based Scoring Function and Its Possible Extensions

As we have mentioned, AMBER ***(41)***, CHARMM ***(42)***, OPLS ***(24)***, and ENCAD ***(43)*** are some examples of the widely used physics-based force fields in protein-folding simulation. These force fields share a lot of commonalities in terms of the underlying physical models used and the mathematical approximations assumed. As an illustration, the AMBER force field, which was first developed under the direction of Professor Peter Kollman, has the following form:

$$V_{\text{total}} = V_{\text{bond}} + V_{\text{angle}} + V_{\text{torsion}} + V_{\text{non-bond}} \tag{1}$$

Here, V_{total} is the total potential energy, V_{bond} is the bond stretching energy, V_{angle} the angle bending energy, and V_{torsion} the angle torsional energy. Together, V_{bond}, V_{angle}, and V_{torsion} are denoted as the bonded interactions terms. $V_{\text{non-bond}}$ is the energy for non-bonded interactions, consisting of a Van der Waals energy term V_{vdW} and an electrostatics term V_{elec}. Other widely used force fields such as CHARMM and OPLS employ similar bonded and non-bonded terms in their formulations, and Eq. 1 is often denoted as the Class I force field.

The bond-stretching energy (*see* **Fig. 1A**) is modeled by treating the bond as an idealized spring and using a simple quadratic function derivable from the Hooke's law.

$$V_{\text{bond}} = k_{\text{bond}}(r - r_{\text{o}})^2 \tag{2}$$

where k_{bond} is the bond-stretching constant, controlling the stiffness of the bond spring, and $(r\text{–}r_{\text{o}})$ is the deviation of the bond length from its equilibrium distance. Unique numerical values for k_{bond} and r_{o} are assigned each pair of atom types.

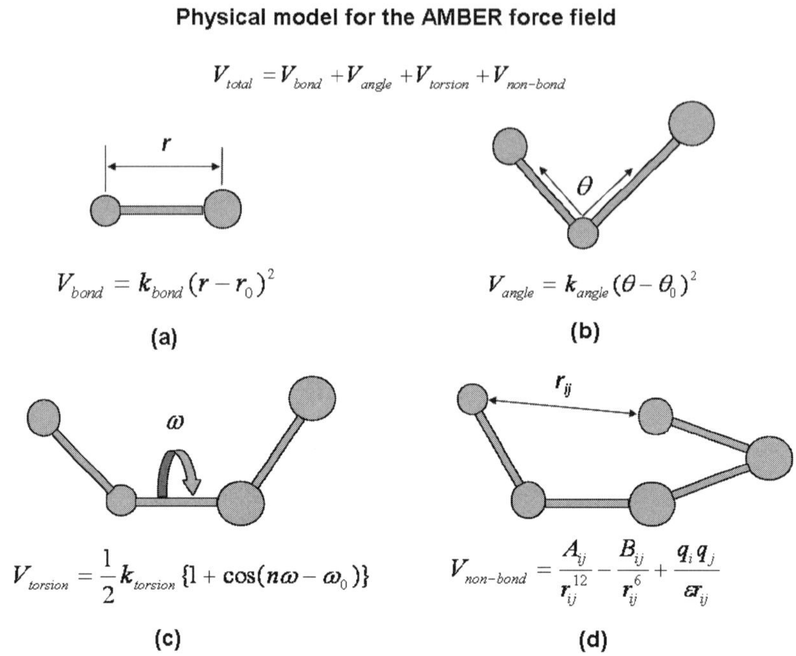

Fig. 1. The physical models for the AMBER molecular mechanics force field. Atoms and bonds are shown. **(A)** The physical model for bond stretching, **(B)** the model for angle bending, **(C)** the model for angle torsional energy, and **(D)** the model for electrostatics and Van der Waals forces.

The angle bending energy (*see* **Fig. 1B**) is similarly modeled by the Hooke's law.

$$V_{\text{angle}} = k_{\text{angle}}(\theta - \theta_{\text{o}})^2 \tag{3}$$

where k_{angle} is the angle bending constant, controlling the stiffness of the angle spring. θ is the angle formed by the atom of interest with its two covalently bonded neighbors, and $(\theta - \theta_{\text{o}})$ is the deviation of the angle from its equilibrium value in radians. Again, unique values for k_{angle} and θ_{o} are determined for each bonded triplet of atom types.

The torsional energy (*see* **Fig. 1C**) is represented by an n-fold periodic function:

$$V_{\text{torsion}} = \frac{1}{2} k_{\text{torsion}} [1 + \cos(n\omega - \omega_0)] \tag{4}$$

Here, the torsional angle ω is the dihedral angle defined by a quartet of bonded atoms, and ω_0 is the reference angle. k_{torsion} is a constant for the

n-fold periodic interaction. n represents the periodicity of the torsional barrier, reflecting the intrinsic symmetry in the dihedral angle for the quartet of the bonded atoms. Unique values of k_{torsion}, n, and ω_0 are assigned to each bonded quartet of atom types. In practice, parameterization of torsional energies also corrects for bonding energy terms unaccounted for by the simple bending and stretching models. Additional torsional energy terms (denoted as "improper torsions" in the literature) can be added to ensure that subtle properties such as chirality and planarity are preserved.

For the non-bonded interactions, AMBER and other commonly used force fields employ a 6–12 Lennard–Jones potential to represent the Van der Waals interactions between two non-bonded atoms, and the Coulomb's law to model the interactions of two charged atoms (*see* **Fig. 1D**):

$$V_{\text{non-bond}} = \left(\frac{A_{ij}}{r_{ij}^{12}} - \frac{B_{ij}}{r_{ij}^{6}}\right) + \left(\frac{q_i q_j}{\varepsilon r_{ij}}\right) \tag{5}$$

The Van der Waals interaction consists of two components, a short-range attractive force that quickly vanishes when the distance between the interacting atoms, r_{ij}, is greater than a few Angstrom and an even shorter-range repulsive force that dominates when r_{ij} is less than the sum of their individual atomic radii. B_{ij} and A_{ij} in Eq. 5 control the attractive and the repulsive components of the steric potential. A_{ij} can be calculated from quantum mechanics considerations or measured from atomic polarizability experiments, and B_{ij} can be calculated from crystallographic data. For the eletrostatics, interacting atoms are treated as point charges of q_i and q_j. The value of the dielectric constant ε accounts for the attenuation of electrostatic interaction by the polar environment. In more sophisticated solvent models, which are discussed later, the constant ε is replaced by a function dependent on r_{ij}. Earlier versions of AMBER had an explicit term to take into account hydrogen bonding. The latest versions incorporate hydrogen-bonding effects into the parameterization of the electrostatic and van der Waals terms, as these two terms are found to be able to sufficiently represent the distance and angle dependencies of hydrogen bonds in molecular mechanics modeling ***(48)***.

Currently, except in the high-resolution refinement stage, idealized backbone and side-chain bond lengths and angles are often used in de novo structure prediction. Hence, the energy associated with the bonded interactions terms V_{bond}, V_{angle}, and V_{torsion} can be regarded as constant. Improvement in structure prediction can conceivably be achieved by enhancing the physical models for the non-bonded terms. For example, one can replace the Van der Waals terms in Eq. 5 by a buffered 14–7 potential ***(49,50)***, by the Morse function ***(51)***, or by the Buckingham–Fowler potential *(52)*. The goal is to reduce the Pauli exclusion barrier so as to allow sufficient sampling of conformations in the neighborhood of the native structure during molecular mechanics or Monte Carlo simulations.

For the electrostatic term, the physical model of fixed charges at atom centers is found to be insufficient to describe charge polarization in the aqueous environment. Examples of the more sophisticated electrostatics models involve generalizing the point charge model with multi-center multi-pole expansion. This can be done through the cumulative atomic multi-pole moment method, the distributed multi-pole analysis, or an atoms-in-molecules-based multi-pole moment method *(53–55)*. Even though these types of model improvement are computationally expensive, several groups have been making significant progress in incorporating polarizable force fields for MD simulation of proteins. For example, see **refs.** ***56–58***.

2.1.2. Protein Structures in Aqueous Environment

Protein structures are formed in the presence of aqueous environment, and therefore, in order for the search of energy-minimized protein conformation to be accurate, the effect of the solvent must be taken into account. Explicit solvent models that simulate individual water molecules [for example, TIPS *(59,60)*, SPC *(61)*, and F3C *(62)*] are too slow to be practicable for protein structure prediction. Truncation of the non-bonded potentials such that interactions beyond a fixed cutoff distance are ignored can improve speed. However, it often leads to undesirable artifacts and reduced accuracy *(63)*. Combining Ewald's approach with fast Fourier transform, Darden and his colleagues have developed the particle mesh Ewald method to describe long-range interactions more efficiently *(64)*. However, direct simulation with explicit water is still highly computational expensive even with this and other advances. On the contrary, the effect of solvation can be modeled implicitly by averaging solvent-solute interaction using mean field formulation and by decomposing the solvation energy into an electrostatic component and a so-called non-polar component, which accounts for everything else. For electrostatics, Poisson–Boltzmann *(65,66)* models extend the simple Coulombic potential by allowing charge distributions within the solute and having separate dielectrics for the solvent and solute. Unfortunately, there are no general analytical solutions for the Poisson–Boltzmann equation for irregular protein shapes and precise numerical solutions (for example, by finite differences using GRASP/Delphi *(67)*) can be very computationally expensive. Faster solutions can be obtained using generalized-Born (GB) approximations *(68)*, which have been incorporated into MD simulations. For the non-polar term, which includes hydrophobic interactions, the energy is usually modeled as a simple linear function of solvent accessible area. The resulting generalized-Born/surface-area (GBSA) models are more accurate than the simple non-bonded interaction terms and can rival knowledge-based functions for scoring small loops in accuracy *(69)*. However, the amount of parameterization involved in GBSA models also rivals that of knowledge-based energies. Recently, other approximate methods for solving the Poisson–Boltzman equation may prove to be as or more accurate with less parameterization *(70)*. Besides the Poisson–Boltzmann and generalized Born-type approaches, another category of implicit models describes the solvent effect in terms of the dielectric screening of electrostatic interaction within the protein molecule. For example, this can be done by defining the dielectric coefficient as a simple function of distance *(71,72)* and as a more detailed function involving solvent-excluded volume *(73)*, the distance of a charge from the protein surface, and the degree of exposure of a charge point to the solvent *(74)*.

In summary, the implicit solvent models are computationally much more efficient than the explicit models. The tradeoff is the inability to represent the detailed interaction structures between the solvent and the solute, which can be essential in determining the overall energy landscape. Furthermore, the lack of polarizability in the continuum solvent treatments precludes a flexible description of charge distributions in the aqueous environment.

2.2. An Overview of the Knowledge-Based Scoring Functions

The physics-based functions are formulated from underlying approximate physical models. In contrast, knowledge-based functions are derivable directly from properties observed in known folded proteins *(75)*. Although the basis of the knowledge-based propensities is still physical, the statistical "black-box" approach to the weighting of physical effects has proved to be more effective than explicitly specifying the form and calculating the coefficients in traditional physics-based energies. As a result, almost all of the most successful de novo structure prediction techniques have both physics-based and knowledge-based components.

The hydrophobic moment *(76)* is an example of a simple heuristic energy function. It is analogous to the physical moment of inertia except that the mass term is replaced by a measure of the hydrophobicity of the residue. Minimization of this function leads to compact structures with hydrophobic residues in the core. In general, any property that is differentially observed in folded proteins and unfolded proteins can be converted into an energy function. Hidden Markov models (HMM), neural nets, support vector machines (SVM), and trial and error have been used to find such properties. A particularly useful class of knowledge-based functions is the pairwise distance preferences ***(11,34,77)***, which reflect proper packing. Consequently, the pairwise distance preference scoring functions can be found in many of the top-performing de novo methods, for example, ROSETTA ***(16)***, FRAGFOLD ***(78)***, TASSER ***(79)***, CABS ***(80)***, and PROTINFO ***(81)***.

2.2.1. Deriving Knowledge-Based Scoring Functions from the Bayesian Probability Formalism

A majority of the knowledge-based scoring functions have their theoretical foundations rooted in the Bayesian (conditional) probability formalism. In such a formalism, we view a given set of conformations for a protein sequence as comprising a subset of correct conformations {C} and a subset of incorrect conformations {I}. Furthermore, we consider a set of conformational properties, which can be any feature of protein structure that differs significantly between the subset of incorrect conformations and the subset of correct conformations. Examples are the preferences of some amino acid subsequences to exhibit certain torsion angles, to form contacts with other amino acid types, and so on. In this subheading, for the purpose of illustration, we focus on the set of inter-atomic distances within a structure $\{d_{ab}^{ij}\}$, where d_{ab}^{ij} is the distance between atoms numbers i and j, of type a and b. We want to determine $P(C|\{d_{ab}^{ij}\})$, the probability that the structure is a member of the "correct" subset, given that it contains the distances $\{d_{ab}^{ij}\}$. A standard way to achieve this is to express $P(C|\{d_{ab}^{ij}\})$ in terms of probabilities derivable from experimental structures, through the Bayes' theorem:

$$P(C|\{d_{ab}^{ij}\}) = P(C) \times \frac{P(\{d_{ab}^{ij}\}|C)}{P(\{d_{ab}^{ij}\})} \tag{6}$$

Here, $P(\{d_{ab}^{ij}\}|C)$ is the probability of observing the set of distances $\{d_{ab}^{ij}\}$ in a correct structure. $P(\{d_{ab}^{ij}\})$ is the probability of observing such a set of distances in any correct or incorrect structure, and $P(C)$ is the probability that any structure picked at random belongs to the correct subset. $P(\{d_{ab}^{ij}\}|C)$ is regarded as a posterior probability in the sense that the underlying population for the probability distribution consists of structures that are already known to belong to the "correct" subset. On the contrary, $P(\{d_{ab}^{ij}\})$ is regarded as a prior probability in the sense that its underlying population is composed of structures whose class memberships have not yet been determined. We should note that both $P(\{d_{ab}^{ij}\}|C)$ and $P(\{d_{ab}^{ij}\})$ are highly difficult to compute, because the input arguments to these probability functions are the multitude of distance variables. A full model capturing the dependency among these variables would be extremely complex and would require a huge amount of training data to determine all the implicit parameters. Hence, to ensure computational feasibility of Eq. 6, one often makes the simplifying, albeit not strictly correct, assumption that the distances are statistically independent of one another, that is:

$$P(\{d_{ab}^{ij}\}|C) = \prod_{i,j} P(d_{ab}^{ij}|C);\ P(\{d_{ab}^{ij}\}) = \prod_{i,j} P(d_{ab}^{ij}) \tag{7}$$

Then, combining Eqs. 6 and 7 gives us

$$P(C|\{d_{ab}^{ij}\}) = P(C)\prod_{i,j}\frac{P(d_{ab}^{ij}|C)}{P(d_{ab}^{ij})} \tag{8}$$

For a given protein sequence, $P(C)$ is a constant independent of conformation and therefore can be omitted because we are only interested in selecting native-like conformations among decoys for a fixed protein sequence. Equation 8 suggests a scoring function S, which is proportional to the negative log conditional probability that the given structure is correct, given a set of distances.

$$S(\{d_{ab}^{ij}\}) = \sum_{i,j} s(d_{ab}^{ij})\,;\ s(d_{ab}^{ij}) = -\log\left(\frac{P(d_{ab}^{ij}|C)}{P(d_{ab}^{ij})}\right) \tag{9}$$

An advantage of using Eq. 9 instead of Eq. 8 as a scoring function is that in the logarithm form, the pitfall of repeated multiplication of small numbers is eliminated, and therefore, it is easier to be implemented on the computer.

One can replace the set of distances $\{d_{ab}^{ij}\}$ with another type of conformational property, say for example $\{m_a^i\}$, where m_a^i represents the value of that conformational property attained by residue number i of amino acid type a. This leads to another scoring function:

$$S(\{m_k\}) = -\sum_k \log\left(\frac{P(m_k|C)}{P(m_k)}\right) \tag{10}$$

To gain an intuitive understanding of the scoring function, we note that if the chosen conformational property does not differ significantly between the subset of incorrect conformations and the subset of correct conformations, then the values of $P(m_k|C)$ and $P(m_k)$ will tend to be close to each other. The resulting score S will always be close to 0 and is not an informative measure for decoy discrimination. On the contrary, if the conformational property is well chosen, that is, it differs significantly between incorrect and correct conformations, then for a native-like structure, $P(m_k|C)$ will tend to dominate $P(m_k)$, yielding a negative (good) score for S. On the contrary, for a non-native structure, the opposite occurs, yielding a positive (bad) score.

2.2.2. Compilation of the Probabilities

Before one can use Eq. 9 as a scoring function, the statistics for the posterior probability $P(d_{ab}^{ij}|C)$ and the prior probability $P(d_{ab}^{ij})$ need to be compiled. To compile the statistics for $P(d_{ab}^{ij}|C)$, we can tabulate the intra-molecular distances observed in a database of experimentally determined conformations. Such a database is usually extracted from the Protein Data Bank (PDB) ***(82,83)***. For example, one can proceed to select all the proteins from the PDB that also appear in the e-value filtered ASTRAL SCOP genetic domain sequence subset list with the threshold e-value set at 10^{-4} ***(84)***. Such an e-value is chosen, so that sampling bias (i.e., including too many homologous proteins) can be avoided. We then evaluate the quantity

$$P(d_{ab}^{ij}|C) \equiv \frac{N(d_{ab})}{\sum_{d} N(d_{ab})} \tag{11}$$

where $N(d_{ab})$ is the number of occurrences of atom types a and b in a distance bin d in the database.

To compile the statistics of the prior probability $P(d_{ab}^{ij})$, we apply a formula similar to Eq. 11. But the question is: What would be an appropriate database from which to tabulate the counts? Samudrala and Moult ***(34)*** argued that methods employed for structure prediction usually produce compact models, whether the result is topologically correct or not. Thus, they consider a good choice of prior distribution to be found in the set of possible compact conformations and assume that averaging over different atom types in experimental conformations is an adequate representation of random arrangements of these atom types in any compact conformation. The probability $P(d_{ab})$ of finding atom types a and b in a distance bin d in any native-like or non-native compact conformation is thus approximated by:

$$P(d_{ab}) = \frac{\sum_{ab} N(d_{ab})}{\sum_{d}\sum_{ab} N(d_{ab})} \tag{12}$$

where $\sum_{ab} N(d_{ab})$ is the total number of contacts between all pairs of atom types in a particular distance bin d, and the denominator is the total number of contacts between all pairs of atom types summed over the distance bins d. The pairwise distance preference function described in **Subheading 2.2.1.**, Eq. 9, together with Eq. 11 and the prior distribution assumption of Eq. 12, is termed the RAPDF in *(34)*. **Figure 2A** highlights the essential components of this scoring function.

Besides the above method of estimating prior distributions, various other approaches have also been suggested. Subramaniam et al. *(85)* assumed that all distances are equally probable, and Avbelj and Moult *(86)* considered the set of distances observed in some random coil model as appropriate. Lu and Skolnick *(87)* employed a quasi-chemical approximation. Alternatively, Zhou and Zhou *(88)* assumed that the residues follow uniform distribution everywhere in the protein and developed a new reference state termed "distance-scaled, finite ideal-gas reference state."

2.2.3. A Pairwise Distance Scoring Function in Continuous Form

The RAPDF scoring function uses discrete distance bins to compile the probability scores. Specifically, contact distances between 0 and 3 Å are grouped into bin 1, 3 and 4 Å into bin 2, 4 and 5 Å into bin 3, and so on up to the 20 Å cutoff. As a result, the score for observing any distance within a bin width is the same for a given pair of atom types. However, the distance preferences between atom types should vary in a continuous manner as the distances between the contacts vary. We can seek a function to interpolate between the scores across the discrete bins such that the score for a given distance can be uniquely defined. Several methods for interpolating discrete points, including linear, polynomial, cubic spline, and band-limited interpolations, have been tested for their efficacy to improve the discriminatory power of RAPDF. The best among the tested methods is band-limited interpolation, derivable from the Fourier Theorems. It assumes that the variation of the log-likelihood scores fluctuates slowly enough such that the scores for any given distance can be exactly reconstructed from the scores across the discrete bins.

Given a pair of atom types a and b at a particular distance, a "continuous" log-likelihood score $s_c(d_{ab})$ can be calculated by interpolating between the scores across the discrete bins of $s(d_{ab})$ through the Shannon's sampling theorem, resulting in a smooth curve *(89)*. (*see* **Fig. 2B** for illustration.) Given an amino acid sequence in a particular conformation, $s_c(d_{ab})$ of all contacts between pairs of atom types at any distance within the 20 Å cutoff is summed to yield the total

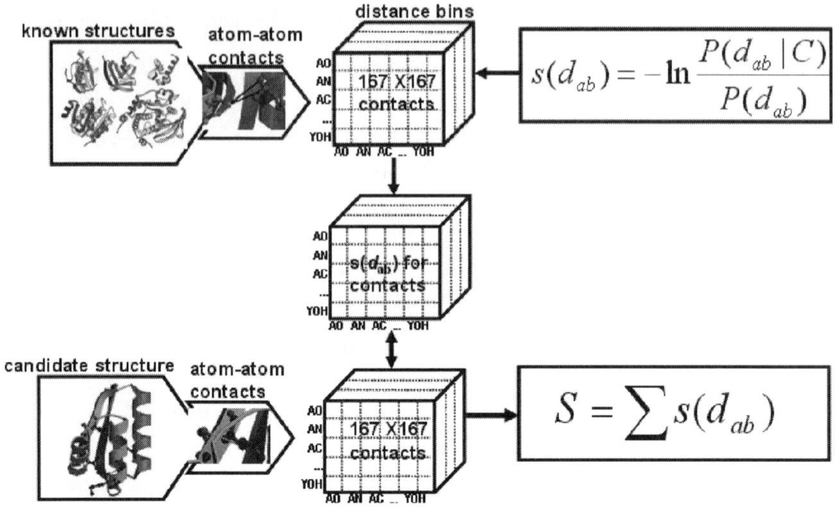

(a)

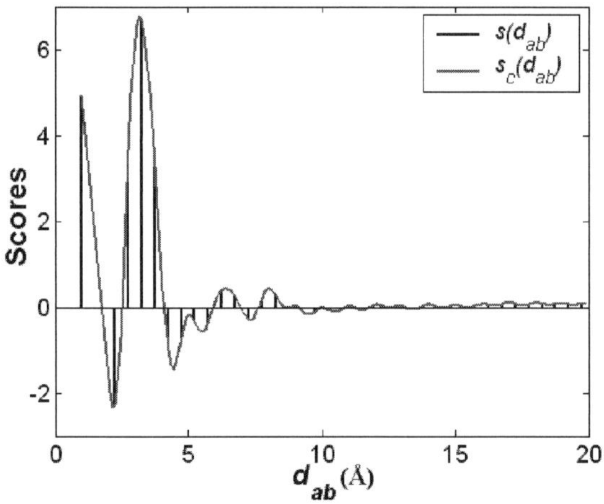

(b)

Fig. 2. The all-atom distance-dependent conditional probability discriminatory function (RAPDF) and its extension, the interpolated RAPDF function. **(A)** The essential feature of the RAPDF scoring function. A matrix giving the log-likelihood scores for pairwise contact among different atom types at various discrete distance bins is computed using a database of known experimental structures. Then, given a candidate ("decoy") structure, appropriate entries in the matrix can be extracted and summed to give a log-likelihood score for the structure. **(B)** The application of band-limited log-likelihood score to evaluate whether the conformation is native-like or not. The interpolated RAPDF (IRAPDF) has been evaluated by various decoy sets. Comparison between the IRAPDF and the RAPDF shows that the band-limited interpolation leads to an improved discriminatory power.

2.3. Neural Network Knowledge-Based Scoring Functions

Rather than predicting whether an entire structure is native-like or not, neural network algorithms are often used to predict the likelihood of occurrence of a certain conformational property for each residue along a given protein sequence. Examples of the properties are the tendencies of an amino acid to be exposed or buried relative to the solvent ***(90–92)***, to be part of the helix, strand, or coil local structures ***(93–95)***, the expected number of contacts a residue makes with other residues ***(96–99)***, and so on. Usually, the conformational property of interest is discretized into a number of states, and a neural network algorithm returns numerical values which correlate with the probabilities of occurrences of those states.

One can combine the neural network algorithms for predicting conformational properties with the Bayesian probability formalism that has been used to construct various knowledge-based functions. This leads to a class of scoring functions that give log-odd scores, indicating whether a given structure is native-like or not, and that have in their core a neural network component. In the following subheadings, we review a standard formulation of the neural network algorithm that is used to predict conformational properties of residues in a protein sequence. We then describe how the neural network and the Bayesian frameworks are combined to form several neural network-based scoring functions.

2.3.1. Neural Network Algorithms for Predicting Local Structures

For concreteness, we consider the prediction of the degree of solvent accessibility of individual residues along a given protein sequence, with the degree discretized into three states: low, medium, and high. The now standard approach, introduced in **ref. *93*** and improved upon in **ref. *94***, uses a feed-forward neural network. The input to the network is a window of sequence

---

Fig. 2. interpolation to the discrete distance bins of the RAPDF function. The score $s_c(d_{ab})$ of a given pair of atom types at any distance within the 20 Å cutoff can be uniquely defined by interpolating across the discrete bins of $s(d_{ab})$. The resulting scoring function is termed as the interpolated RAPDF (IRAPDF).

profile corresponding to a consecutive sequence of residues. Such a windowed sequence profile can be obtained by following a procedure described in **ref. *94***. The protein sequence of interest is employed as input to PSI-BLAST ***(100)***, which generates a position-specific scoring matrix (PSSM) associated with that sequence. The PSSM consists of $20 \times M$ entries, where M being the length of the sequence, and each entry in a column gives the log-likelihood for one of the twenty possible amino acid substitutions for the residue position of interest. The standard logistic transform is then applied to each entry of the PSSM, so that these values are rescaled to the 0–1 range, appropriate to serve as neural network inputs. The neural network itself can consist of one or more hidden layers, and its output layer comprises three output units, representing the low, medium, and high solvent accessibility states, respectively. Training of the network is done with back-propagation ***(101)***, using the database of experimentally determined protein structures we have already described in **Subheading 2.2.2.** Given a window of sequence profile of the residue of interest (i.e., the sequence profile of the residue as well as those of the neighboring residues), the resulting neural network returns a numerical value in each output unit correlating with the probability with which the residue assumes the corresponding state.

2.3.2. Combining the Neural Network Algorithms with the Bayesian Probability Formalism

To describe how one combines the Bayesian and the neural network frameworks to construct new scoring functions, for concreteness, suppose once again that the conformational property of interest is the degree of solvent accessibility. Using the language of the preceding subheadings, we want to calculate the probability that a given structure belongs to the subset of correct structures, given the associated conformational string $\{q_a^i\}$. Here, $q_a^i \in \{l, m, h\}$, where l represents low solvent accessibility state, m medium, and h high, i is the residue number, and a is the amino acid type. A scoring function described in Eq. 10 now takes the following form:

$$S(\{q_a^i\}) = -\sum_i \log\left[\frac{P(q_a^i|C)}{P(q_a^i)}\right] \quad (13)$$

$P(q_a^i|C)$ is simply the (posterior) probability of residue i taking on a particular solvent accessibility state q_a^i in a native structure. With an additional processing step involving the nearest-neighbor approach of Yi and Lander ***(102)*** to be discussed in detail in the next subheading, this probability can be estimated by using the neural network algorithm previously described. $P(q_a^i)$, on the contrary, is the (prior) probability that the residue is observed to assume the solvent accessibility state q_a^i in any native-like or non-native structure. It can be estimated using the formula

$$P(q_a) \equiv \frac{N(q_a)}{\sum_{q \in \{l,m,h\}} N(q_a)} \tag{14}$$

where $N(q_a)$ is the number of occurrences of the amino acid type a taking on the solvent accessibility state q in some database of structures, and $\sum_{q \in \{l,m,h\}} N(q_a)$ is the total number of occurrences of the amino acid type a in that database. Again, the question is: What is an appropriate database from which to tabulate the counts? We can use the same approach adopted by Samudrala and Moult in **ref.** ***34***, arguing that the set of possible compact conformations is a good choice of prior distribution. Then, the database to use will simply be the database of the experimentally determined structures. Alternatively, we can employ a database of decoy structures. Such a database can be created by applying a de novo conformational space sampling protocol to generate n decoy structures (for example, $n = 10$) for each protein sequence that appears in the database of the experimentally determined structures and then gathering the resulting decoys.

We note that as $P(q_a^i|C)$ is estimated by the neural network algorithm with a window of sequence profile as its input, the influence of the neighbors of residue i on its conformation is automatically taken into account. Thus, the posterior probability that residue i assumes a particular conformation is calculated in the context of its surrounding environment. In contrast, the probability distribution $P(q_a)$ is compiled on a "single-residue" basis. Thus, $P(q_a)$ can be viewed as the tendency of the amino acid type a to adopt a certain conformation averaged over the various types of neighborhood environments.

For further illustration, we generate a neural network-based Bayesian scoring function for each of the following conformational properties: the virtual torsion angle, the virtual bending angle, and the degree of solvent accessibility. The virtual torsion angle and the virtual bending angle are calculated by the DSSP program ***(103)***. Specifically, given a residue i of interest, the virtual torsion angle for i is the dihedral angle defined by the C_α atoms of residues $i-1$, i, $i+1$, and $i+2$. The virtual bending angle is the bending angle defined by the C_α atoms of residues $i-2$, i, and $i+2$. Solvent accessibility is the residue water exposed surface in Å^2. To implement the scoring functions, the virtual torsion angle are manually divided into two discrete states, whereas the virtual bending angle and the degree of solvent exposure are each manually divided into three discrete states.

2.3.3. Training and Post-Processing of the Neural Network

The Stuttgart Neural Network Simulator ***(104)*** is a versatile and convenient tool to configure and train the neural networks for predicting the various conformational properties. The network configurations follow the description given in **Subheading 2.3.1.** The input layer receives a window of sequence profile. The window size typically ranges from 1 to 17 consecutive residues. The network has a single hidden layer and an output layer of two or three units representing two or three discrete states. See **Fig. 3** for an illustration.

We divide the database of experimentally determined structures into two equal subsets *A* and *B*, which are alternately used as the training and the test sets. The neural network training is done in batch mode using standard back-propagation, and the cycle of batch-mode training is repeated until the test error reaches a minimum. We note that two neural networks are obtained at the conclusion of the training—one (denoted as NN_A) trained with subset *A* and tested with subset *B* and another one (denoted as NN_B) trained with subset *B* and tested with subset *A*.

Given a residue of interest together with its windowed sequence profile, it is desired to extract from NN_A and NN_B the posterior probabilities with which the residue assumes each of the three states, say in the case of solvent accessibility prediction (two states in the case of virtual torsion angle prediction and three states in the case of virtual bending angle prediction). To this end, the nearest-neighbor approach of Yi and Lander ***(102)*** is employed: The output layer of NN_A gives a 3-tuple vector (s_{lA}, s_{mA}, s_{hA}). The closeness of this vector with respect to vectors corresponding to all instances in the test set can be calculated through the Euclidean measure

$$\left((s_{lA} - s_{lA}^{g})^2 + (s_{mA} - s_{mA}^{g})^2 + (s_{hA} - s_{hA}^{g})^2\right)^{1/2} \tag{15}$$

where g stands for instance g in the test set. The k-nearest neighbors [e.g., the closest 5% of all instances in the test set with respect to (s_{lA}, s_{mA}, s_{hA})] are then determined, and the actual solvent accessibility states of those nearest neighbors are tabulated, yielding the counts (c_{lA}, c_{mA}, c_{hA}). The same procedure is repeated with NN_B. The probability that the residue of interest takes on each of the three states is thus estimated by

$$P(s_q) = \frac{c_{qA} + c_{qB}}{\sum_{r \in \{l,m,h\}} c_{rA} + c_{rB}} \tag{16}$$

where q stands for low, medium, or high accessibility state. Equation 16 supplies the posterior probabilities required in Eq. 13 for score calculation.

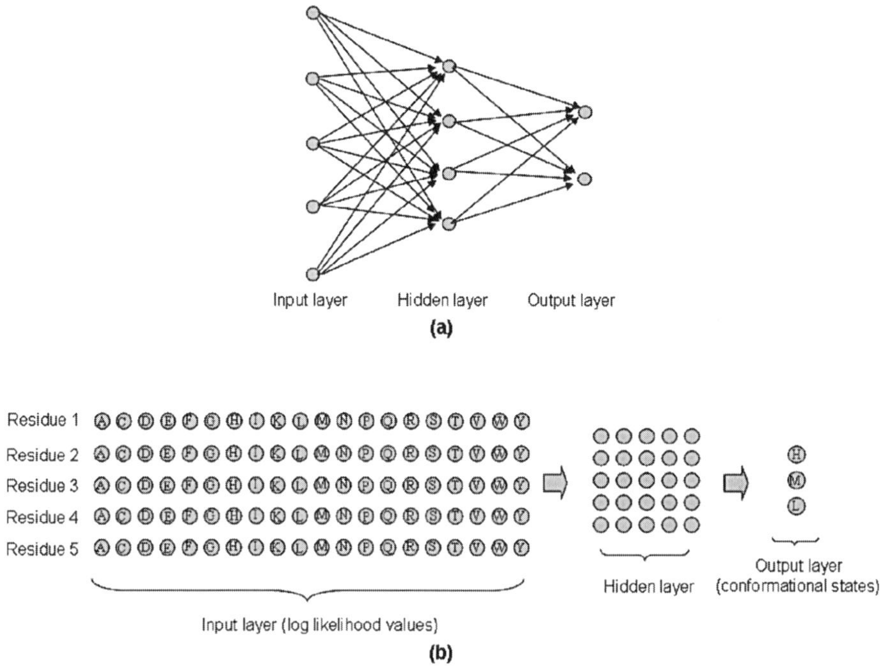

Fig. 3. Schematic diagrams of the neural networks used to predict conformational property given a sequence profile. **(A)** A fully connected neural network with input (5 units), hidden (4 units), and output (2 units) layers. Every unit in the input layers is connected with every unit in the hidden layers. The same holds true for the hidden and the output layers. **(B)** The typical size of a neural network we use for constructing the knowledge-based functions. In this example, the window size of the input sequence profile is five residues. Each residue provides twenty input units, representing the log-likelihood values for the twenty possible amino acid substitutions for that residue position. The hidden layer consists of 25 units. The output layer has three units. In the case of solvent accessibility prediction, these output units correspond to low, medium, and high solvent accessibility states, respectively. The input and the hidden layers, and the hidden and the output layers, are fully connected as in **(A)**, but for simplicity, the connections are not shown.

2.3.4. Decoy Sets and Evaluation of the Knowledge-Based Scoring Functions

One evaluates the usefulness of a scoring function by examining the ability of the scoring function to distinguish native-like conformations from non-native ones. This is achieved through generating test decoy sets and testing the performance of the function on those sets. There are various approaches to generate test decoys. For example, they can be created by sampling discrete-state models starting from a native conformation ***(105)***, having amino acid sequences with known folds mounted onto different folds ***(106,107)***, and using crystal structures of various resolutions ***(85)***. Databases of test decoy sets have been created to enable the evaluation of scoring functions on multiple types of decoys ***(108–110)***. An approach most relevant to evaluating scoring functions for de novo structure prediction is to create test decoys through de novo conformational space sampling. A typical de novo conformational space sampling protocol consists of an MCSA search procedure guided by a set of energy functions, with move set based on lattice models ***(111,112)***, fragment substitution ***(113,114)***, or continuous torsional distributions ***(81)***.

There are several commonly used measures for evaluating the usefulness of scoring functions. The $\log P_{B1}$ measure is the log probability of selecting the lowest C_α root mean square deviation (RMSD) conformation in a test decoy set, calculated with the formula

$$\log P_{B1} = \log_{10}\left(\frac{R_i}{n}\right) \tag{17}$$

Here, R_i is the C_α RMSD rank of the best scoring conformation in the test set of n decoys. The $\log P_{B10}$ measure is the log probability of selecting the lowest C_α RMSD conformation among the top-10 best-scoring conformations, that is, instead of using the RMSD rank of the best-scoring conformation, the best RMSD rank achieved among the top-10 best-scoring conformations is used as R_i in Eq. 17. The CC measure is the correlation coefficient between the C_α RMSDs and the scores generated by the scoring function. The enrichment ratio measure is the fraction enrichment of the top 10% lowest RMSD conformations in the top 10% best scoring conformations. Specifically, after a scoring function is applied to a test decoy set, we count the number of decoys (denoted as a), which are in the top 10% in terms of both their scores and their C_α RMSDs relative to the native structure. The expected number in a random distribution is 10% × 10% × (number of decoys in the set) (denoted as b). The enrichment ratio is a/b. A value above 1 indicates enrichment over the random distribution. The four evaluation measures are illustrated in an example in **Fig. 4**.

To examine the utility of the knowledge-based scoring functions in decoy discrimination, we apply both the RAPDF and the neural network-based functions to 41 test decoy sets of varying quality generated with de novo conformational space sampling. Each decoy set contains approximately 10,000 decoy conformations. Table 1 summarizes the PDB identifiers and the SCOP

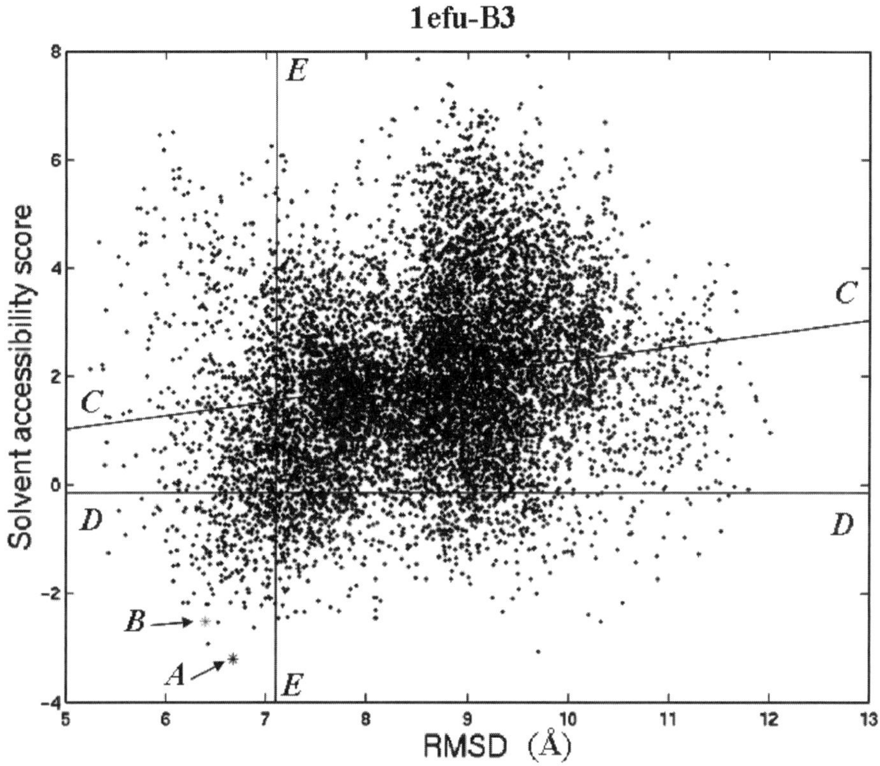

Fig. 4. Measures for evaluating scoring functions. Log P_{B1} is the log probability of selecting the lowest C_α RMSD conformation in a test decoy set (point *A*), which is −1.42 in this example. Log P_{B10} is the log probability of selecting the lowest C_α RMSD conformation among the top 10 best-scoring conformations in a test decoy set (point *B*), which is −1.76 in this example. The correlation coefficient between the C_α RMSDs and the scores is equal to the slope of line *C-C* and has the value of 0.25 in the present case. Line *D-D* represents the top 10% score cutoff for the decoy set. By counting the number of decoys below this line, which are also within the top 10% RMSD cutoff (left of line *E-E*), and dividing this number by the expected value for a random distribution, an enrichment ratio of 2.7 is obtained. Different measures are needed dependent on the specific purposes and roles of the scoring functions.

classifications of the 41 protein sequences used in generating the test decoy sets. Also included is the C_α RMSD of the best decoy relative to the corresponding native structure in each test set. Among them, fifteen test decoy sets have their best structures below 6 Å C_α RMSD relative to their native conformations.

Table 1
List of the Protein Sequences Used in Generating the Test Decoy Sets

Protein	SCOP classifications	Length	Minimum RMSD
1b0n-A2	a.35.1.3 (A:1–68)	68	2.729
1b33-N	d.30.1.1 (N:)	67	7.349
1b34-A	b.38.1.1 (A:)	80	7.943
1b4b-A	d.74.2.1 (A:)	71	5.506
1b79-A	a.81.1.1 (A:)	102	5.29
1ck9-A	d.79.3.1 (A:)	104	7.661
1ctf	d.45.1.1 (–)	68	4.37
1dgn-A	a.77.1.1 (A:)	89	4.482
1dj8-A	a.57.1.1 (A:)	79	5.092
1dtj-A	d.51.1.1 (A:)	74	4.902
1e68-A	a.64.2.1 (A:)	70	3.794
1eai-C	g.22.1.1 (C:)	61	6.914
1edz-A2	c.58.1.2 (A:3–148)	146	9.277
1efu-B3	a.5.2.2 (B:1–54)	54	5.247
1ev0-A	d.71.1.1 (A:)	58	6.641
1f53-A	b.11.1.4 (A:)	84	9.123
1fc3-A	a.4.6.3 (A:)	119	8.184
1fmt-A1	b.46.1.1 (A:207–314)	108	7.385
1g6e-A	b.11.1.6 (A:)	87	7.891
1g7d-A	a.71.1.1 (A:)	106	5.867
1goi-A1	b.72.2.1 (A:447–498)	52	6.111
1gut-A	b.40.6.1 (A:)	67	6.459
1h5p-A	b.99.1.1 (A:)	95	8.223
1h8a-C1	a.4.1.3 (C:87–143)	57	2.941
1ijy-A	a.141.1.1 (A:)	122	7.916
1ira-Y1	b.1.1.4 (Y:1–101)	101	8.317
1iwg-A1	d.58.44.1 (A:38–134)	97	5.7
1jju-A3	b.1.18.14 (A:274–351)	78	6.614
1jos-A	d.52.7.1 (A:)	100	5.302
1jyg-A	a.60.11.1 (A:)	69	3.471
1k2y-X2	c.84.1.1 (X:155–258)	104	6.889
1ktz-B	g.7.1.3 (B:)	106	8.586
1l9l-A	a.64.1.1 (A:)	74	4.041
1msp-A	b.1.11.2 (A:)	124	9.932
1n69-A	a.64.1.3 (A:)	78	6.753
1qu6-A1	d.50.1.1 (A:1–90)	90	8.597
1rie	b.33.1.1 (–)	127	9.548
1sra	a.39.1.3 (–)	151	8.781

1sro	b.40.4.5 (–)	76	6.031
2igd	d.15.7.1 (–)	61	6.508
7gat-A	g.39.1.1 (A:)	66	7.248

Each row lists the Protein Data Bank (PDB) identifier of the sequence, the SCOP classification, the length of the protein sequence, and the C_α RMSD of the best decoy structure relative to the native conformation in the test decoy set. Each test decoy set contains $\sim 10,000$ decoys. Fifteen test decoy sets have their best structures below 6 Å C_α RMSD relative to their corresponding native conformations. Twenty-four test decoy sets have their best structures below 7 Å C_α RMSD relative to their corresponding native conformations.

Twenty-four decoy sets have their best structures below 7 Å C_α RMSD relative to their native conformations, and so on. For illustration purpose, we employ the enrichment ratio measure to evaluate the scoring functions. The results are displayed in **Fig. 5**. From the figure, we observe that the RAPDF function gives uniform performance for decoy discrimination across decoy sets of different quality, whereas the neural network-based scoring functions tend to perform better for decoy sets with better quality.

2.4. Some Other Knowledge-Based Scoring Functions in the Recent Literature

In the formulation of the RAPDF scoring function as well as of the other pairwise distance preference functions described in **refs.** ***11,77,87*** and ***(88)***, the solvation effect is not explicitly modeled. However, as we have previously discussed, as protein folding occurs in the aqueous environment, a careful accounting of the solvent effect is important in determining the native conformation. In this regard, McConkey et al. ***(115)*** quantify contact surfaces of atoms by integrating the solvent accessible surface and the inter-atomic contacts into one quantity and construct an all-atom contact potential based on the contact preferences of 167 residue-specific atom types with 168 possible contact types (167 possible atom contact types and one solvent contact). They demonstrate that this all-atom contact potential delivers satisfactory performance for distinguishing native conformations from decoy structures.

Another possible approach to augment the pairwise distance preference scoring functions is by considering various multi-body geometric properties. In **ref.** ***116***, a four-body SNAPP potential involving the tiling of protein structures with tetrahedra having the center of mass of each amino acid side-chain at each vertex is introduced. This formulation results in 8855 possible tetrahedron types with the corresponding log-likelihoods computed from structural databases. It is found that the SNAPP potential is accurate in predicting the

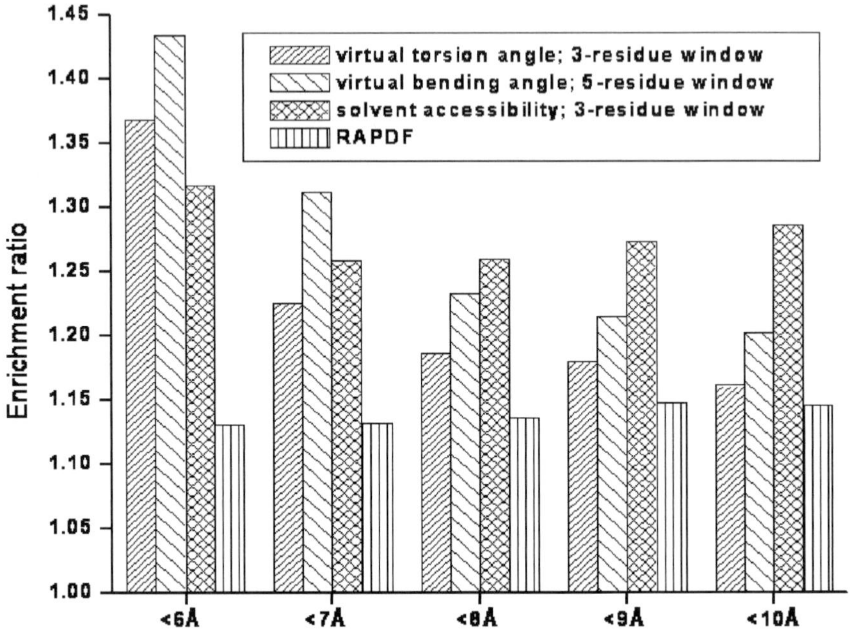

Fig. 5. Performances of the various knowledge-based scoring functions. The functions are evaluated using the average enrichment ratios on test decoy sets of varying quality. For example, the first four bars indicates the average enrichment ratios attained by the individual functions for the test decoy sets that contain structures of less than 6 Å C_α RMSD relative to the native conformations. The following scoring functions are examined in the figure: a neural network-based virtual torsion angle scoring function with a three-residue window; a neural network-based virtual bending angle scoring function with a five-residue window; a neural network-based solvent accessibility scoring function with a three-residue window; and the all-atom distance-dependent conditional probability function.

effects of hydrophobic core mutations. A similar four-body scoring function derived through the Delauney tessellation of side-chain centroids of amino acids is shown to be able to distinguish native conformation from partially unfolded and deliberately misfolded structures ***(117)***. On the basis of the work of Professor Banavar and his colleagues, Ngan et al. ***(118)*** construct a three-body knowledge-based potential involving the radii of curvature formed among triplets of residues in protein conformations. The resulting residue-triplet function is shown to be of utility in discriminating native-like conformations from non-native structures. Finally, Li et al. ***(119)*** introduce a knowledge-based scoring function based on the edge simplices from the alpha shape of the protein structure. Formally, their statistical alpha contact potential is a two-body scoring function, and their definition of contact is when atoms from non-bonded residues share a Voronoi edge, with the edge at least partially contained in the body of the protein. This formulation has the benefit of avoiding spurious contact between two residues when a third residue is between them. The authors have shown that the alpha contact potential performs comparably with other atom-based potentials, while requiring fewer parameters.

In summary, the construction of a knowledge-based scoring function involves the following steps: (1) selection of a conformational property that differs between native-like and non-native structures; (2) compilation of the posterior probability distributions of this conformational property by direct counting or through statistical techniques such as neural network, based on a database of experimentally determined structures; (3) derivation of the prior probability distributions based on a database of decoy structures or through simplifying assumptions such as the averaging-over-atom-types argument of Samudrala and Moult *(34)*, the quasi-chemical approximation of Lu and Skolnick *(87)*, or the uniform distribution argument of Zhou and Zhou *(88)*; and (4) formation of the log-odd scores from the prior and posterior probabilities. Step 1 is perhaps the most critical step and is largely dependent on one's insights into the physical and chemical processes involved in protein folding and by trial and error. In step 2, the selection of appropriate statistical techniques is heavily influenced by the size and quality of the available data set, because these factors have a direct impact on determining whether certain statistical assumptions (e.g., the conditional independence assumption in Eq. 7) are needed.

2.5. The Design of Decoy Filters

As we have discussed, conformational search algorithms produce a multitude of candidate conformations. Various scoring functions can be combined into a filter to distill this vast collection of decoys, to retain those that are native-like. An approach to constructing such a filter is to assign weights to the different scoring functions, such that the resulting linear combination of the scores gives the overall quantitative assessment of a decoy structure of interest. The weights used in the linear combination can be derived by performing logistic regression on test decoy sets. Specifically, native-like decoys (determined by a suitably chosen C_α RMSD cutoff) in each test set are labeled as belonging to class 1, and the rest labeled as class 0. The normalized scores for an individual decoy become the independent variables (x_j; $j = 1 \ldots k$; $k =$ the total number of score types), whereas its associated class label forms the dependent variable (p), which are then used to fit the following equation to obtain the weights w_js:

$$\log\left(\frac{p}{1-p}\right) = \alpha + w_1 x_{1,i} + \ldots + w_k x_{k,i} \quad (18)$$

Here, α is a constant representing the intercept. i ranges from 1 to N, and N is the total number of decoys. Normalization of a scoring function can be achieved by subtracting its mean and dividing by its standard deviation, where the mean and the standard deviation are computed over all decoys within a test set, or by replacing the raw score of a decoy with its rank and then dividing by the total number of decoys in the test set. Techniques such as leave-one-out cross-validation and forward and backward stepwise regression can be applied to determine which independent variables are helpful in assessing the accuracy of a given decoy structure and which can be discarded. Essentially, functions describing useful orthogonal characteristics of protein native conformations will receive large weights, whereas those that are less useful or containing overlapping information will have smaller or zero weights. Finally, alternative approach to performing logistic regression is also possible, for example, by replacing it with machine-learning techniques such as the neural network or SVM. The decision is again influenced by the size and quality of the available test data.

2.6. Further Enhancement of Decoy Selection Through Conformer Clustering and High-Resolution Refinement

Conformer clustering and high-resolution refinement are often used as additional steps in the decoy selection process to further refine the set of native-like conformations retained by the decoy filter. The idea of conformer clustering is based on the following observation: Conformers with correct folds are in general similar to other conformers with correct folds. On the contrary, it is unlikely that multiple conformers share the same mistake, and therefore, conformers with incorrect folds are in general dissimilar to each other as well as to conformers with correct folds. Hence, the conformers that are most similar to the others, that is, those at the cluster centers of the conformational distribution, will tend to be the correct ones. Various metrics are used to describe the conformational distribution, including pairwise RMSD, pairwise RMSD with cutoffs, and number of neighbors ***(16,120)***. Heuristic schemes such as k-mean clustering, visual inspection following dimensionality reduction, and iterative sampling ***(121)*** can be used to locate these cluster centers.

Figure 6 illustrates the performance of a conformer-clustering algorithm [the density score function available in the RAMP package ***(122)***] in distinguishing

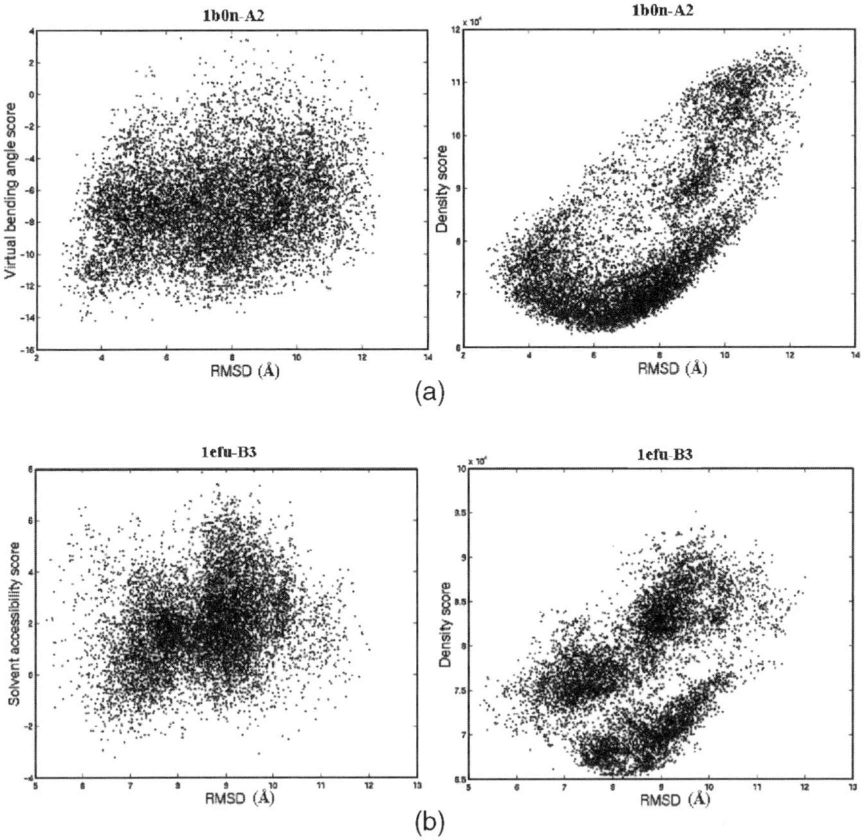

Fig. 6. The comparison of some knowledge-based scoring functions and the density score function in discriminating decoys. In **(A)**, the virtual bending angle scoring function is compared to the density score function, whereas in **(B)**, the solvent accessibility scoring function is compared to the density score function. The diagrams show that the density score function produces improved correlation between the C_α RMSDs and the scores in both cases, suggesting that conformer clustering is useful as a complementary step in decoy selection.

native-like structures from non-native conformations. Compared with the neural network-based virtual bending angle and solvent accessibility scoring functions, the density score function produces results that show improved correlation between the C_α RMSDs and the generated scores. This observation suggests that applying conformer clustering in addition to using scoring functions as filter can enhance the overall ability to select native-like structures from decoys.

The goal of high-resolution refinement is to further optimize the remaining candidate structures that have passed through the decoy filtering and conformer clustering stages. The optimization is carried out by making small perturbations to a candidate structure guided by a highly detailed energy potential. One of the most notable methods is that of Misura et al., which has been shown to be effective in the Sixth Critical Assessment of Techniques for Protein Structure Prediction (CASP-6) ***(123,124)***. It involves applying perturbations to backbone and side-chain torsion angles using an all-atom force field. The force field consists of a standard 6–12 Lennard–Jones potential for Van der Waals packing, the implicit solvation model of Lazaridis and Karplus describing dielectric screening ***(73)***, and a new orientation-dependent hydrogen bonding term ***(125)***. The hydrogen-bonding term is derived based on observed geometrical parameters of hydrogen bonds in high-resolution crystal structures of proteins. Using this combined physics-based and knowledge-based function as part of their prediction protocol, Bradley et al. have reported success in high-resolution structure prediction of less than 1.5 Å for protein domain of less than 85 residues ***(124)***.

A summary of the scoring functions discussed in this chapter can be found in Table 2. We should note that there are other means to guide conformational search and decoy filtering besides using scoring functions. For example, filtering schemes based on contact order ***(126)*** and beta sheet topology ***(127)*** have been found to be beneficial in enriching the ensemble quality of decoy structures.

3. Discussion and Conclusion

A main objective of the structural genomic initiatives, spurred by large-scale genome sequencing efforts, is to determine as many protein folds as possible. The need to determine protein structures rapidly and inexpensively in turn leads to an increased interest in computational protein structure prediction, the two main approaches of which being homology modeling and de novo structure prediction.

The key components in de novo protein structure prediction are conformational space sampling and decoy selection. Scoring functions are employed in both the conformational sampling stage and the decoy selection stage. In the first stage, a selected combination of scoring functions approximates the energy landscape of the conformational space, and conformational search algorithms generate trajectories leading to the landscape minima, whereas in the second stage, another set of possibly different scoring functions are used as filter to

Table 2
A list of the scoring functions discussed in Section 2

Scoring function	Subheading	Usage	Description
Class I force field	2.1.1.	Conformational space search	Physics-based force field modeling bonded and non-bonded interactions among atoms
RAPDF	2.2.1.	Conformational space search/decoy filtering	Knowledge-based potential describing atom–atom distance preferences
IRAPDF	2.2.3.	Conformational space search/decoy filtering	Continuous version of the RAPDF function
Neural network knowledge-based functions	2.3.	Conformational space search/decoy filtering	Incorporation of neural network into the Bayesian probability framework to describe various conformational properties
Atom–atom contact scoring function	2.4.	Conformational space search/decoy filtering	Knowledge-based atom–atom contact preference function taking solvent accessibility into account
SNAPP potential	2.4.	Conformational space search/decoy filtering	A four-body knowledge-based function describing tiling of protein structures with tetrahedra
Four-body contact scoring function	2.4.	Conformational space search/decoy filtering	A four-body knowledge-based function based on Delauney tessellation of side chain
Residue triplet scoring function	2.4.	Conformational space search/decoy filtering	A three-body knowledge-based function based on the radii of curvature formed among triplets of residues

(Continued)

Table 2
(Continued)

Scoring function	Subheading	Usage	Description
Alpha contact potential	2.4.	Conformational space search/decoy filtering	A two-body knowledge-based function based on edge simplices from the alpha shape of the protein structure
Structure refinement potential of Misura et al.	2.6.	High-resolution refinement	A combined physics- and knowledge-based function modeling Van der Waals interaction, solvent effects, and hydrogen bonding

Each row gives the name of the scoring function, the subheading in which it is discussed, its usage, and a brief description of its components.

retain a collection of the native-like structures. Conformer clustering and high-resolution refinement can also be used as additional steps to further refine this collection. In this chapter, we have studied some examples of the physics-based and knowledge-based scoring functions. For the physics-based approach, the Class I force field and its extensions as well as solvation modeling were discussed. For the knowledge-based approach, we studied the Bayesian probability formalism and used it to derive the RAPDF ***(34)***. In addition, we detailed the construction of the neural network-based Bayesian scoring functions. The Bayesian probability formalism was combined with the neural network methodology to construct various types of log-likelihood scoring functions. Then, we described some of the new knowledge-based scoring functions from in the recent literature. These functions extend the pairwise distance preference scoring functions in various ways, for example, by explicitly modeling the solvent effects and by considering multi-body geometric arrangements and interactions. Finally, we briefly discussed conformer clustering and described a detailed energy potential used for high-resolution refinement. In general, because of the weaknesses of solvent and electrostatic modeling, simulations attempting to fold proteins de novo from physics-based scoring functions alone do not perform satisfactorily. The statistical models that are used to construct knowledge-based functions provide added flexibilities over direct physical modeling, and as a result, most of the successful de novo structure prediction protocols have both physics-based and knowledge-based components.

Scoring function design remains a very difficult problem. None of the existing physics-based and knowledge-based functions can faithfully reproduce the true energy landscape of the protein conformational space, and none of them can consistently and reliably select native-like conformations from non-native structures for a broad spectrum of proteins. The difficulty is mainly because the physical and statistical models considered so far in the literature cannot well approximate the quantum mechanical character of intra-molecular and solvent-protein interactions. Furthermore, scoring functions describing truly orthogonal characteristics of protein native conformations are difficult to discover, especially for the knowledge-based functions that are the sum of many constituent effects. Thus, it is of practical interest to continue developing various types of new scoring functions, to exploit their differences, and to capture the cumulative effect of incremental enrichments. Fortunately, the increase in the size of the PDB together with increased computational power means that the construction of more sophisticated knowledge-based scoring functions are now possible. More realistic electrostatics and solvation models are also being developed, increasing the capabilities of the physics-based force fields. These advances will play important roles to improving the state of the art of protein folding simulation and de novo structure prediction.

Acknowledgments

We thank Drs. Enoch Huang and Britt Park for their earlier edition on scoring functions for de novo protein structure prediction and the anonymous reviewer for the many helpful suggestions. This work is supported in part by a Searle Scholar Award, NSF Grant DBI-0217241, an NSF CAREER award, and NIH Grant GM068152 to R.S. and the University of Washington's Advanced Technology Initiative in Infectious Diseases.

References

1. Brenner, S., Levitt, M. (2000) Expectations from structural genomics. *Protein Sci.*, **9**, 197–200.
2. Brenner, S.E. (2001) A tour of structural genomics. *Nat. Genet.*, **210**, 801–809.
3. Burley, S.K. (2000) An overview of structural genomics. *Nat. Struct. Biol.*, **7 (Suppl)**, 932–934.
4. Heinemann, U., Illing, G., Oschkinat, H. (2001) High-throughput three-dimensional protein structure determination. *Curr. Opin. Biotech.*, **12**, 348–354.
5. Bonneau, R., Baker, D. (2001) Ab initio protein structure prediction: progress and prospects. *Annu. Rev. Biophys. Biomol. Struct.*, **30**, 173–189.

6. Anfinsen, C.B., Haber, E., Sela, M., White, F.H., Jr. (1961) The kinetics of formation of active ribonuclease during oxidation of the reduced polypeptide chain. *Proc. Natl. Acad. Sci. U. S. A.*, **47**, 1309–1314.
7. Doolittle, R. (1981) Similar amino acid sequences: chance or common ancestry? *Science*, **214**, 149–159.
8. Sander, C., Schneider, R. (1991) Database of homology-derived protein structures and the structural meaning of sequence alignment. *Proteins*, **9**, 56–68.
9. Murzin, A., Bateman, A. (1997) Distance homology recognition using structural classification of proteins. *Proteins*, **29S**, 105–112.
10. Bowie, J., Luthy, R., Eisenberg, D. (1991) Method to identify protein sequences that fold into a known three-dimensional structure. *Science*, **253**, 164–170.
11. Jones, D., Taylor, W., Thornton, J. (1992) A new approach to protein fold recognition. *Nature*, **258**, 86–89.
12. Moult, J., Fidelis, K., Zemla, A. Hubbard, T. (2003) Critical assessment of methods of protein structure prediction (CASP): round V. *Proteins*, **53**, 334–339.
13. Moult, J., Fidelis, K., Rost, B., Hubbard, T., Tramontano, A. (2005) Critical assessment of methods of protein structure prediction (CASP) – round 6. *Proteins*, **61**, 3–7.
14. Lee, J., Liwo, A., Ripoll, D., Pillardy, J., Scheraga, J. (1999) Calculation of protein conformation by global optimization of a potential energy function. *Proteins*, **S3**, 204–208.
15. Samudrala, R., Xia, Y., Huang, E., Levitt, M. (1999) Ab initio protein structure prediction using a combined hierarchical approach. *Proteins*, **S3**, 194–198.
16. Simons, K., Bonneau, R., Ruczinski, I., Baker, D. (1999) Ab initio structure prediction of CASP3 targets using ROSETTA. *Proteins*, **S3**, 171–176.
17. Samudrala, R., Xia, Y., Levitt, M., Huang E.S. (1999) A combined approach for ab initio construction of low resolution protein tertiary structures from sequence, in *Proceedings of the Pacific Symposium on Biocomputing* (Altman, R. B., Dunker, A.K., Hunter, L., Klein, T.E., Lauderdale, K., eds.), World Scientific Press, Singapore, pp. 505–516.
18. Samudrala, R., Levitt, M. (2002) A comprehensive analysis of 40 blind protein structure predictions. *BMC Struct Biol*, **2**, 3–18.
19. Moult, J., Hubbard, T., Bryant, S.H., Fidelis, K., Pedersen, J.T. (1997) Critical assessment of methods of protein structure prediction (CASP): round II. *Proteins*, **29**, 2–6.
20. Moult, J., Hubbard, T., Fidelis, K., Pedersen, J.T. (1999) Critical assessment of methods of protein structure prediction (CASP): round III. *Proteins*, **37**, 2–6.
21. Moult, J., Fidelis, K., Zemla, A., Hubbard, T. (2001) Critical assessment of methods of protein structure prediction (CASP): round IV. *Proteins*, **45**, 2–7.
22. Brooks, B., Bruccoleri, R., Olafson, B., States, D., Swaminathan, S., Karplus, M. (1983) CHARMM: a program for macromolecular energy, minimization, and dynamics calculations. *J. Comp. Chem.*, **4**, 187–217.

23. Weiner, S., Kollman P., Nguyen, D., Case, D. (1986) An all atom force field for simulations of proteins and nucleic acids. *J. Comp. Chem.*, **7**, 230–252.
24. Jorgensen, W., Tirado-Rives, J. (1988) The OPLS potential function for proteins. Energy minimisations for crystals of cyclic peptides and crambin. *J. Amer. Chem. Soc.*, **110**, 1657–1666.
25. MacKerell, A.D., Jr., Bashford, D., Bellott, M., Dunbrack, R.L., Jr., Evanseck, J.D., et al. (1998) All-atom empirical potential for molecular modeling and dynamics studies of proteins. *J. Phys. Chem. B*, **102**, 3586–3616.
26. Cornell, W.D., Cieplak, P., Bayly, C.I., Gould, I.R., Merz, K.M., Jr., Fergusson, D.M., Spellmeyer, D.C., Fox, D.C., Caldwell, J.W., Kollman, P.A. (1995) A second generation force field for the simulation of proteins and nucleic acids. *J. Amer. Chem. Soc.*, **117**, 5179–5197.
27. Nemethy, G., Gibson, K.D., Palmer, K.A., Yoon, C.N., Paterlini, G., Zagari, A., Rumsey, S., Scheraga, H.A. (1992) Energy parameters in peptides: improved geometrical parameters and non-bonded interactions for use in the ECEPP/3 algorithm, with application to proline-containing peptides. *J. Phys. Chem.*, **96**, 6472–6484.
28. Wodak, S., Rooman, M. (1993) Generating and testing protein folds. *Curr. Opin. Struct. Biol.*, **3**, 247–259.
29. Sippl, M. (1995) Knowledge based potentials for proteins. *Curr. Opin. Struct. Biol.*, **5**, 229–235.
30. Gilis, D., Rooman, M. (1996) Stability changes upon mutation of solvent-accessible residues in proteins evaluated by database-derived potentials. *J. Mol. Biol.*, **257**, 1112–1126.
31. Jernigan, R.L., Bahar I. (1996) Structure-derived potentials and protein simulations. *Curr. Opin. Struct. Biol.*, **6**, 195–209.
32. DeBolt, S.E., Skolnick, J. (1996) Evaluation of atomic level mean force potentials via inverse refinement of protein structures: atomic burial position and pairwise non-bonded interactions. *Protein Eng.*, **8**, 637–655.
33. Zhang, C., Vasmatzis, G., Cornette, J.L., DeLisi, C. (1997) Determination of atomic desolvation energies from the structures of crystallised proteins. *J. Mol. Biol.*, **267**, 707–726.
34. Samudrala, R., Moult, J. (1998) An all-atom distance-dependent conditional probability discriminatory function for protein structure prediction. *J. Mol. Biol.*, **275**, 895–916.
35. Huang, E.S., Samudrala, R., Park, B.H. (2000) Scoring functions for ab initio protein structure prediction. *Methods Mol. Biol.*, **143**, 223–245.
36. Hartree, D.R. (1957) *The Calculation of Atomic Structure*. John Wiley & Sons, New York.
37. Hohenberg, P., Kohn, W. (1964) Inhomogeneous electron gas. *Phys. Rev.*, **136**, 864.

38. Kauzmann, W. (1959) Some factors in the interpretation of protein denaturation. *Adv. Protein Chem.*, **14**, 1–64.
39. Dill, K.A. (1990) Dominant forces in protein folding. *Biochemistry*, **29**, 7133–7155.
40. Morozov, A.V., Kortemme, T., Tsemekhman, K., Baker, D. (2004) Close agreement between the orientation dependence of hydrogen bonds observed in protein structures and quantum mechanical calculations. *Proc. Natl. Acad. Sci. U. S. A.*, **101**, 6946–6951.
41. Weiner, P.K., Kollman P.A. (1981) AMBER: Assisted model building with energy refinement. A general program for modeling molecules and their interactions. *J. Comp. Chem.*, **2**, 287–303.
42. Brooks, B.R., Bruccoleri, R.E., Olafson, B.D., States, D.J., Swaminathan, S., Karplus, M. (1983) CHARMM: a program for macromolecular energy, minimization, and dynamics calculations. *J. Comp. Chem.*, **4**, 187–217.
43. Levitt, M., Hirshberg, M., Sharon, R., Daggett, V. (1995) Potential energy function and parameters for simulations of the molecular dynamics of proteins and nucleic acids in solution. *Comp. Phys. Comm.*, **91**, 215–231.
44. Levitt, M. (1983) Molecular dynamics of native protein. I. Computer simulation of trajectories. *J. Mol. Biol.*, **168**, 595–617.
45. Daggett, L.P., Sacaan, A.I., Akong, M., Rao, S.P., Hess, S.D., Liaw, C., Urrutia, A., Jachec, C., Ellis, S.B., Dreessen J, et al. (1995) Molecular and functional characterization of recombinant human metabotropic glutamate receptor subtype 5. *Neuropharmacology*, **34**, 7133–7155.
46. Levitt, M. (1983) Protein folding by restrained energy minimization and molecular dynamics. *J. Mol. Biol.*, **170**, 723–764.
47. Brunger, A.T., Clore, G.M., Gronenborn, A.M., Karplus, M. (1986) Three-dimensional structure of proteins determined by molecular dynamics with interproton distance restraints: application to crambin. *Proc. Natl. Acad. Sci. U. S. A.*, **83**, 3801–3805.
48. Ferguson, D.M., Kollman, P.A. (1991) Can the Lennard-Jones 6-12 function replace the 10–12 form in molecular mechanics calculations? *J. Comput. Chem.*, **12**, 620–626.
49. Halgren, T.A. (1992) Representation of van der Waals (vdW) interactions in molecular mechanics force fields: potential form, combination rules, and vdW parameters. *J. Am. Chem. Soc.*, **114**, 7827–7843.
50. Halgren, T.A. (1996) Merck molecular force field. I. Basis, form, scope, parameterization, and performance of MMFF94. *J. Comput. Chem.*, **17**, 490–519.
51. Hart, J.R., Rappe, A.K. (1992) van der Waals functional forms for molecular simulations. *J. Chem. Phys.*, **97**, 1109–1115.
52. Buckingham, A.D., Fowler, P.W. (1985) A model for the geometries of van der Waals complexes. *Can. J. Chem.*, **63**, 2018.

53. Sokalski, W.A., Shibata, M., Ornstein, R.L., Rein, R. (1993) Point charge representation of multicenter multipole moments in calculation of electrostatic properties. *Theor. Chim. Acta*, **85**, 209–216.
54. Stone, A.J. (1981) Distributed multipole analysis, or how to describe a molecular charge distribution. *Chem. Phys. Lett.*, **83**, 233–239.
55. Kosov, D., Popelier, P.L.A. (2000) Atomic partitioning of molecular electrostatic potentials. *J. Phys. Chem. A*, **104**, 7339–7345.
56. Cieplak, P., Caldwell, J., Kollman, P. (2001) Molecular mechanical models for organic and biological systems going beyond the atom centered two body additive approximation: aqueous solution free energies of methanol and N-methyl acetamide, nucleic acid base, and amide hydrogen bonding and chloroform/water partition coefficients of the nucleic acid bases. *J. Comput. Chem.*, **22**, 1048–1057.
57. Kaminski, G.A., Stern, H.A., Berne, B.J., Friesner, R.A., Cao, Y.X., Murphy, R.B., Zhou, R., Halgren, T.A. (2002) Development of a polarizable force field for proteins via ab initio quantum chemistry: first generation model and gas phase tests. *J. Comput. Chem.*, **23**, 1515–1531.
58. Ren, P., Ponder, J.W. (2003) Polarizable atomic multipole water model for molecular mechanics simulation. *J. Phys. Chem. B*, **107**, 5933–5947.
59. Jorgensen, W.L. (1981) Transferable intermolecular potential functions for water, alcohols, and ethers. Application to liquid water. *J. Am. Chem. Soc.*, **103**, 335–340.
60. Jorgensen, W.L., Chandrasekhar, J., Madura, J.D., Impey, R.W., Klein, M.L. (1983) Comparison of simple potential functions for simulating liquid water. *J. Chem. Phys.*, **79**, 926–935.
61. Berendsen, H.J.C., Grigera, J.R., Straatsma, T.P. (1987) The missing term in effective pair potentials. *J. Phys. Chem.*, **91**, 6269–6271.
62. Levitt, M., Hirshberg, M., Sharon, R., Laidig, K.E., Daggett, V. (1997) Calibration and testing of a water model for simulation of the molecular dynamics of proteins and nucleic acids in solution. *J. Phys. Chem. B*, **101**, 5051–5061.
63. York, D.M., Darden, T., Pedersen, L.G. (1993) The effect of long-range electrostatic interactions in simulations of macromolecular crystals: a comparison of the Ewald and truncated list methods. *J. Chem. Phys.*, **99**, 8345–8348.
64. Darden, T., York, D., Pedersen, L. (1993) Particle mesh Ewald: an N*log(N) method for Ewald sums in large systems *J. Chem. Phys.*, **98**, 10089–10092.
65. Gouy, M. (1910) Sur la constitution de la charge électrique a la surface d'un électrolyte. *Journ. Phys.*, **9**, 457–468.
66. Gilson, M.K., Honig, B. (1988) Calculation of the total electrostatic energy of a macromolecular system: solvation energies, binding energies, and conformational analysis. *Proteins*, **4**, 7–18.
67. Nicholls, A., Honig, B. (1991) A rapid finite difference algorithm, utilizing successive over-relaxation to solve the Poisson-Boltzmann equation. *J. Comp. Chem.*, **12**, 435–445.

68. Bashford, D., Case, D.A. (2000) Generalized Born models of macromolecular solvation effects. *Annu. Rev. Phys. Chem.*, **51**, 129–152.
69. de Bakker, P.I.W., DePristo, M.A., Burke, D.F., Blundell, T.L. (2003) Ab initio construction of polypeptide fragments: accuracy of loop decoy discrimination by an all-atom statistical potential and the AMBER force field with the generalized born solvation model. *Proteins*, **51**, 21–40.
70. Fogolari, F., Brigo, A., Molinari, H. (2003) Protocol for MM/PBSA molecular dynamics simulations of proteins. *Biophys. J.*, **85**, 159–166.
71. Warshel, A., Levitt, M. (1976) Theoretical studies of enzymic reactions – dielectric, electrostatic and steric stabilization of carbonium-ion in reaction of lysozyme. *J. Mol. Biol.*, **103**, 227–249.
72. Gelin, B.R., Karplus, M. (1979) Side-chain torsional potentials: effect of dipeptide, protein, and solvent environment. *Biochemistry*, **18**, 1256–1268.
73. Lazaridis, T., Karplus, M. (1999) Effective energy function for proteins in solution. *Proteins*, **35**, 133–152.
74. Mallik, B., Masunov, A., Lazaridis, T. (2002) Distance and exposure dependent effective dielectric function. *J. Comp. Chem.*, **23**, 1090–1099.
75. Moult, J. (1997) Comparison of database potentials and molecular mechanics force fields. *Curr. Opin. Struct. Biol.*, **7**, 194–199.
76. Eisenberg, D., Weiss, R.M., Terwillinger, T.C. (1982) The helical hydrophobic moment: a measure of the amphiphilicity of a helix. *Nature*, **299**, 371–374.
77. Sippl, M.W., S. (1992) Detection of native-like models for amino acid sequences of unknown three-dimensional structure in a database of known protein conformations. *Proteins*, **13**, 258–271.
78. Jones, D.T. (2001) Predicting novel protein folds by using FRAGFOLD. *Proteins*, **45**, 127–132.
79. Zhang, Y., Skolnick, J. (2004) Tertiary structure predictions on a comprehensive benchmark of medium to large size proteins. *Biophys. J.*, **87**, 2647–2655.
80. Boniecki, M., Rotkiewicz, P., Skolnick, J., Kolinski, A. (2003) Protein fragment reconstruction using various modeling techniques. *J. Comput. Aided Mol. Des.*, **17**, 725–738.
81. Hung, L.H., Ngan, S.C., Liu, T., Samudrala, R. (2005) PROTINFO: new algorithms for enhanced protein structure predictions. *Nucleic Acids Res.*, **33**, W77–W80.
82. Westbrook, J., Feng, Z., Chen, L., Yang, H., Berman, H.M. (2003) The Protein Data Bank and structural genomics. *Nucleic Acids Res.*, **31**, 489–491.
83. Bourne, P.E., Addess, K.J., Bluhm, W.F., Chen, L., Deshpande, N., Feng, Z., Fleri, W., Green, R., Merino-Ott, J.C., Townsend-Merino, W., Weissig, H., Westbrook, J., Berman, H.M. (2004) The distribution and query systems of the RCSB Protein Data Bank. *Nucleic Acids Res.*, **32**, D223–D225.
84. Chandonia, J.M., Hon, G., Walker, N.S., LoConte, L., Koehl, P., Levitt, M., Brenner, S.E. (2004) The ASTRAL compendium in 2004. *Nucleic Acids Res.*, **32**, D189–D192.

85. Subramaniam, S., Tcheng, D.K., Fenton, J. (1996) Knowledge-based methods for protein structure refinement and prediction, in *Proceedings of the Fourth International Conference on Intelligent Systems in Molecular Biology* (States, D., Agarwal, P., Gaasterland, T., Hunter, L. & Simth, R., eds.), AAAI Press, Menlo Park, CA, pp. 218–229.
86. Avbelj, F., Moult, J. (1995) Role of electrostatic screening in determining protein main chain conformational preferences. *Biochemistry*, **34**, 755–764.
87. Lu, H., Skolnick, J. (2001) A distance-dependent atomic knowledge-based potential for improved protein structure selection. *Proteins*, **44**, 223–232.
88. Zhou, H., Zhou, Y. (2002) Distance-scaled, finite ideal-gas reference state improves structure-derived potentials of mean force for structure selection and stability prediction. *Protein Sci.*, **11**, 2714–2726.
89. Oppenheim, A.V., Schafer, R.W., Buck, J.R. (1999) *Discrete-Time Signal Processing*, 2nd ed. Prentice Hall, Upper Saddle River, NJ.
90. Rost, B., Sander, C. (1994) Conservation and prediction of solvent accessibility in protein families. *Proteins*, **20**, 216–226.
91. Ahmad, S., Gromiha, M.M. (2002) NETASA: neural network based prediction of solvent accessibility. *Bioinformatics*, **18**, 819–824.
92. Kim, H., Park, H. (2004) Prediction of protein relative solvent accessibility with support vector machines and long-range interaction 3D local descriptor. *Proteins*, **54**, 557–562.
93. Rost, B., Sander, C. (1993) Prediction of protein secondary structure at better than 70% accuracy. *J. Mol. Biol.*, **232**, 584–599.
94. Jones, D.T. (1999) Protein secondary structure prediction based on position-specific scoring matrices. *J. Mol. Biol.*, **292**, 195–202.
95. Cuff, J.A., Barton, G.J. (1999) Application of enhanced multiple sequence alignment profiles to improve protein secondary structure prediction. *Proteins*, **40**, 502–511.
96. Lund, O., Frimand, K., Gorodkin, J., Bohr, H., Bohr, J., Hansen, J., Brunak, S. (1997) Protein distance constraints predicted by neural networks and probability density functions. *Protein Eng.*, **10**, 1241–1248.
97. Pollastri, G., Baldi, P., Fariselli, P., Casadio, R. (2002) Prediction of coordination number and relative solvent accessibility in proteins. *Proteins*, **47**, 142–153.
98. Olmea, O., Valencia, A. (1997) Improving contact predictions by the combination of correlated mutations and other sources of sequence information. *Fold Des.*, **2**, S25–32.
99. Fariselli, P., Casadio, R. (1999) Neural network based predictor of residue contacts in proteins. *Protein Eng.*, **12**, 15–21.
100. Altschul, S.F., Madden, T.L., Schaffer, A.A. (1997) Gapped BLAST and PSI-BLAST: a new generation of protein database search programs. *Nucleic Acids Res.*, **25**, 3389–3402.

101. Rumelhart, D.E., Hinton, G.E., Williams, R.J. (1986) Learning representations by back-propagating errors. *Nature*, **323**, 533–536.
102. Yi, T.-M., Lander, E.S. (1993) Protein secondary structure prediction using nearest-neighbor methods. *J. Mol. Biol.*, **232**, 1117–1129.
103. Kabsch, W., Sander, C. (1983) Dictionary of protein secondary structure: pattern recognition of hydrogen-bonded and geometrical features. *Biopolymers*, **22**, 2577–2637.
104. Zell, A., Mamier, G., Vogt, M., et al. (2005) The SNNS users manual version 4.1. Available at http://www-ra.informatik.uni-tuebingen.de/snns.
105. Park, B., Levitt, M. (1996) Energy functions that discriminate x-ray and near native folds from well-constructed decoys. *J. Mol. Biol.*, **266**, 831–846.
106. Novotny, J., Bruccoleri, R., Karplus, M. (1984) An analysis of incorrectly folded protein models. Implications for structure predictions. *J. Mol. Biol.*, **177**, 787–818.
107. Holm, L., Sander, C. (1992) Evaluation of protein models by atomic solvation preference. *J. Mol. Biol.*, **225**, 93–105.
108. Samudrala, R., Levitt, M. (2000) Decoys 'R' Us: a database of incorrect conformations to improve protein structure prediction. *Protein Sci.*, **9**, 1399–1401.
109. Tsai J., B., R., Morozov, A.V., Kuhlman, B., Rohl, C.A., Baker, D. (2003) An improved protein decoy set for testing energy functions for protein structure prediction. *Proteins*, **53**, 76–87.
110. Park, B.H., Huang, E.S., Levitt, M. (1997) Factors affecting the ability of energy functions to discriminate correct from incorrect folds. *J. Mol. Biol.*, **266**, 831–846.
111. Hinds, D.A., Levitt, M. (1992) A lattice model for protein structure prediction at low resolution. *Proc. Natl. Acad. Sci. U. S. A.*, **89**, 2536–2540.
112. Park, B., Levitt, M. (1995) The complexity and accuracy of discrete state models of protein structure. *J. Mol. Biol.*, **249**, 493–507.
113. Simons, K.T., Kooperberg, C., Huang, E., Baker, D. (1997) Assembly of protein tertiary structures from fragments with similar local sequences using simulated annealing and Bayesian scoring functions. *J. Mol. Biol.*, **268**, 209–225.
114. Hung, L.H., Samudrala, R. (2003) PROTINFO: secondary and tertiary protein structure prediction. *Nucleic Acids Res.*, **31**, 3296–3299.
115. McConkey, B.J., Sobolev, V., Edelman, M. (2003) Discrimination of native protein structures using atom-atom contact scoring. *Proc. Natl. Acad. Sci. U. S. A.*, **100**, 3215–3220.
116. Carter, C.W., Jr., LeFebvre, B.C., Cammer, S.A., Tropsha, A., Edgell, M.H. (2001) Four-body potentials reveal protein-specific correlations to stability changes caused by hydrophobic core mutations. *J. Mol. Biol.*, **311**, 625–638.
117. Krishnamoorthy, B., Tropsha, A. (2003) Development of a four-body statistical pseudo-potential to discriminate native from non-native protein conformations. *Bioinformatics*, **19**, 1540–1548.

118. Ngan, S.-C., Inonye, M.T, Samudrala, R. (2006) A knowledge-based scoring function based on residue triplets for protein structure prediction. *Protein Eng.*, **19**, 187–193.
119. Li, X., Hu, C., Liang, J. (2003) Simplicial edge representation of protein structures and alpha contact potential with confidence measure. *Proteins*, **53**, 792–805.
120. Wang, K., Fain, B., Levitt, M., Samudrala, R. (2004) Improved protein structure selection using decoy-dependent discriminatory functions. *BMC Struct. Biol.*, **4**, 8.
121. Zhang, Y., Skolnick, J. (2004) SPICKER: a clustering approach to identify near-native protein folds. *J. Comput. Chem.*, **25**, 865–871.
122. Samudrala, R. (2006). RAMP Howto. Available at http://software.compbio.washington.edu/ramp/ramp.html
123. Misura, K.M.S., Baker, D. (2005) Progress and challenges in high-resolution refinement of protein structure models. *Proteins*, **59**, 15–29.
124. Bradley, P., Misura, K.M.S., Baker, D. (2005) Toward high-resolution de novo structure prediction for small proteins. *Science*, **309**, 1868–1871.
125. Kortemme, T., Morozov, A.V., Baker, D. (2003) An orientation-dependent hydrogen bonding potential improves prediction of specificity and structure for proteins and protein-protein complexes. *J. Mol. Biol.*, **326**, 1239–1259.
126. Bonneau, R., Ruczinski, I., Tsai, J., Baker, D. (2002) Contact order and ab initio protein structure prediction. *Protein Sci.*, **11**, 1937–1944.
127. Bradley, P., Malmstrom, L., Qian, B., Schonburn, J., Chivian, D., Kim, D.E., Meiler, J., Misura, K.M., Baker D. (2005) Free modeling with Rosetta in CASP6. *Proteins*, **61**, 128–134.

11

Protein–Protein Docking: Overview and Performance Analysis

Kevin Wiehe, Matthew W. Peterson, Brian Pierce, Julian Mintseris, and Zhiping Weng

Summary

Protein–protein docking is the computational prediction of protein complex structure given the individually solved component protein structures. It is an important means for understanding the physicochemical forces that underlie macromolecular interactions and a valuable tool for modeling protein complex structures. Here, we report an overview of protein–protein docking with specific emphasis on our Fast Fourier Transform-based rigid-body docking program ZDOCK, which is consistently rated as one of the most accurate docking programs in the Critical Assessment of Predicted Interactions (CAPRI), a series of community-wide blind tests. We also investigate ZDOCK's performance on a non-redundant protein complex benchmark. Finally, we perform regression analysis to better understand the strengths and weaknesses of ZDOCK and to suggest areas of future development for protein-docking algorithms in general.

Key Words: Protein–protein docking; ZDOCK; RDOCK; Fast Fourier Transform; benchmark; CAPRI; shape complementarity; electrostatics; desolvation energy; regression analysis.

1. Introduction

Protein–protein interactions play a central role in biochemistry. This can be seen in cell-signaling cascades, enzyme catalysis, the immune response by means of antibody–antigen interactions, and the large-scale motions of organisms. These interactions are also implicated in many diseases.

From: *Methods in Molecular Biology, Vol. 413: Protein Structure Prediction, Second Edition*
Edited by: M. Zaki and C. Bystroff © Humana Press Inc., Totowa, NJ

While experimental techniques such as yeast two-hybrid system and mass spectrometry are able to determine the existence of protein–protein interactions, the structure of the macromolecular complex of two interacting proteins can provide additional information about their interaction, such as the specific residues involved in the interaction and the degree of conformational change undergone by the proteins upon binding.

X-ray crystallography and nuclear magnetic resonance have provided us with the structures of many complexes, but numerous structures still remain unsolved because of time and experimental limitations. This leads to a need for computational methods to understand the nature of protein–protein interactions, one of which is protein–protein docking.

This chapter is divided into three sections. The first section provides an overview of protein–protein docking and describes some of the available algorithms for docking. The second describes the ZDOCK suite of programs in detail, and the third describes an analysis of the performance of ZDOCK.

1.1. Protein–Protein Docking: An Overview

Protein–protein docking is defined as the prediction of the structure of two proteins in a complex, given only the structure of the interacting proteins. The "docking problem" can be broken down into two types of docking: bound docking, in which a complex is separated and reassembled, and unbound docking, where the structure of the complex is found from the individually solved structures of the interacting proteins. Obviously, bound docking has little applicable value, but it is often used for testing and verification purposes.

Unbound docking is much more difficult than bound docking because the proteins involved can change conformation upon binding. A study of conformational changes in protein complexes *(1)* showed that while the general model for protein–protein recognition is an induced fit model where the proteins must change conformation in order to bind, the amount of conformational change was small enough such that binding could be modeled as a "lock-and-key" mechanism as a first approximation. This allows for successful docking results even when there are noticeable changes in the conformation of the interacting proteins. This "rigid-body" approximation has been invaluable in the advancement of the protein–protein docking field. However, modeling induced fit by flexible docking remains a central challenge, and a large portion of current docking research is focused in this area.

There are two main challenges in the development of methods for protein–protein docking. The first is the construction of a scoring function that allows for the discrimination between correct or near-correct predictions and incorrect predictions. The second is the development of an algorithm that quickly searches and scores all possible orientations of the proteins to be docked. The most obvious way to dock two proteins would be simulate the molecular dynamics, as this would allow the complex to reach its native state with time. Unfortunately, the computational power necessary for such a simulation makes this currently intractable.

Protein–protein docking is often carried out in two stages. The initial stage treats the proteins as rigid bodies, allowing for an efficient search of the six-dimensional (6-D) space (three dimensions of translational freedom and three dimensions of rotational freedom). The 6-D space is searched for regions of high shape and biochemical complementarity, using a "soft" scoring function that allows for some clashes between atoms. A critical component of docking research has been the development of novel techniques for increasing the speed of the search. One of the most popular methods is the Fast Fourier Transform (FFT) *(2)*, used in ZDOCK *(3)*, FTDock *(4)*, and GRAMM *(5)* to search translational space and in HEX *(6)* to search angular space. Other search methods that have been used include representing the proteins using grids of bits *(7)*, Monte Carlo sampling *(8,9)*, genetic algorithms *(10)*, and geometric hashing *(11)*.

Many docking algorithms have a refinement and re-ranking stage. This involves making small changes to the highest-scoring predictions from the initial stage using techniques such as 6-D rigid-body movements, molecular dynamics, and the clustering of similar predictions. Often, a more advanced scoring function, designed to increase the rank of near-native structures and decrease the rank of false positives, is introduced. This allows for a more descriptive approximation of biochemical properties such as desolvation free energy, electrostatics, and hydrogen bonding. **Table 1** provides a list of current docking methods, along with their methodologies.

1.2. Measuring the Accuracy of Predicted Complexes

Once a prediction has been created, it is useful to evaluate it in a quantitative fashion. This is most often done using root mean square deviation (RMSD) between the atoms (using all atoms, backbone atoms, or Cα atoms) of the prediction and the complex. This is done by first aligning the predicted structure with the crystallized complex in a manner that minimizes RMSD. RMSD between the predicted (p) and actual (a) Cα atoms is calculated as follows (with n being the total number of atoms):

$$RMSD = \sqrt{\frac{1}{n}\sum_{i=1}^{n}\left\{[p_x(i) - a_x(i)]^2 + [p_y(i) - a_y(i)]^2 + [p_z(i) - a_z(i)]^2\right\}} \quad (1)$$

Two of the most often used metrics for measuring the accuracy of a predicted structure are interface RMSD (iRMSD) and ligand RMSD (lRMSD). iRMSD

Table 1
A Summary of Docking Tools

Name	Method	Server/DL[a]	Reference
AutoDOCK	Flexible docking using Monte Carlo search and incremental construction	N/Y	***12***
BiGGER	Global search using bit mapping, rescored with multiple filters	N/N	***7***
ClusPro	Docking with DOT/ZDOCK, clustering	Y/N	***13***
DOCK	Global search with grid-based energy function, flexible docking with random search and incremental construction	N/Y	***14***
DOT	FFT global search using shape complementarity and electrostatics	Y/N	***15***
FTDock	FFT rigid-body search	N/Y	***4***
GRAMM	FFT with clustering and rescoring	Y/Y	***5***
HADDOCK	Rigid-body minimization, semi-flexible simulated annealing, rescoring using biochemical data	N/Y	***16***
HEX	Fourier correlation of spherical harmonics, explicit translational search	Y/N	***6***
ICM	Docking by combining pseudo-Brownian potential and torsional steps with local gradient minimization	N/N[b]	***8***
PatchDock/SymmDock	Geometric hashing and pose-clustering to score shape complementarity	Y/N	***17***
RosettaDock	Optimization of side-chain conformation with rigid- body Monte Carlo minimization	N/Y	***9***
ZDOCK	FFT search using shape complementarity, desolvation, and electrostatics. Refinement and re-ranking with RDOCK	Y/Y	***18***

FFT, Fast Fourier Transform.
[a] Availability of download to academic users.
[b] Browser can be downloaded; docking component must be purchased.

is defined as the Cα RMSD of those residues having at least one atom within a distance cutoff of the interacting partner; lRMSD is calculated by superposing the receptor of the predicted structure with the known structure, performing the same transformation on the ligand, and calculating the Cα RMSD of the ligand. An advantage of using iRMSD is that unlike lRMSD, it is not affected by conformational change in domains that do not include the binding site.

Often, a prediction is classified as a "hit" if the iRMSD and lRMSD are below a threshold. Unfortunately, this hard cutoff does not take into account many nuances. Another method of evaluating the accuracy of docking predictions is the fraction of native and non-native contacts (f_{nat} and $f_{\text{non-nat}}$). Contacts are defined as residue pairs with less than 5Å distance between the receptor and ligand. f_{nat} is a measure of the number of contacts correctly predicted, and $f_{\text{non-nat}}$ measures the number of incorrectly predicted contacts. $f_{\text{non-nat}}$ serves as an indication of atomic clash between the interface residues in the predicted complex and also as a proxy for conformational change, as residues may move into the interface upon binding.

1.3. The Critical Assessment of Predicted Interactions Experiment

The CAPRI (Critical Assessment of Predicted Interactions) experiment was created to compare the performance of docking algorithms of various groups ***(19)***. CAPRI was modeled after Critical Assessment of Structural Prediction (CASP), which started in 1994 to compare the performance of protein-folding algorithms ***(20)***.

CAPRI is a blind competition, so the participating groups do not receive the complex structure until after all predictions have been made. Each group submits 10 predictions, ranked by confidence. The predictions are then evaluated based on various factors and assigned a score [incorrect, acceptable (one star), medium (two stars), and high (three stars)] based on their accuracy. The CAPRI metrics for these scores are described by the Boolean expressions below:

$$\begin{aligned}
High &= (f_{nat} \geq 0.5) \cap [(lRMSD \leq 1.0) \cup (iRMSD \leq 1.0)] \\
Medium &= \{[(f_{nat} \geq 0.3) \cap (f_{nat} < 0.5)] \cap [(lRMSD \leq 5.0) \cup (iRMSD \leq 2.0)]\} \cup \\
&\quad [(f_{nat} \geq 0.5) \cap (lRMSD > 1.0) \cap (iRMSD > 1.0)] \\
Acceptable &= \{[(f_{nat} \geq 0.1) \cap (f_{nat} < 0.3)] \cap [(lRMSD \leq 10.0) \cup (iRMSD \leq 4.0)]\} \cup \\
&\quad [(f_{nat} \geq 0.3) \cap (lRMSD > 5.0) \cap (iRMSD > 2.0)] \qquad (2)
\end{aligned}$$

We have made predictions for all CAPRI targets, and **Table 2** summarizes our performance. As an example, **Fig. 1** shows the close resemblance between

Table 2
ZDOCK/RDOCK Performance in the CAPRI Experiment, Rounds 1–5

Target	Protein complex	Accuracy[a]	Contact %[b]	Rank[c]	Success rate[d] (%)
1	Hpr kinase-HPr	Incorrect	7.7	9	3.19
2	Rotavirus VP6-Fab	Medium	96.1	3	2.29
3	Hemagglutinin-Fab HC63	Incorrect	59.7	7	2.19
4	Alpha-amylase-camelide AMD10	Incorrect	0.0	1	0.0
5	Alpha-amylase-camelide AMB7	Incorrect	6.3	2	0.0
6	Alpha-amylase-camelide AMD9	Incorrect	27.6	5	8.96
7	T cell Receptor V(BETA)2-exotoxin A1	Medium	83.8	1	16.91
8	Nidogen-laminin	Medium	47	1	7.45
9	LicT homodimer	Incorrect	8	2	0.20
10	TBEV trimer	Incorrect	11	3	0.97
11	Cohesin-dockerin (model)	Acceptable	13	1	9.12
12	Cohesin-dockerin	High	84	1	10.95
13	SAG1-FAB	High	87	1	7.18
14	MYPT-PP1	High	53	8	16.06
18[e]	GH11 Xylanase-TAXI	Medium	91	1	2.15
19	Ovine Prion-FAB	Medium	57	8	4.52

CAPRI, Critical Assessment of Predicted Interactions.

[a] Accuracy, as scored by the CAPRI evaluation team based on interface root mean square deviation (RMSD), ligand RMSD, and percentage of correct contacts predicted.

[b] Percentage of correct interface residue contact pairs predicted.

[c] Rank, as assigned by ZDOCK team, of best prediction out of the 10 submission for that target.

[d] A metric used to evaluate the success of predictions across all groups in CAPRI ***(21)***.

[e] Targets 15–17 were canceled.

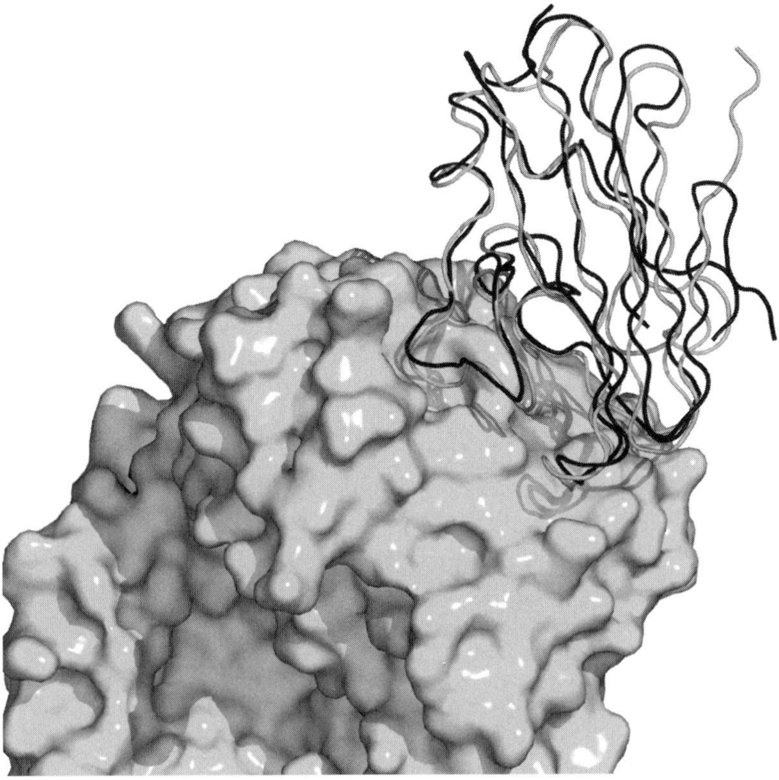

Fig. 1. Prediction of the structure of the SAG1–antibody complex [Critical Assessment of Predicted Interactions (CAPRI) Target 13]. The antibody of the prediction was superposed onto the crystal structure; the predicted SAG1 is in gray loops, whereas the crystal structure SAG1 is shown in black loops (the antibody is shown using surface representation). The non-binding domain of the SAG1 molecule is not shown. Pymol *(22)* was used to generate this figure.

our predicted structure and the crystal structure for Target 13 (SAG1–antibody complex).

1.4. A Benchmark for Protein–Protein Docking

In order to provide the docking community with a standard set of test cases to test docking algorithms, we developed two protein–protein docking benchmarks. The first benchmark, Benchmark 1.0 *(23)*, contained 59 test cases, consisting of 22 enzyme–inhibitor complexes, 19 antibody–antigen complexes, 11 other complexes, and 7 difficult complexes. Of these complexes, 31 are unbound–unbound, and 28 are bound–unbound. A number of groups have used this benchmark to test the performance of their docking algorithms ***(9,24–26)***.

A newer version of the docking benchmark, Benchmark 2.0 ***(27)***, has been created. It includes 84 test cases and was designed to focus on unbound–unbound test cases. Structural classification of proteins (SCOPs) ***(28)*** was used to avoid redundancy in the benchmark. This benchmark is classified by docking difficulty, based on the amount of conformational change undergone by the interacting proteins. Complexes classified as rigid and medium fall into the realm of rigid-body docking, whereas complexes classified as difficult would require algorithms that explicitly search backbone conformations.

2. The ZDOCK/RDOCK/M-ZDOCK Approach

2.1. ZDOCK: An FFT-Based Initial Stage Docking Algorithm

ZDOCK is an initial-stage docking algorithm that uses an FFT to find the three-dimensional (3-D) structure of a protein complex. The ZDOCK algorithm optimizes three parameters: shape complementarity, electrostatics, and desolvation free energy.

ZDOCK takes Protein Data Bank (PDB) ***(29)*** files as input. The larger of the two interacting proteins is considered the receptor (R), whereas the smaller of the two is considered the ligand (L). These PDB files are first parsed through the supplied program *mark_sur*, which measures the amount of accessible surface area (ASA) of each atom using a water probe of radius 1.4 Å. If an atom has an ASA of more than 1 $Å^2$, it is marked as a surface atom. *mark_sur* also marks the atom type for each atom in the structure, based on the 18 atom types based on atomic contact energy (ACE) ***(30)***. For any given rotational orientation, the L and R are both discretized onto a 3-D grid of size $N \times N \times N$ with a spacing of 1.2 Å. N must be large enough such that the grid can cover the sum of the maximal spans of R and L, plus 1.2 Å, and it is often set at 128.

2.1.1. The Fast Fourier Transform

As previously mentioned, the FFT is a popular method for quickly searching 3-D translational space. A diagram of the general FFT docking approach is detailed in **Fig. 2**. The search is performed by randomly perturbing both the receptor and ligand to avoid starting from a near-native state, and then discretizing them into discrete functions [$R(x, y, z)$ and $L(x, y, z)$ for the receptor and ligand, respectively] onto separate 3-D grids. ZDOCK searches rotational space explicitly by rotating the ligand in either 15° or 6° steps, which result in 3600 and 54,000 total angles, respectively. For each angle, only the

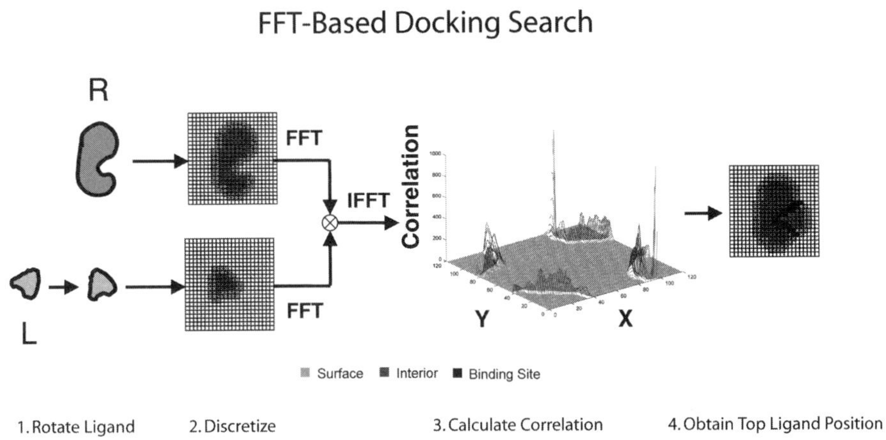

Fig. 2. The steps involved in a Fast Fourier Transform (FFT)-based docking search. For each ligand rotation, it is discretized and this discretization is then correlated with the discretized receptor to obtain the top-scoring ligand position. These steps are repeated to cover all ligand rotations in three dimensions, if necessary. In the case of ZDOCK, this involves 3600 iterations for 15° sampling and 54,000 iterations for 6° sampling.

top-scoring translation is found. To find the highest-scoring translation, we performed a cross-correlation. The correlation for a particular x, y, z translation (i, j, k) is found by taking the complex conjugate of the one of the functions, offsetting the grids, and multiplying the overlapping grid points together, with the sum of these products representing the score for that translation.

$$S(i, j, k) = \sum_{x,y,z} L^*(x+i, y+j, z+k)\, R(x, y, z) \quad (3)$$

Cross-correlations can be performed globally in a single step by working in the frequency domain. This is done using the Discrete Fourier Transform (DFT) and Inverse Fourier Transform (IFT):

$$S(i, j, k) = \frac{1}{N^3} IFT\left\{IFT\left[L(x, y, z)\right]^* DFT\left[R(x, y, z)\right]\right\} \quad (4)$$

The FFT is a method for computing the DFT and IFT efficiently. Each FFT is $O(\log_2(N^3))$, whereas the multiplication of the grids is $O(N^3)$. Therefore, using the FFT to perform the translational search reduces the computational complexity of the search from $O(N^6)$ to $O(N^3 \log_2(N^3))$.

ZDOCK uses a combination of three physical and biochemical properties to describe ligand and receptor: shape complementarity, desolvation free energy, and electrostatics.

2.1.2. Shape Complementarity

The physical basis for shape complementarity comes from the van der Waals (vdW) potential. Atoms are subject to an attractive force at long distances, and a repulsive force at short distances, caused by the overlap of electronic orbitals. Most often, this is approximated by the Lennard–Jones 6–12 potential, shown below:

$$V_{\mathrm{L-J}} = \frac{A}{r^{12}} - \frac{B}{r^{6}} \tag{5}$$

The r^6term represents the attractive energy, whereas the r^{12} term represents the repulsive energy. The minimum of the vdW potential is found at the sum of the vdW radii, which can be thought of as the effective sizes of the interacting atoms.

Early versions of ZDOCK used a shape complementarity function known as grid-based shape complementarity (GSC) ***(3)***. Here, two discrete functions, R_{GSC} (GSC function for the receptor) and L_{GSC} (GSC function for the ligand), are used to describe the geometric characteristics of the two proteins as follows:

$$R_{\mathrm{GSC}} = \begin{cases} 1 & \text{solvent-accessible surface} \\ 9i & \text{solvent-excluding surface} \\ 9i & \text{core} \\ 0 & \text{open space} \end{cases}$$

$$L_{\mathrm{GSC}} = \begin{cases} 0 & \text{solvent-accessible surface} \\ 1 & \text{solvent-excluding surface} \\ 9i & \text{core} \\ 0 & \text{open space} \end{cases} \tag{7}$$

The solvent-excluding surface layer is defined by the grid points marked as surface atoms by *mark_sur*, whereas the core is defined as the atoms not on the surface. The solvent-accessible surface layer is an additional layer of grid points surrounding the surface of the protein.

The current version of ZDOCK uses a complementarity function known as pairwise shape complementarity (PSC) ***(31)***. PSC is composed of a favorable term and a penalty term. The favorable term calculates the number of atom pairs between R and L within a distance cutoff D, whereas the penalty component of PSC is proportional to the number of overlapping grid points between

R and L, much like GSC. Whereas the GSC function results in grid spaces with purely real or imaginary values, the PSC function is complex. L_{PSC} and R_{PSC} are shown below.

$$\Re\left[L_{psc}\right] = \begin{cases} 1 & \text{if nearest grid point to ligand atom} \\ 0 & \text{otherwise} \end{cases}$$

$$\Re\left[R_{psc}\right] = \begin{cases} \text{Number of receptor atoms within D} = +\text{vdW radius of nearest atom} & \text{open space} \\ 0 & \text{otherwise} \end{cases}$$

$$\Im\left[L_{PSC}\right] = \Im\left[R_{PSC}\right] = \begin{cases} 3 & \text{solvent-excluding surface} \\ 9 & \text{core} \\ 0 & \text{open space} \end{cases} \tag{8}$$

The use of PSC rather than GSC for scoring shape complementarity was shown to greatly increase the number of near-native predictions for Benchmark 1.0 during initial stage docking ***(31)***.

2.1.3. Desolvation Free Energy and Electrostatics

ACE ***(30)*** is used by ZDOCK to estimate desolvation free energy. ACE is defined as the change in free energy resulting from the breaking of two atom–water contacts and the formation of an atom–atom contact and a water–water contact. This is also referred to as the hydrophobic effect, which is known to play a critical role in protein–protein binding. ZDOCK introduces two discrete functions, L_{DE} and R_{DE}, to describe the desolvation energy of the ligand and receptor:

$$\Re\left[L_{DE}\right] = \Re\left[R_{DE}\right] = \begin{cases} \text{PSC} + \text{ACE scores of all nearby atoms} & \text{open space} \\ 0 & \text{otherwise} \end{cases}$$

$$\Im\left[L_{DE}\right] = \Im\left[R_{DE}\right] = \begin{cases} 1 & \text{if nearest grid point to atom} \\ 0 & \text{otherwise} \end{cases} \tag{9}$$

The electrostatics energy term for ZDOCK can be expressed as a correlation between the electric potential generated by the receptor with the charges of the ligand atoms. ZDOCK adopts the Coulombic formula used by Gabb et al. ***(4)*** but incorporates partial charges using the CHARMM19 parameters from the CHARMM molecular mechanics program ***(32)***.

2.1.4. ZDOCK Scoring Function

There are two ZDOCK versions that use PSC to describe shape complementarity: ZDOCK 2.1 simply uses PSC as the scoring function, whereas ZDOCK

2.3 uses a linear combination of the shape complementarity-electrostatics score and the desolvation score. ZDOCK 2.3 incorporates PSC and electrostatics into single complex functions ($R_{\mathrm{PSC+ELEC}}$ and $L_{\mathrm{PSC+ELEC}}$) to improve computation time. These functions are described below:

$$\Re\left[L_{\mathrm{PSC+ELEC}}\right]=\Re\left[R_{\mathrm{PSC+ELEC}}\right]=\begin{cases}3.5 & \text{solvent-excluding surface}\\ 3.5^{2} & \text{core}\\ 0 & \text{open space}\end{cases}$$

$$\Im\left[R_{\mathrm{PSC+ELEC}}\right]=\begin{cases}\beta^{*}\ \text{electric potential of all R atoms} & \text{open space}\\ 0 & \text{otherwise}\end{cases}$$

$$\Im\left[L_{\mathrm{PSC+ELEC}}\right]=\begin{cases}-1^{*}\ \text{atom charge} & \text{grid point closes to ligand atom}\\ 0 & \text{otherwise}\end{cases} \tag{10}$$

2.2. RDOCK: Refining ZDOCK Predictions

The refinement stage of protein docking with ZDOCK is carried out using an algorithm known as RDOCK *(33)*. Because of the soft scoring function in ZDOCK, many of the top-scoring predictions are false positives (not near-native). RDOCK refines these output structures through energy minimization. This is carried out in three steps, using CHARMM *(32)*.

1. Removal of clashes by minimization of vdW and internal energies.
2. Minimization of total (Coulombic electrostatics, vdW, internal) energy, constraining non-hydrogen atoms, and keeping ionic side chains in their neutral states.
3. Minimization of total energy with no restrictions.

Once energy minimization has been performed, the minimized structures are re-ranked. Any complexes that still exhibit clashes (those that have vdW energy of 10 kcal/mol or greater) after minimization are discarded. Electrostatics and desolvation energy for the complexes are calculated using CHARMM and ACE, respectively. The RDOCK scoring function, $\Delta G_{\mathrm{binding}}$, is a linear combination of desolvation score (ΔG_{ACE}) and electrostatic energy (ΔE_{elec}).

$$\Delta G_{\mathrm{binding}}=\Delta G_{\mathrm{ACE}}+0.9^{*}\Delta E_{\mathrm{ELEC}} \tag{11}$$

2.3. M-ZDOCK: Symmetric Multimer Docking with ZDOCK

The ZDOCK algorithm has been modified to predict the structure of C_n multimer complexes, in which two or more identical proteins interact, resulting in a ring-shaped complex. M-ZDOCK *(34)* reconstructs the multimer based on the optimal position of two adjacent monomers in a single plane. This leads to a reduction in computational time due to the reduced search space, as well as an increase in performance when compared with docking C_n multimers with ZDOCK.

2.4. ZDOCK Performance on Benchmark 2.0

ZDOCK was tested against version 2.0 of the docking benchmark using ZDOCK 2.3 and ZDOCK 2.1, with 6° and 15° angular sampling.

2.4.1. Prediction Evaluation

To evaluate the structure predictions produced by ZDOCK, we used the RMSD of the interface Cα atoms. Interface Cα atoms were identified by selecting residues that had any atom within 10 Å of the other molecule in the bound complex. A hit was defined as a prediction with an iRMSD $\leq$ 2.5 Å.

Two measures are defined to evaluate the average performance of a docking algorithm over the entire benchmark. *Success rate* is defined as the percentage of test cases that have a hit in the top *N* predictions. *Average hit count* is the number of hits for all test cases in the top *N* predictions, divided by the number of test cases.

2.4.2. Running ZDOCK

Several considerations were taken before and while running ZDOCK. To remove bias from the starting positions (the Benchmark 2.0 unbound test cases are by default aligned to the bound proteins, to facilitate the evaluation of predicted structures), we used a different random seed to rotate the ligand for each case. In addition, the antibodies (apart from the camelid 1KXQ) had most of their non-complementarity-determining region (non-CDR) loops blocked to avoid false-positive predictions. The CDRs of the antibodies were identified using their sequences (loops L1, L2, L3, H1, H2, and H3) and by examination of the structures (loops L4 and H4 and the N-termini).

2.4.3. Success Rate and Hit Count

Figure 3 shows the success rate for ZDOCK when run against all rigid-body cases from Benchmark 2.0. It can be seen that ZDOCK 2.3 performs better overall than ZDOCK 2.1 in terms of success rate. This is because the scoring function used in ZDOCK 2.3 is better at discriminating hits against incorrect predictions across the benchmark. Also, for both ZDOCK 2.1 and ZDOCK 2.3, the 15° sampling has a higher success rate than the 6° sampling. This indicates that for more predictions (i.e., finer sampling), there are more false positives introduced that reduce the rank of the first hit in some of the test cases. However, the 6° sampling is superior with regard to the number of hits, indicated by the hit count plot.

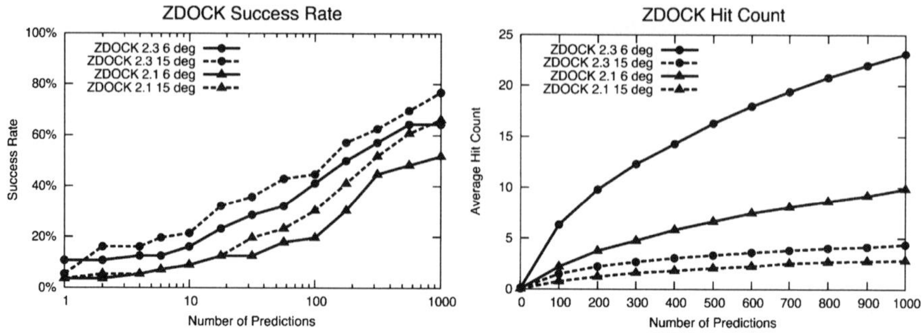

Fig. 3. ZDOCK success rate (left) and hit count (right) for the rigid-body test cases of Benchmark 2.0, for $N = 1$ through 1000 predictions. The success rate is the percentage of test cases with a hit in the top N, whereas the hit count is the number of hits for all test cases within the top N divided by the number of cases. Hits are defined as predictions that have an interface root mean square deviation (RMSD) ≤ 2.5 Å, as described in the text.

Also in **Fig. 3** is the average hit count for the four ZDOCK modes tested. In this plot, it is clear that ZDOCK 2.3 with 6° sampling is the best, followed by ZDOCK 2.1 with 6° sampling. The greater number of hits produced by ZDOCK 2.3 with 6° sampling make this the best option for following up with re- ranking and refinement of the top predictions (e.g., $N = 2000$), as suggested by us earlier ***(33)***.

2.4.4. ZDOCK Performance by Test Case Category

Figure 4 gives the success rate curves for ZDOCK across the three types of test cases in Benchmark 2.0: Enzyme/Inhibitor, Antibody/Antigen, and Others (the latter is defined as those cases that fall into neither of the first two categories).

ZDOCK has the best success rate for the Antibody/Antigen test cases, with the success rate at 1000 predictions with 95% success for ZDOCK 2.3 15° sampling. This may be partly because approximately half of the Antibody/Antigen cases use bound forms of the antibody; thus the interface conformational change of these cases is on average smaller. In addition, the search space is reduced by blocking the non-CDR portions of the antibody.

The Enzyme/Inhibitor cases did not match the Antibody/Antigen cases in terms of success rate at 1000 predictions, although for the top predictions (i.e., small N), ZDOCK performed better for this category. Most notable is

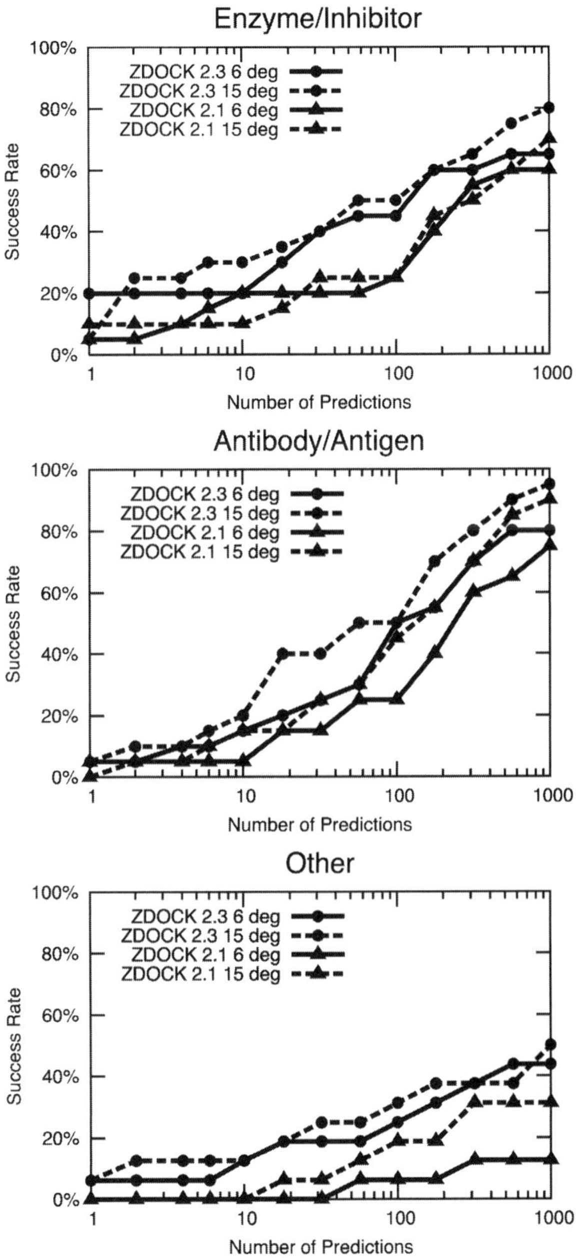

Fig. 4. ZDOCK success rate for Benchmark 2.0 cases, broken down by category: Enzyme/Inhibitor, Antibody/Antigen, and Other.

the 20% success for ZDOCK 2.3 15° sampling for the top prediction (four of 20 cases). This may be due to the PSC scoring function, which when combined with desolvation and electrostatics (as in ZDOCK 2.3) is well suited to identify the pocket-shaped binding sites on the enzymes.

ZDOCK did not perform quite as well on the Others test cases; this was also seen when ZDOCK was run against Benchmark 1.0 Others test cases. Of the four ZDOCK options tested, both sampling levels of ZDOCK 2.3 performed better than ZDOCK 2.1. In fact, at $N = 1000$, ZDOCK 2.3 still performed better than ZDOCK 2.1, whereas for Enzyme/Inhibitor and Antibody/Antigen cases, the ZDOCK 2.1 15° sampling performed better than ZDOCK 2.3 6° sampling. This trend may indicate that shape complementarity (which is the only scoring metric used for ZDOCK 2.1) is less important (versus electrostatics and desolvation) for the Others cases than for the Enzyme/Inhibitor and Antibody/Antigen cases.

2.5. Docking Overview: Summary

Protein–protein docking has evolved to the point where it is possible to predict the structures of many protein complexes based on their unbound proteins. This is demonstrated above using a protein-docking benchmark and the rigid-body-docking algorithm ZDOCK. However, based on the success rate plots of **Fig. 3**, it is evident that not all cases are successfully predicted within the top few thousand docking predictions, and for a few cases, no hits are found. What leads to this variation in docking success across a set of cases? The final section of this chapter takes an in-depth look at how various properties of proteins impact the ability of docking to successfully predict the complex structure.

3. The Relationships Between ZDOCK Performance and Protein Complex Characteristics

The performance of ZDOCK is dependent on both the accuracy of the energy function and the comprehensiveness of the search algorithm. Both of these are in turn dependent on the many physicochemical characteristics of the protein–protein complex that ZDOCK is attempting to predict. For example, in any particular complex, the exact shape of the protein–protein interface will undoubtedly have an effect on how high shape complementarity is scored in the energy function. Protein–protein complexes with planar interfaces may prove to be the most challenging for ZDOCK. Thus, it is important to examine how ZDOCK performs with respect to differing interface shapes in order to gauge the effectiveness of the shape complementarity term. Knowing how

ZDOCK performs with respect to a vast array of different protein–protein complex characteristics provides an understanding of what types of complexes ZDOCK can be expected to excel in predicting. It also can help lead to more focused improvements in the development of protein docking by identifying the strengths and weaknesses of the algorithm. In addition, it may be possible to extend the conclusions drawn from such an examination to other FFT-based docking algorithms.

3.1. Near-Native Prediction Definitions

In order to objectively and systematically evaluate the performance of the ZDOCK algorithm, it is necessary to compare the near-native docking orientations produced by ZDOCK to the space of orientations available given a particular complex within the rigid-body FFT framework. While the fields of protein structure prediction and docking commonly make use of "decoys" to evaluate algorithm performance, here we adopt an alternative approach. We estimate the space of potential near-native conformations using a newly designed program called HitFinder. This space is reasonably limited under the assumption of rigid-body docking, and therefore focus was placed on the 64 rigid-body cases from the protein-docking benchmark *(27)*.

Using the core framework of the ZDOCK algorithm, HitFinder maps the complex components onto a 1.2 Å grid and uses a 6° Euler angle set *(18)* to perform FFT search for orientations that would represent near-native hits. HitFinder iterates over the same set of angles and translations as ZDOCK but uses a simple RMSD filter instead of a docking scoring function. For every potential ligand-docking orientation where the ligand overlaps with the native ligand orientation, the docking orientation is retained for further processing if the ligand Cα RMSD is less than or equal to 10 Å. Following this initial search, potential docking hits are further defined using a more nuanced protocol based on the CAPRI prediction accuracy criteria. As in CAPRI, these hit definitions rely on the combination of RMSD and native contact fraction criteria. Here two kinds of hits are classified: high quality and medium quality. They are defined by the following Boolean relationships:

$$\text{High-quality hits} = \left[iRMSD \leq \left(iRMSD_{\text{superposed unbound complex}} + 1\,\text{Å}\right)\right] \cap (f_{nat} > 0.5) \cap (f_{non\text{-}nat} < 0.5) \quad (12)$$

$$\text{Medium-quality hits} = \left[iRMSD \leq \left(iRMSD_{\text{superposed unbound complex}} + 1\,\text{Å}\right)\right] \cap (f_{nat} > 0.3) \cap (f_{non\text{-}nat} < 0.7) \quad (13)$$

Because HitFinder does not include the shape complementarity functions that are normally a part of the ZDOCK algorithm, there is no control over potential ligand/receptor clash for those orientations where they come too close to each other. Therefore, this study uses the more strictly defined space of high-quality hits (or three-star hits; **Eq. 2**) as a guide and eliminates all hits with clash significantly greater than the average three-star hit. Clashes are defined as the number of interface contacts within 3 Å. All docking orientations with a clash total greater than the mean number of clashes for the three-star hits plus 2 standard deviations are eliminated. Finally, if an orientation meets all the required hit criteria, it is labeled a "potential hit" and all such structures are recorded for a complex.

3.2. Measuring ZDOCK Success

To examine the success of ZDOCK, a metric for protein–protein docking accuracy is needed. Measuring ZDOCK accuracy per complex could be accomplished by merely counting the number of medium- or high-quality hits the algorithm achieves out of a certain number of predictions. However, because the number of potential hits is inherent to each particular protein–protein complex (*see* **Fig. 5A and B**), this measure would not reflect precisely how well ZDOCK performs. As an example, if complex A has 100 potential hits and complex B has 1000 potential hits, ZDOCK's accuracy is not equivalent if it finds one hit for both complexes. Complex B is easier to predict because it possesses some characteristics that allow for a greater number of hits possible. Further discussion as to what characteristics these may be will follow in Section 3. It may make sense to simply take the percentage of hits predicted out of the number of potential hits as a metric for docking accuracy. In this metric, ZDOCK makes successful predictions at a 1% rate for complex A and only a 0.1% rate for complex B and thus clearly performs better on complex A. Yet there is a flaw to this measure as well. As explained previously, the ZDOCK algorithm only keeps the highest-scoring translation for every rotation angle searched. This means that if multiple hits exist in the same rotational angle, ZDOCK will at best only select one of them. In the example of complex A versus complex B, complex A has 100 potential hits, but hypothetically could have 99 in one rotational angle. In that case, the highest number of hits an optimal ZDOCK search could find would only be 2. Thus, accuracy as defined as a percentage of potential hits would reach the upper limit at 2%. It is necessary then to introduce another definition, that of the "hit angle." A hit angle is defined as any rotational angle in a ZDOCK search that has at least one translation that results in a potential hit (*see* **Fig. 5C and D**). Using this

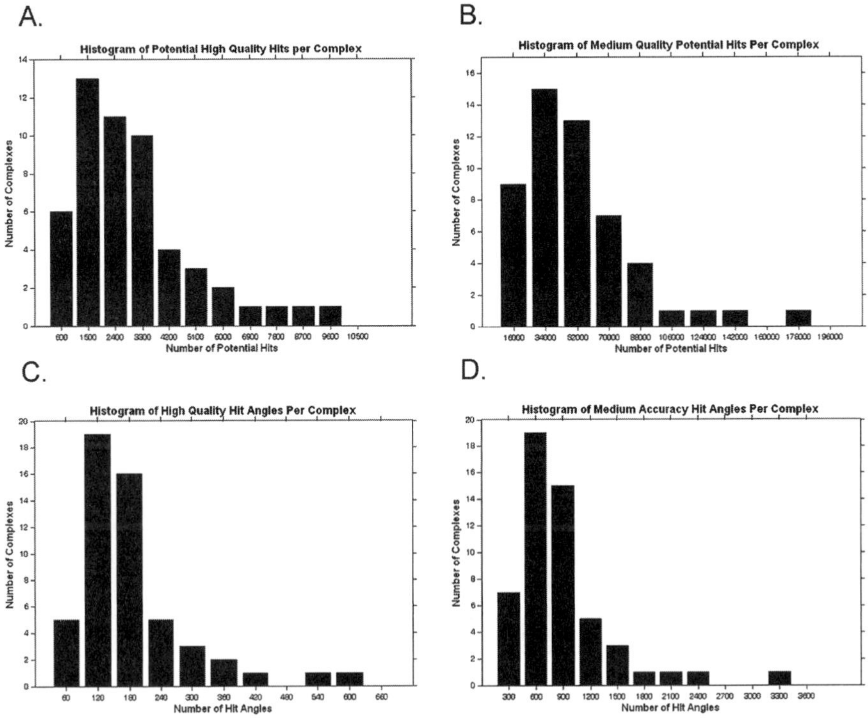

Fig. 5. Medium- and high-quality potential hits and hit angles.

definition, we found complex A has two hit angles because it has 100 potential hits but only two rotational angles in which those potential hits can be found. Finally, the accuracy of ZDOCK performance can robustly be measured as the percentage of hits predicted out of the possible number of hit angles. In the above example, if ZDOCK finds one hit for complex A and there are only two hit angles possible, the accuracy rate on that particular complex is 50%. Thus, for complex A, ZDOCK is operating at half of its maximum performance level. The distribution of accuracy rates for medium-quality and high-quality hits are shown in **Fig. 6**.

If the accuracy rate of ZDOCK on complex B is much lower than the 50% of complex A, it leads to questions of why ZDOCK performs better on complex A. What are the characteristics of complex A that make it more suitable for creating good ZDOCK predictions? What are the characteristics of complex B that are associated with ZDOCK missing many good predictions? In the next section, these questions are examined for many complexes with myriad

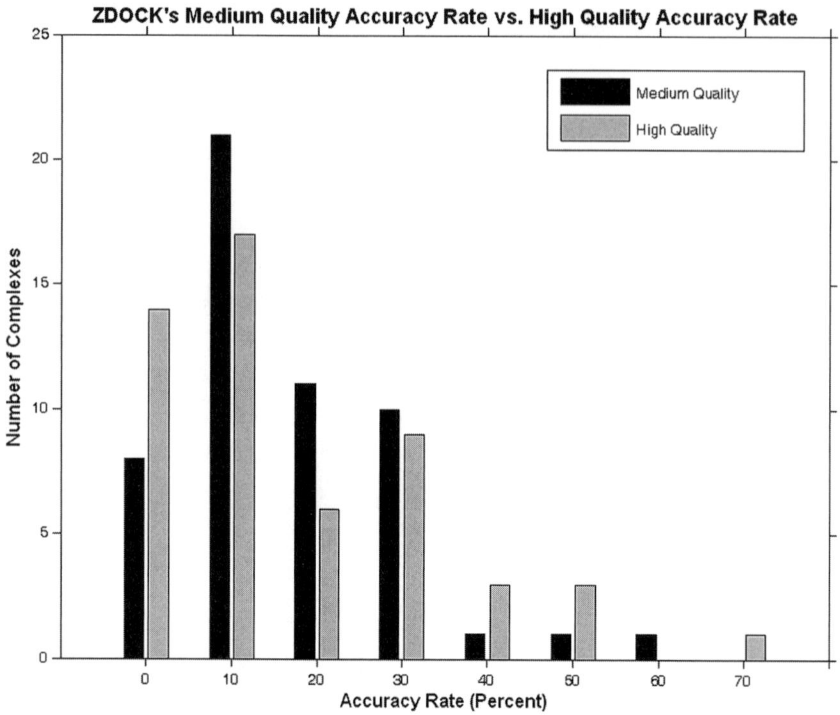

Fig. 6. ZDOCK Performance as measured by accuracy rate of medium- and high-quality predictions. Accuracy rate varies substantially across the 53 complexes of the regression data set.

physicochemical characteristics by employing regression analysis. The goal of the analysis is to get a better understanding about what types of attributes lead to successes and failures in protein–protein docking predictions with ZDOCK.

3.3. Regression Analysis of ZDOCK Performance

To begin to examine by regression analysis which types of attributes of protein–protein complexes are important to the success of ZDOCK, it is necessary to have a large data set of complexes. Fortunately, the protein–protein benchmark represents the largest data set of protein–protein complexes in the docking field and includes 84 such transient complexes. Some paring down of that original data set is required in order for the study to control for factors already known to affect docking results. Previously, 20 of the 84 complexes have been characterized as undergoing large conformational change upon binding. These complexes were removed from this study in order to focus on docking performance in rigid-body cases. In addition, 11 benchmark complexes are antibody–antigen complexes in which the antibody structure is only solved in the complex. These "unbound–bound" complexes were removed in order to best represent true docking performance. The remaining 53 complexes comprise the data set used in the regression analysis (*see* **Table 3**).

A comprehensive list of protein–protein complex attributes is needed to establish what characteristics influence the performance of ZDOCK. One hundred twelve such attributes were used in the regression analysis, some of which are closely related. For brevity, only the general attributes are included in **Table 4**.

3.4. Simple Linear Regression Approach

Simple linear regression was first employed to determine which single attributes could be associated with ZDOCK performance. Accuracy rate is used as the response variable in the regression. Accuracy rate can be broken down according to the two categories of hits: high and medium quality. Also, accuracy

Table 3
Protein Data Bank Codes in the Regression Analysis Data Set

1A2K	1E6E	1HE8	1RLB
1AHW	*1E6J*	1HIA	1SBB
1AK4	1E96	1I4D	1TMQ
1AKJ	1EAW	*1JPS*	1UDI
1AVX	1EWY	1KAC	*1VFB*
1AY7	1EZU	1KLU	*1WEJ*
1B6C	1F34	1KTZ	2BTF
1BUH	1F51	1KXP	2MTA
1BVK	1FC2	1MAH	2PCC
1BVN	1FQJ	1ML0	2SIC
1CGI	1GCQ	*1MLC*	2SNI
1D6R	1GHQ	1PPE	*2VIS*
1DFJ	1HE1	1QA9	7CEI
1DQJ			

Antibody–antigen complexes are in italics.

Table 4
General Attributes of Protein–Protein Complexes

Attributes
Surface
Shape
Volume
Weight
Curvature
Interface size
Side-chain conformational change
Backbone conformational change
Number of interface hydrogen bonds
Number of native complex clashes
Hydrophobic character of interface
Polar character of interface
Charged character of interface

Of the 112 attributes used in the regression analysis, only the general attributes are listed here. Most attributes are expanded to include separately their values for complex, receptor, and ligand as well as the unbound and bound states of each.

rate is dependent on the number of predictions made, and this analysis uses all hits from the top 54,000 predictions, corresponding to one prediction per rotational angle at 6° sampling density.

As expected for an intricate system such as protein–protein docking, most of the simple linear regression models in this analysis fail to establish good relationships between single independent protein complex attributes and the outcomes investigated. Only simple linear regression on the accuracy rate for medium-quality predictions resulted in predictors with highly significant correlations ($p < 0.001$). Curvature of the interface has the strongest correlation ($R^2 = 0.36$) with medium-quality accuracy rate (*see* **Fig. 7**). It is a positive linear relationship, and thus ZDOCK performance tends to increase as curvature of the interface also increases. Interface curvature is calculated by first fitting a plane to the atoms of the interface. The RMSD from this plane is the curvature score *(35)*. The ZDOCK scoring function relies heavily on shape complementarity for computing the energy of predicted complexes, and thus the importance of that energy term to the performance of ZDOCK is apparent from this correlation.

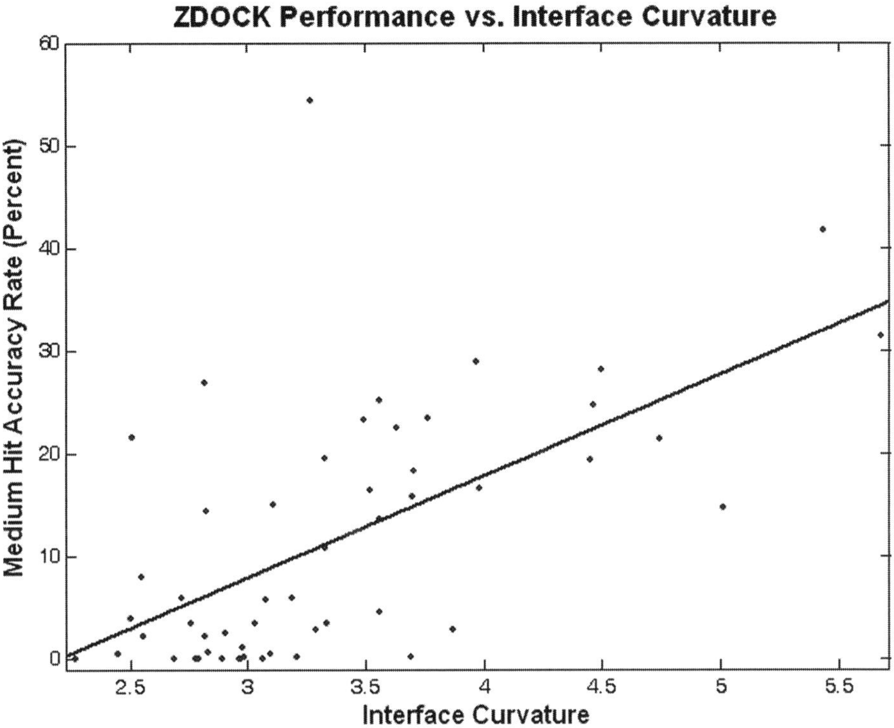

Fig. 7. ZDOCK performance versus interface curvature of the test case. Interface curvature is the strongest correlated predictor in simple linear regression for medium-quality accuracy rate ($R^2 = 0.36$).

The four remaining predictors that showed statistically meaningful correlations all were related to the size of the interface. The strongest correlation among these ($R^2 = 0.30$) was the difference in accessible solvent area between the complex and its constituents, referred to as dASA. The linear relationship between interface size and ZDOCK performance is positive, meaning ZDOCK performs well on protein complexes with larger interfaces. Scores representing larger interfaces are more statistically significant than scores representing smaller interfaces, which lead to better discriminatory power of the algorithm and hence better docking performance. In addition, it suggests that the sensitivity to a few bad contacts is lowered in larger interfaces because they make up a smaller percentage of the overall interface. By contrast, in smaller interfaces, one or two mispositioned side chains could proportionately contribute enough high energy to the overall docking score to sufficiently lower the rank of a near-native structure such that it is not included in the final prediction set.

3.5. Multiple Linear Regression Approach

Whereas simple linear regression is an important first look at which single characteristics of protein complexes are relevant to the performance of ZDOCK, a more comprehensive approach should involve the employment of multiple linear regression analysis. Finding the relationship of combined attributes to ZDOCK sampling and accuracy gives a better indication of what to expect in terms of successes and failures depending on the type of complex involved in the prediction. In multiple linear regression, it is important to avoid over-fitting the data caused by using a small ratio of outcome variables to predictor variables. Therefore in this study, only sets of four attributes were considered for the regression with 53 complexes. It was computationally tractable to do the regression on all permutations of four attributes and thus avoid the pitfalls associated with a stepwise regression approach.

3.5.1. Medium Quality Predictions

Multiple linear regression analysis was computed for the response variables of accuracy rate for medium- and high-quality hits with four predictors. For medium-quality accuracy rate, the four attributes with the highest correlation (R^2 adjusted = 0.53) were: curvature of the interface, size of the ligand interface relative to the size of the ligand, ligand side-chain conformational change, and the hydrophobicity of atoms that are completely buried upon binding (*see* **Table 5**).

The inclusion of curvature of interface in the top correlated set of attributes suggests the importance of shape complementarity just as it did in the simple linear regression for medium-quality accuracy rate. It is possible to exactly determine how important interface curvature or any other predictor is to the overall correlation by looking at the coefficients of partial determination for the regression model. A coefficient of partial determination in this analysis measures the proportionate reduction in variation in ZDOCK performance when a particular predictor is included in the regression model. With the above four attributes, the coefficient of partial determination for the inclusion of interface curvature in the regression model is 0.41. This explains quantitatively that interface curvature accounts for a 41% reduction in the regression error when it is added to the three-attribute model of interface hydrophobicity, ligand side-chain conformational change, and ligand interface size relative to the size of the ligand. Thus, interface curvature is highly important to the multiple linear relationship between these four predictors and ZDOCK performance.

Table 5
The Highest Correlated Regression Models

Attribute	Coefficient	Partial Coefficientof Determination
Medium-quality accuracy rate (R^2-*adjusted: 0.53*)		
Interface Curvature	0.58	0.41
Interface Hydrophobicity	−0.34	0.21
% of Ligand in Interface	0.27	0.13
Ligand Side-chain Change	−0.20	0.08
High-quality accuracy rate (R^2-*adjusted: 0.41*)		
% of Ligand in Interface	0.50	0.24
Native Complex Close Contacts	0.39	0.21
Ligand Side-chain Change	−0.33	0.16
Complex Shape	0.36	0.14

Hydrophobicity, with a coefficient of partial determination of 0.21, is the second most important predictor in this regression model. Hydrophobicity was characterized for atoms in the interface that are completely buried upon binding using an atom-typing scheme ***(36)*** representing three categories: polar, hydrophobic, and charged. Unexpectedly, in this regression model, the relationship between ZDOCK performance and complexes with interfaces with a large amount of hydrophobic atoms buried is negative. Although the correlation is weak, simple linear regression of ACE score versus medium accuracy rate confirms this inverse relationship (*see* **Fig. 8**). Previous analysis ***(3)*** on an earlier test-case data set found a positive relationship between hydrophobicity and ZDOCK performance. However, the earlier data set included several homodimer test cases that were not included in the current benchmark. Homodimers are known to have strong hydrophobic interfaces, and their absence in the current benchmark explains the loss of a positive correlation between hydrophobicity and ZDOCK performance.

ZDOCK uses a 6-Å cutoff for defining the interface for calculating the desolvation energy of the prediction. In the multiple regression analysis, the relationship of hydrophobicity and ZDOCK performance is most significant when the interface is limited to the atoms that become buried upon complex formation. Although the results were surprising that hydrophobicity is negatively correlated with ZDOCK performance, it underscores a potential area for improvement in the ZDOCK algorithm. Calculating the desolvation energy of just the buried atoms instead of using a 6 Å contact radius may better represent the role of the hydrophobic effect in protein–protein binding and consequently increase the accuracy of ZDOCK.

The size of the ligand interface relative to the size of the ligand is the third most important attribute in the highest correlated regression model for medium-quality accuracy rate. The relationship is positive and for ligands in which the interface represents a large proportion of the total size, ZDOCK performance increases for this regression model. From a probability standpoint, this certainly makes sense as the greater the ratio between ligand interface size and ligand size, the higher the probability any docking prediction can be considered near native.

The final attribute of the regression model is a measure of how much side-chain conformational change occurs in the ligand interface. Specifically, it is calculated by determining the percentage of ligand interface residues that differ in rotamer type between the unbound and bound states. Rotamers were defined using the Dunbrack rotamer libraries *(37)*. Most of the conformational change that occurs in side chains does not result in large structural differences such as in

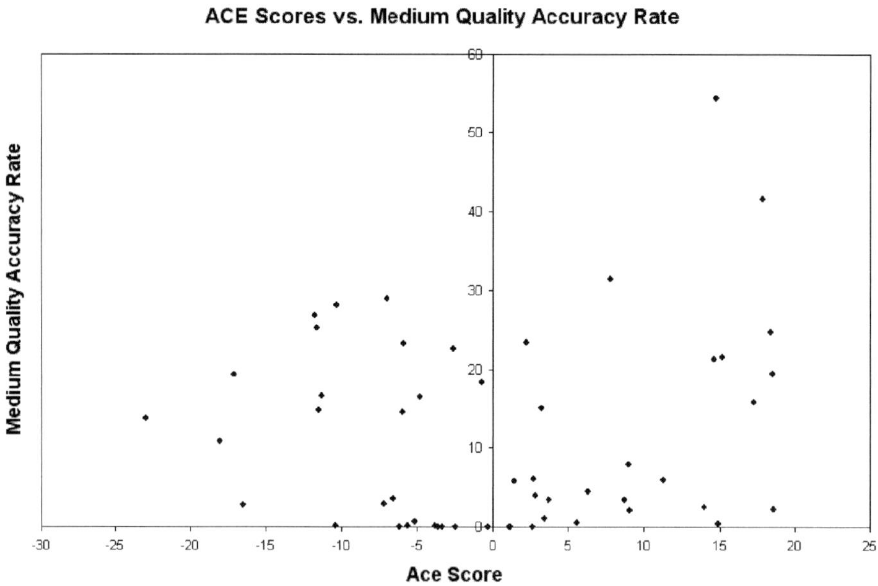

Fig. 8. Medium-quality accuracy rate shows a very weak but positive correlation with the atomic contact energy (ACE) score of the native complex interface. ACE scores decrease as interface hydrophobicity increases and medium-quality accuracy rate is therefore negatively correlated with interface hydrophobicity.

the complexes with backbone conformational change that were removed in the creation of the data set. However, even small differences in side-chain positions can cause large inaccuracies in the calculation of the scoring function especially within the vdW terms. Because ZDOCK does not attempt to move side chains during docking, interfaces with more side chains in different positions than in their unbound state will cause an inaccurate representation of the true bound interface and thus ZDOCK performance will suffer. Side-chain search is an actively pursued area in protein–protein docking research, and from the results of this regression analysis, it is understandable why accurate placement of side chains is a vital part of making successful docking predictions.

3.5.2. High-Quality Predictions

In comparison to the ability of ZDOCK to produce medium–quality predictions, there may exist a different set of characteristics of protein complexes that associate with ZDOCK's ability to generate high-quality predictions.

To this end, all regression models with four predictors were run using the high-quality accuracy rate as the response variable. The highest correlated model (R^2 adjusted $= 0.40$) included the following four attributes: complex shape, size of the ligand's interface relative to the size of the ligand, ligand side-chain movement, and number of close contacts in the native complex (*see* **Table 6**). Whereas two of these attributes are the same as in the medium-quality accuracy rate regression, two are different and will be explored further in this section.

The inclusion of native complex close contacts in the regression model was a surprising result, and even more unexpected was that the relationship between the number of close contacts and accuracy rate in the model was positive. Close contacts were calculated as all intermolecular atomic contacts less than 3 Å in the native complex structure. The positive relationship means that in the highest correlated model, ZDOCK performance is higher in complexes with many close contacts. It would seem that close contacts occur more often in larger interfaces and at least partly explain the positive relationship based on the aforementioned reasons why larger interfaces are preferred for better ZDOCK performance. However, there is no strong correlation between the two attributes of native complex close contacts and interface size ($R^2 = 0.25$). Thus, it may instead be that a complex with many close contacts represents a tightly packed interface. This would suggest once again the importance of the shape complementarity term in the ZDOCK energy function and in particular the necessity for a well struck balance between the vdW repulsion and attraction parameters.

Table 6
Intercorrelation of Attributes in the Regression Models for Medium- and High-Quality Accuracy Rates (R^2 values)

Medium–Quality Accuracy Rate				
Attribute	Interface Curvature	Interface Hydrophobicity	% of Ligand in Interface	Ligand Side-chain Change
Interface Curvature	1	–	–	–
Interface Hydrophobicity	0.03	1	–	–
% of Ligand in Interface	0.09	0.005	1	–
Ligand Side-chain Change	0.0001	0.009	0.02	1

High-Quality Accuracy Rate				
Attribute	% of Ligand in Interface	Native Complex Close Contacts	Ligand Side-chain Change	Complex Shape
% of Ligand in Interface	1	–	–	–
Native Complex Close Contacts	0.05	1	–	–
Ligand Side-chain Change	0.03	0.006	1	–
Complex Shape	0.28	0.01	0.02	1

Complex shape is the final attribute of the highest correlated regression model for high-quality accuracy rate. Complex shape is measured using the radius of gyration of the bound receptor and ligand. In this regression model, ZDOCK performance tends to increase with elongated complex shapes. The most commonly elongated complex shapes in the data set are the Antibody/Antigen cases, and removing these from the regression model reduces the coefficient of partial determination for this characteristic by more than half (0.14–0.06). The diminishing importance of complex shape when antibody–antigens are excluded suggests a relationship between ZDOCK's high-quality accuracy rate and whether or not the complex is an antibody–antigen. Antibody–antigen complexes are known to be high-affinity binders and perhaps ZDOCK's performance correlates well with binding affinity as such complexes would require very low energy conformations that simple scoring functions such as ZDOCK's could find with greater success. Unfortunately, accurate binding affinity data for each complex in the data set are not available to proceed further with such an analysis.

The coefficients of partial determination for the high-quality accuracy rate regression model for four predictors show more balance in the importance of the attributes than in the medium-quality accuracy rate model (*see* **Table 6**) Ligand interface size relative to ligand size and number of native complex clashes contribute almost equally to the reduction of regression error in the variation with coefficients of partial determination of 0.24 and 0.21, respectively. Ligand side-chain movement and complex shape were slightly less important with coefficients of 0.16 and 0.14, respectively.

3.6. Regression Analysis Conclusion

The relationships between complex characteristics and high-quality performance and medium-quality performance for ZDOCK are clearly similar especially with shape complementarity, side-chain conformational change, and the ratio of ligand interface size to ligand size. However, the difference in the two types of performance seems to be in how much each attribute contributes relative to the others. Shape complementarity, in the form of interface curvature, is ZDOCK's dominating discriminating force in medium-quality predictions. Yet, for high-quality predictions, it is clearly not as important and more attributes are equally as necessary. Understanding the differences in how ZDOCK performs with varying levels of prediction quality could allow for a future strategy of tweaking the parameters of the scoring function to fit a user's goals depending on what level of precision they require. Given the results of the regression analysis, it may be possible to target improvements to ZDOCK that would sacrifice high-quality performance for an increased amount of medium-quality predictions. Conversely, if only high-quality predictions are required, the quantity of medium level predictions could be sacrificed for a small amount of high-quality predictions.

Regression analysis is a good tool for finding the underlying relationships between characteristics of protein–protein complexes and ZDOCK performance. With this knowledge, it is possible to get a better idea of when and why ZDOCK makes successful predictions. Through this analysis, the shared importance of shape complementarity, side-chain conformational change, and interface size in ZDOCK's ability to predict high- and medium-quality protein complex structures is readily apparent.

In addition, understanding the relationships between each attribute in a comprehensive characterization of protein–protein complexes and how ZDOCK performs gives insight into where best to make future improvements to the algorithm. Advancements in side-chain search and an approach for scoring only the buried interface atoms in the desolvation energy calculations are some possible avenues of pursuit for further ZDOCK development.

Acknowledgments

We are grateful to the Scientific Computing Facilities at Boston University and the Advanced Biomedical Computing Center at NCI, NIH for support in computing. This work was funded by NSF grants DBI-0133834 and DBI-0116574.

References

1. Betts, M.J. and M.J. Sternberg. An analysis of conformational changes on protein-protein association: implications for predictive docking. *Protein Eng*, 1999, **12**(4): p. 271–83.
2. Katchalski-Katzir, E., et al. Molecular surface recognition: determination of geometric fit between proteins and their ligands by correlation techniques. *Proc Natl Acad Sci USA*, 1992, **89**(6): p. 2195–9.
3. Chen, R. and Z. Weng. Docking unbound proteins using shape complementarity, desolvation, and electrostatics. *Proteins*, 2002, **47**(3): p. 281–94.
4. Gabb, H.A., R.M. Jackson, and M.J. Sternberg. Modelling protein docking using shape complementarity, electrostatics and biochemical information. *J Mol Biol*, 1997, **272**(1): p. 106–20.
5. Vakser, I.A. Protein docking for low-resolution structures. *Protein Eng*, 1995, **8**(4): p. 371–7.
6. Ritchie, D.W. and G.J. Kemp. Protein docking using spherical polar Fourier correlations. *Proteins*, 2000, **39**(2): p. 178–94.
7. Palma, P.N., et al. BiGGER: a new (soft) docking algorithm for predicting protein interactions. *Proteins*, 2000, **39**(4): p. 372–84.
8. Abagyan, R., M. Totrov, and D. Kuznetsov. ICM – a new method for protein modeling and design – applications to docking and structure prediction from the distorted native conformation. *J Comput Chem*, 1994, **15**(5): p. 488–506.
9. Gray, J.J., et al. Protein-protein docking with simultaneous optimization of rigid-body displacement and side-chain conformations. *J Mol Biol*, 2003, **331**(1): p. 281–99.
10. Gardiner, E.J., P. Willett, and P.J. Artymiuk. Protein docking using a genetic algorithm. *Proteins*, 2001, **44**(1): p. 44–56.
11. Fischer, D., et al. A geometry-based suite of molecular docking processes. *J Mol Biol*, 1995, **248**(2): p. 459–77.
12. Morris, G.M., et al. Automated docking using a Lamarckian genetic algorithm and an empirical binding free energy function. *J Comput Chem*, 1998, **19**(14): p. 1639–62.

13. Comeau, S.R., et al. ClusPro: an automated docking and discrimination method for the prediction of protein complexes. *Bioinformatics*, 2004, **20**(1): p. 45–50.
14. Kuntz, I.D., et al. A geometric approach to macromolecule-ligand interactions. *J Mol Biol*, 1982, **161**(2): p. 269–88.
15. Mandell, J.G., et al. Protein docking using continuum electrostatics and geometric fit. *Protein Eng*, 2001, **14**(2): p. 105–13.
16. Dominguez, C., R. Boelens, and A.M. Bonvin. HADDOCK: a protein-protein docking approach based on biochemical or biophysical information. *J Am Chem Soc*, 2003, **125**(7): p. 1731–7.
17. Schneidman-Duhovny, D., et al. PatchDock and SymmDock: servers for rigid and symmetric docking. *Nucleic Acids Res*, 2005, **33**(Web Server issue): p. W363–7.
18. Chen, R., L. Li, and Z. Weng. ZDOCK: an initial-stage protein-docking algorithm. *Proteins*, 2003, **52**(1): p. 80–7.
19. Janin, J., et al. CAPRI: a critical assessment of predicted interactions. *Proteins*, 2003, **52**(1): p. 2–9.
20. Moult, J., et al. Critical assessment of methods of protein structure prediction (CASP) –round 6. *Proteins*, 2005, **61 Suppl 7**: p. 3–7.
21. Vajda, S. Classification of protein complexes based on docking difficulty. *Proteins*, 2005, **60**(2): p. 176–80.
22. Delano, W.L. *The PyMOL Molecular Graphics System*, 2002.
23. Chen, R., et al. A protein-protein docking benchmark. *Proteins*, 2003, **52**(1): p. 88–91.
24. Kozakov, D., et al. Optimal clustering for detecting near-native conformations in protein docking. *Biophys J*, 2005, **89**(2): p. 867–75.
25. Duan, Y., B.V. Reddy, and Y.N. Kaznessis. Physicochemical and residue conservation calculations to improve the ranking of protein-protein docking solutions. *Protein Sci*, 2005, **14**(2): p. 316–28.
26. Tovchigrechko, A. and I.A. Vakser. Development and testing of an automated approach to protein docking. *Proteins*, 2005, **60**(2): p. 296–301.
27. Mintseris, J., et al. Protein-Protein Docking Benchmark 2.0: an update. *Proteins*, 2005, **60**(2): p. 214–6.
28. Murzin, A.G., et al. SCOP: a structural classification of proteins database for the investigation of sequences and structures. *J Mol Biol*, 1995, **247**(4): p. 536–40.
29. Berman, H.M., et al. The Protein Data Bank. *Nucleic Acids Res*, 2000, **28**(1): p. 235–42.
30. Zhang, C., et al. Determination of atomic desolvation energies from the structures of crystallized proteins. *J Mol Biol*, 1997, **267**(3): p. 707–26.
31. Chen, R. and Z. Weng. A novel shape complementarity scoring function for protein-protein docking. *Proteins*, 2003, **51**(3): p. 397–408.
32. Brooks, B.R., et al. CHARMM: a program for macromolecular energy, minimization, and dynamics calculations. *J Comput Chem*, 1983, **4**: p. 187–217.

33. Li, L., R. Chen, and Z. Weng. RDOCK: refinement of rigid-body protein docking predictions. *Proteins*, 2003, **53**(3): p. 693–707.
34. Pierce, B., W. Tong, and Z. Weng. M-ZDOCK: a grid-based approach for Cn symmetric multimer docking. *Bioinformatics*, 2005, **21**(8): p. 1472–8.
35. Laskowski, R.A. SURFNET: a program for visualizing molecular surfaces, cavities, and intermolecular interactions. *J Mol Graph Model*, 1995, **13**(5): p. 323–30, 307–8.
36. Mintseris, J. and Z. Weng. Optimizing protein representations with information theory. *Genome Inform Ser Workshop Genome Inform*, 2004, **15**(1): p. 160–9.
37. Dunbrack, R.L., Jr. and M. Karplus. Backbone-dependent rotamer library for proteins. Application to side-chain prediction. *J Mol Biol*, 1993, **230**(2): p. 543–74.

12

Molecular Dynamics Simulations of Protein Folding

Angel E. Garcia

Summary

I illustrate the use of the replica exchange molecular dynamics (REMD) algorithm to study the folding of a small (57 amino acids) protein that folds into a three-helix bundle, protein A. The REMD is a trivially parallel method that uses multiple copies of the system of interest to study the canonical ensemble equilibrium properties. Each replica represents a different thermodynamic state, usually at different temperatures. This method enhances the configurational sampling of proteins and allows us to study folding in simulations that are much shorter than the folding timescale for the system at ambient temperature. I show that using REMD and the Amber force field, I can obtain stable configurations of protein A whose backbone root mean square distance (RMSD) is within 0.17 nm of the nuclear magnetic resonance (NMR)-determined structure without biasing the system toward the folded structure. The simulations are done in explicit solvent and starting from nearly extended configurations. This calculation shows that currently available force fields and enhanced sampling methods perform reasonably well in describing the folded structure of small proteins.

Key Words: Molecular dynamics; thermodynamics; folding; hydration; enhanced sampling methods; replica exchange.

1. Introduction

The use of all-atom molecular dynamics (MD) simulations to study protein folding has been limited by the accuracy of the force fields and the inability of current simulation methods and computers to sample the large configurational space of proteins. These two problems are interrelated because the inability to sample configurational space limits the development and validation of force

From: *Methods in Molecular Biology, vol. 413: Protein Structure Prediction, Second Edition*
Edited by: M. Zaki and C. Bystroff © Humana Press Inc., Totowa, NJ

fields. Much progress has been made over the last few years on both aspects of this problem. The use of enhanced sampling methods has enabled the study of the folding/unfolding kinetics and thermodynamics of small proteins, mini proteins, and peptides. Two main methods have emerged as the most powerful: replica exchange MD (REMD) ***(1–3)*** and replica dynamics (RD) and its variations ***(4–6)***. REMD is best suited for studying protein thermodynamics, whereas RD is more suitable for studying kinetics (dynamics). Both methods rely on the use of multiple computer processors and the simulation of multiple copies of the system of interest. REMD, also known as parallel tempering, simulates multiple copies of the system, where each copy samples a different temperature, and configurations at different temperatures are exchanges in a Monte Carlo move. Variations of this method have generated different states of the system with different force fields, volumes, or perturbation potentials ***(7–9)***. This method is best suited for studying the thermodynamics of the folding/unfolding transition. Given that the force fields have not been fully calibrated and validated, it might be possible that structural transitions will occur at higher or lower temperatures than observed experimentally. If one is interested in the properties of the system at one T (or over a narrow range of T's), then REMD may not be the method of choice. REMD has been shown to produce the thermodynamics of protein folding for systems as large 46 amino acids, as well as for peptides and mini proteins. The kinetics of the folding/unfolding transition can be obtained from REMD simulations but not in a straightforward manner. The REMD method parallelizes trivially, because replicas need to communicate only their current temperature and energy during exchange attempts. Typical calculations using REMD have used eight to 82 replicas.

In the RD method, as described by Voter, multiple copies of the system are simulated with all copies starting from an identical configuration (or energy basin), but different velocities. In the simulation, there is a dephasing period where the systems are thermalized without escaping the basin. During the production period, the system is monitored for transitions from one energy basin to another. Once a transition is detected, all copies are re-started from this configuration until another transition is detected. This method accelerates the dynamics of the system proportionally to the number of replicas simulated. The use of this method for heterogeneous protein-solvent systems has proven difficult. However, alternatives to the method, which rely on assumptions about the kinetics of the system, have been implemented with remarkable success. RD simulations have been able to reproduce the experimentally observed timescales for folding of peptides, proteins, mini proteins, and RNA oligomers. This method parallelizes trivially, given that little or no communication is needed between replicas. The RD method has been used with tens of thousands of replicas using the Folding@home distributed computing platform at Stanford ***(5–6)***.

In this chapter, I will describe in detail how REMD is implemented in a biomolecular system. I am particularly interested in the simulation of folding of proteins and peptides, using models that include the solvent explicitly. The REMD method has also been used with implicit solvent models ***(10)***, as well as with knowledge-based energy functions ***(11)***.

2. Molecular Dynamics Simulations

MD simulations of biomolecular systems typically use a semi-empirical force field that has been parameterized with quantum mechanical and solvation free energies calculations of model solutes ***(12–14)***. Descriptions of force fields can be found in the literature ***(12,15)***. The potential energy function commonly used is approximated using the following equation:

$$U(X) = \sum_{bonds} k_b (l - l_0)^2 + \sum_{bondangles} k_{theta} (\theta - \theta_0)^2 + \sum_{dihedrals} \sum_{n} V_n [1 + \cos(n\phi + \gamma)] + \sum_{(i<j)} 4\varepsilon_{ij} \left[\left(\frac{\sigma_{ij}}{r_{ij}}\right)^{12} - \left(\frac{\sigma_{ij}}{r_{ij}}\right)^{6} \right] + \sum_{(i<j)} \frac{q_i q_j}{r_{ij}}, \quad (1)$$

where there are bonding energy terms (bond stretching, bond bending, and rotations around bonds), and non-bonding terms (van der Waals and Coulomb interactions). Here, X denotes the configuration of the system, l_0 the equilibrium bond lengths, θ_0 the equilibrium bond angles, γ the dihedral angle phases for each Fourier component n, q_i the partial charge of atom i, and σ_{ij} the van der Waals contact distance for atoms i and j. This parameterization of the energy does not include bond or atomic polarization effects, as the values of the atoms' partial charges are fixed. This force field, although classical in nature and the result of many approximations, is already quite complicated to solve numerically by simulation. The inclusion of additional terms (polarizability, bond formation and breaking, non-linear bond bending and stretching, mixed bending and stretching terms, and angular dependent hydrogen bonding, to name a few) will significantly limit the ability to perform simulations. I prefer to simulate systems with this kind of force field and add additional terms when proven necessary, rather than trying to include all effects at once and being unable to solve a simple folding problem.

To simulate the folding/unfolding of a small protein or peptide, I prefer to simulate the system in explicit solvent (water) and under periodic boundary conditions, at constant temperature, volume, or pressure, and with a fixed number of particles in the system. These ensembles are called constant NTP or NTV ensembles. The use of periodic boundary conditions and treating Coulomb interactions with Ewald sums are the simplest choices that avoid non-physical behavior due to cutoffs, boundaries, etc. To perform simulations at constant T and P, I must artificially couple the system to a heat bath, because MD conserves energy and naturally simulates a constant NVE ensemble. In our calculations, I use the Nose–Hoover temperature-coupling ***(16,17)*** and the Parinello–Rahman pressure-coupling algorithms.

A small biomolecular system of M atoms is solvated by approximately $N_w = 3M$ to $10M$ water molecules. Typical sizes for solvated systems are from 2k to 20k atoms. By including the solvent explicitly, one can better account for hydrophobic interactions and for the balance between the gain in system entropy that drives the spontaneous folding and the reduction in configurational entropy of the folded chain. The complexity of the calculation scales linearly with M for the bonded energy terms. The non-bonded energy terms scale as $N \ln N$ for Coulomb interactions, if approximate methods for Ewald sums are used ***(18)***, and order N for van der Waals interactions when a cutoff is used. The energy and force evaluations needed for performing an MD calculation are time consuming. For example, an approximately 60 amino acid protein (900 atoms) in water (6000 water molecules) takes 22 h per nanosecond of simulation on a single processor. Considering that the folding time of this protein is in the microsecond timescale, it will take more than 10^4 CPU days to compute folding trajectories.

2.1. The Replica Exchange Algorithm

Let us assume that I want to simulate the folding thermodynamics of a biomolecular system with energy function $U(X)$, described in **Eq. 1**, where X represents the configuration of the system. The replica exchange (RE) is an extension of the Metropolis Monte Carlo (MC) simulation. The main idea is to simulate multiple copies (replicas) of the system, with each replica sampling a different thermodynamic state. For simplicity, I will assume that I am sampling different temperatures and that the replicas sample the same energy function and have identical composition and volume. All these conditions can be relaxed when needed. In the RE, one can perform two kinds of moves sequentially:

1. changes in configurations by MC or MD, and
2. MC exchanges between configurations at different temperatures.

In what follows, I assume that I am simulating R replicas of the system, with a distribution of target temperatures (T_1, ..., T_i, T_j, ..., T_R), coordinates represented by $X = (x_1, \ldots, x_m, x_n, \ldots, x_R)$, and energies ($E_1$, ..., E_i, E_j, ..., E_R). At fixed time intervals, systems labeled i and j, with temperatures T_i and T_j, respectively, can exchange temperatures, such that system i changes to temperature T_j and system j changes to temperature T_i. The transition probability for exchanges must satisfy detailed balance,

$$W(X)\, w(X, X') = W(X')\, w(X', X)\,, \tag{2}$$

where $W(X)$ is the weighting factor for the state X and $w(X, X')$ is the transition probability of exchanging system X by system X'. $W(X)$ is given by the product of the Boltzmann factors for each of the R replicas,

$$W(X) = \exp\left(-\sum_{i=1}^{R} \beta_i E_i\right), \tag{3}$$

with $\beta_i = 1/RT_i$, $E_i = E_{\text{kin}\,i} + U_i$, where $E_{\text{kin}\,i}$ is the kinetic energy and U_i is the potential energy of the corresponding replica. This gives

$$\frac{w(X, X')}{w(X', X)} = \exp(-\Delta)\,,$$

where

$$\Delta = (\beta_j - \beta_i)(E_i - E_j)\,. \tag{4}$$

These transition probabilities between configuration X and X' are implemented using the Metropolis criterion,

$$W(X, X') = \min(1, \exp(-\Delta))\,. \tag{5}$$

In MD simulations, the kinetic energy is related to the temperature of the system, and therefore, when exchanging configurations, their E_{kin} must be also changed. The approach first introduced by Sugita and Okamoto ***(1)*** is to scale the particles momenta uniformly by $\sqrt{T_i/T_j}$, such that the kinetic energy terms in the Boltzmann factor cancel out and then the exchange transition probability for REMD has the same form as in the MC RE calculations,

$$\Delta = (\beta_j - \beta_i)(U_i - U_j) = \Delta\beta\Delta U, \tag{6}$$

where $\Delta\beta = \beta_j - \beta_i$ and $\Delta U = U_i - U_j$. In a parallel implementation of this method, processors communicate only when exchanges are attempted. Because the number of time integration steps between exchange attempts is much greater than unity, the communication requirements of this method are minimal, resulting in near-linear scaling of speedup with processor number.

In what follows, I will discuss practical issues about the implementation of REMD. Two parameters that must be determined to apply REMD are the temperature distribution ($T_1, \ldots T_j, \ldots, T_R$) and the number of replicas, R, to include in the simulation. These two issues are related, because the number of replicas is going to be determined using the range of temperatures one wishes to cover in the simulation. For large systems, the energy changes with temperature are large. Therefore, only replicas at similar temperatures can exchange, although exchanges among all pairs can be attempted. The temperatures are usually distributed exponentially, but changes between neighboring temperatures are chosen such that a reasonable exchange rate (10–25%) is obtained. The exchange rate and temperatures must be chosen such that one can ensure that all replicas span the whole range of temperatures. In some instances, one may select two neighboring temperatures that are too far apart, and, as a consequence, the replicas will remain within a limited range of temperature values, thus limiting the efficiency of the method. The exchange rates can be monitored along the equilibration stage of the simulation, and temperature differences between neighbors can be increased or reduced if the exchange rates are too large, or slow, respectively.

The number of replicas, R, depends on the temperature range that the simulations will cover and the size (number of atoms) of the system *(7)*. The enhancement in sampling is more efficient when the temperatures are distributed over a broad range that covers and exceeds the temperatures in which the system exhibits conformational changes of interest (e.g., folding/unfolding; helix/coil transitions). At the highest temperatures, the system should have very short correlation times and fast conformational changes. In a card game, the highest temperature runs correspond to the shuffling of the cards. The high temperature runs will allow the system to lose memory of the current state, jump over energy barriers, and sample new regions of space. The low temperature replicas will tend to occupy the lowest energy regions of space sampled. They will heat up with thermal fluctuations, or when other replicas find a lower energy state. In the last case, the replica that finds the lower energy basins will cool down. This is similar to a self-regulated simulated annealing schedule.

The assignment of temperatures for different replicas can be obtained from the system energy distribution as a function of temperature. The acceptance of an exchange attempt depends on the potential energy of the two configurations and the temperature of the replicas. The probability of having a configuration with energy E at temperature T is denoted by $P(E, T)$. The exchange acceptance rate will be given by

$$R_{\text{acc}}(T_1, T_2) = \int_{-\infty}^{\infty} dx\, P(x, T_1) \left[\int_{-\infty}^{x} dy\, P(y, T_2) + \int_{x}^{\infty} dy\, P(y, T_2) \exp(\Delta\beta(x-y)) \right], \tag{7}$$

where $T_1 < T_2$ are the temperature of the two replicas being exchanged. The first integral comes from the acceptance of exchanges when $y \leq x$; that is, when the T_2 replica has lower or equal potential energy than the T_1 replica (the exchange is always accepted). The second integral comes from the acceptance of exchanges when $x > y$. A good approximation for the potential energy distribution is a Gaussian distribution that is determined by the first two moments of the potential energy distribution: the average $< E(T) >$ and the variance $\sigma = \sigma_E(T)^2 = \langle (E(T) - < E(T) >)^2 \rangle$. From these two moments, one can write the histogram of observed energies as

$$P(E, T| < E(T) >, \sigma(T)) = \frac{1}{\sqrt{2\pi}\sigma(T)} \exp\left(\frac{-(E - < E(T) >)^2}{2\sigma^2(T)} \right). \tag{8}$$

Within this approximation, the acceptance rate will be given by

$$\begin{aligned} R_{\text{acc}}(T_1, T_2) = {} & \frac{1}{2}\left[1 + \operatorname{erf}\left(\frac{< E_2 > - < E_1 >}{\sqrt{2\sigma_1^2 + 2\sigma_2^2}}\right)\right] \\ & + \frac{1}{2}\left[\exp\left(\Delta\beta(< E_2 > - < E_1 >) + \left(\frac{\Delta\beta}{2}\right)^2 (2\sigma_1^2 + 2\sigma_2^2)\right)\right. \\ & \left. \operatorname{erfc}\left(\frac{\Delta\beta(\sigma_1^2 + \sigma_2^2) + < E_2 > - < E_1 >}{\sqrt{2\sigma_1^2 + 2\sigma_2^2}}\right)\right]. \end{aligned} \tag{9}$$

In this equation, $< E > = < E(T) >$ and $\sigma = \sigma(T)$. Using this formula, and given $< E(T) >$ and $\sigma(T)$ as a function of T, one can iteratively solve for T_{i+1} given T_i, for a given acceptance rate. **Figure 1** shows the histogram distribution, the average energy, and variance of the energy as a function T

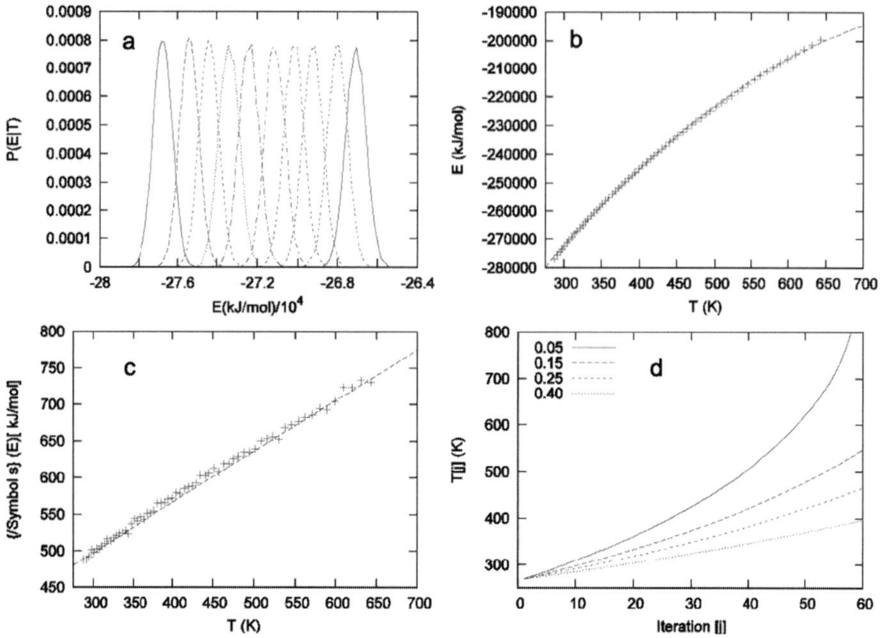

Fig. 1. **(A)** Potential energy histograms sampled at various T's. **(B)** First ($< E >$) and **(C)** second (σ) moments of the potential energy as a function of T for the 57 amino acid protein A in water. **(D)** Optimal replica temperatures obtained with **Eq. 9** for exchange acceptance rates of 0.05, 0.15, 0.25, and 0.40.

for a broad range of T obtained for protein A in water. In practice, one will perform short (1 ns) simulations of the system at constant NTV over a few temperatures covering a broad range of temperatures and fit the averages of $< E(T) >$ and $\sigma(T)$ as a function T. I fit $\overline{E}$ to a quadratic function and $\sigma(T)$ to a linear function of T. **Figure 1A** shows the potential energy histograms at various T's, with T difference between histograms of 3–4 K. The figure also shows the average energy and standard deviation of the potential energy as a function of T. **Figure 1D** shows the curves of T_j as a function of the iteration index j for various acceptance rates ($R_{acc} = 0.15, 0.25, 0.40$). The higher the chosen acceptance rate, the larger the number of replicas needed to span the same T range. For lower exchange acceptance rates, one need fewer replicas, but at some limit, the system will behave as independent runs and will not benefit from the RE algorithm. The exchange rate can be chosen to be different for different T regions. One can chose low exchange rate for low T and high exchange rate for high T and obtain longer constant temperature segments for low temperature replicas, which could be used to analyze the dynamics of the system.

2.2. Application of REMD to Fold a Three-Helix Bundle Protein

I simulated the folding of amino acids 1–57 of protein A. Protein A folds into a three-helix bundle—probably the simplest protein fold. Protein A is a model protein for determining the protein-folding kinetics. The structure of protein A has been determined by nuclear magnetic resonance (NMR), and amino acids 11–56 are well ordered ***(19)***. Amino acids 1–11 are disordered and their structure cannot be resolved. Previous calculations on protein A have simulated the fragment 11–56. However, experiments on protein A have shown that the fragment without amino acids 1–11 is not stable in solution ***(20)***. Extensive protein engineering and ϕ-value analysis have been performed in protein A ***(21)***. These experiments show that folding occurs predominantly through an ensemble where helix II forms first, followed by helix III and helix I. Previous calculations on folding ***(22,23)*** and unfolding ***(24)*** are not in agreement with each other. Here I report ab initio folding of protein A from an extended conformation. Previous all-atom, explicit solvent, simulations of protein folding have been done on smaller peptides. A detailed analysis of the folding mechanism is outside the scope of this chapter and will be presented elsewhere.

Here, I show recent results of a simulation of protein A (amino acids 1–57) using the AMBER94 force field ***(12)*** and 6583 TIP3P ***(25)*** water molecules as a solvent. The simulation is starting from an artificially prepared compact coil structure. This structure is obtained by starting from an extended conformation and performing a high T (400 K) MD simulation in vacuum. The resulting partially collapsed structure has a (CA C N O) Root Mean Square Distance (RMSD) of 0.77 nm from the NMR structure ***(19)***. This collapsed structure lacks any regular secondary structure elements and serves as the starting configuration for the solvated system. A cubic box, 6.72 nm on the side, of TIP3P water molecules is equilibrated at 300 K and 1 atmosphere (atm) for 5 ps. This water box is used to solvate the partially collapsed structure described above by putting the protein at the center of the box and deleting all water molecules within 0.28 nm from the protein. The resulting system has 6583 water molecules and 896 protein atoms, for a total of 20,645 atoms. This system is partially equilibrated for 10 ps at 330 K and 1 atm. The resulting system is further equilibrated at 300 K and 1 atm for 2 ns. The equilibrated system is contained in a cube with 5.97 nm on the side. The final configuration of this system is used as the initial configuration for 52 replicas simulated at constant volume over a temperature range of 275–630 K. These replicas are simulated for 5 ns. From these simulations, I calculate the first and second moments of the potential energy of the system as a function of temperature. These moments are used to determine the temperatures and number of replicas to be studied in the production run. For the production part of the simulation, I use 64 replicas over a temperature range of 287–643 K. The temperature spacing between each of the replicas is chosen such that the energy distributions overlap sufficiently and state exchange attempts are (on average) accepted with a 25% probability. **Figure 1A** shows that the Gaussian approximation for the energy histograms is accurate when one use a quadratic in T interpolation for the first moment and a linear interpolation for the second moment of the potential energy. The temperature assignment for the replicas obtained using this Gaussian approximation and **Eq. 9** is shown in **Fig. 1D**.

All simulations in explicit solvent are conducted with the following parameters. The electrostatic interactions are treated with the smooth particle mesh Ewald summation *(18)* with a real space cutoff of 0.9 nm and a $50 \times 50 \times 50$ mesh with fourth order interpolation for the reciprocal lattice contribution. The Ewald convergence factor is set to $3.47\,\text{nm}^{-1}$. Corrections for the Lennard Jones due to cutoff are taken into account in the pressure and energy calculations. The time-step used in the MD steps is 2 fs, and a Nose–Hoover *(16,17)* thermostat is used with a time coupling of 0.5 ps. For bonds within the protein, constraints are applied using SHAKE *(26)*, and the water constraints are solved using SETTLE *(27)*. Simulations are carried out using GROMACS 3.2 modified by us to perform REMD using the Amber force fields. The simulations use the AMBER94 force field *(12)* without any modifications. The REMD simulation is conducted for 30 ns. Using our GROMACS implementation, the simulations, take about 22 h per nanosecond when distributed over 64 processors of our Opteron 2.2 GHz Linux-Cluster.

Figure 2A shows a time-evolution of the CA-C-N-O-backbone RMSD with respect to the NMR structure for one of the replicas. **Figure 2B** shows the time series for the temperature of this replica. Note that as the system folds, the T decreases and then fluctuates around 325 K. After 12 ns, this replica samples conformations that are within 0.2 nm RMSD from the NMR structure. The best fit between a sample configuration and the NMR structure is 0.17 nm. **Figure 3** shows a set of sampled conformations along the folding trajectory. Note that the RMSD does not follow a monotonic decrease toward the folded state but follows a path where the structure is as close as 0.5 nm and then increases to 1.2 nm before falling into the folding basin. During the first 30 ns of REMD,

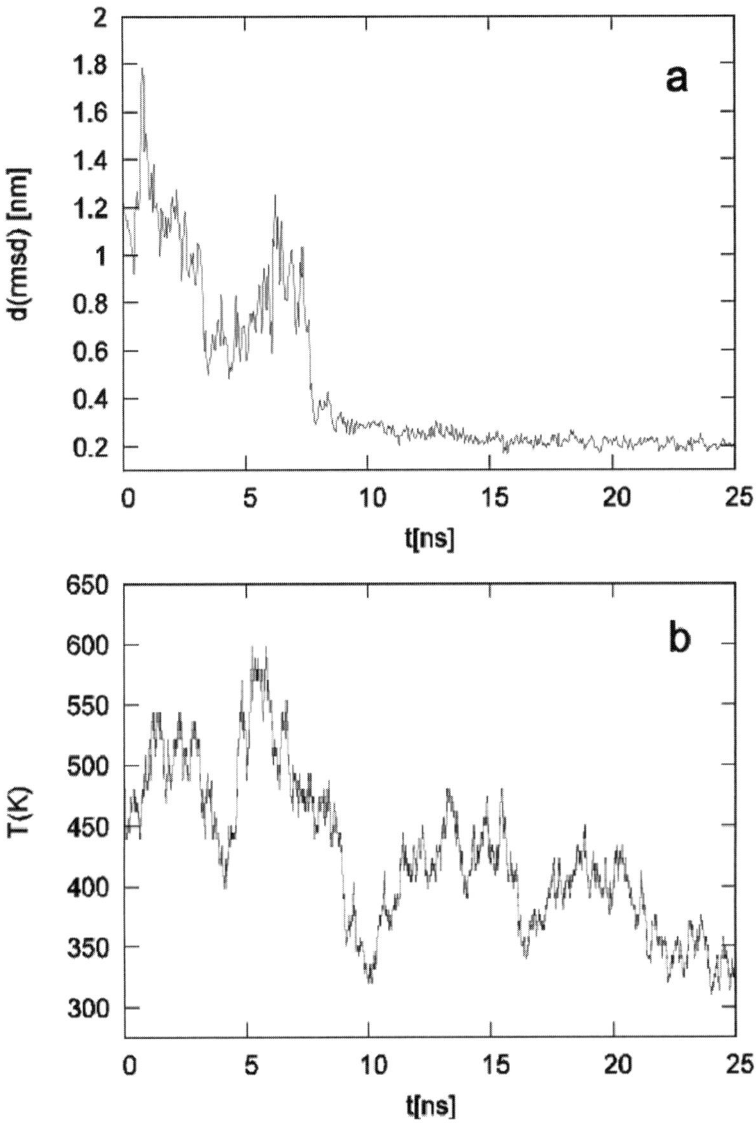

Fig. 2. (**A**) Backbone RMSD, in nm, as a function of time (in nanoseconds) between one of the simulated replicas and the nuclear magnetic resonance (NMR) structure ***(19)***. (**B**) Temperature (in K) of the same replica as a function of time. The plot shows that the protein reaches the folded state basin at high T, and then T gradually decreases after the RMSD has reached below 0.2 nm. This figure illustrates that the force field used in the simulation adequately samples the folded state as a low free energy basin.

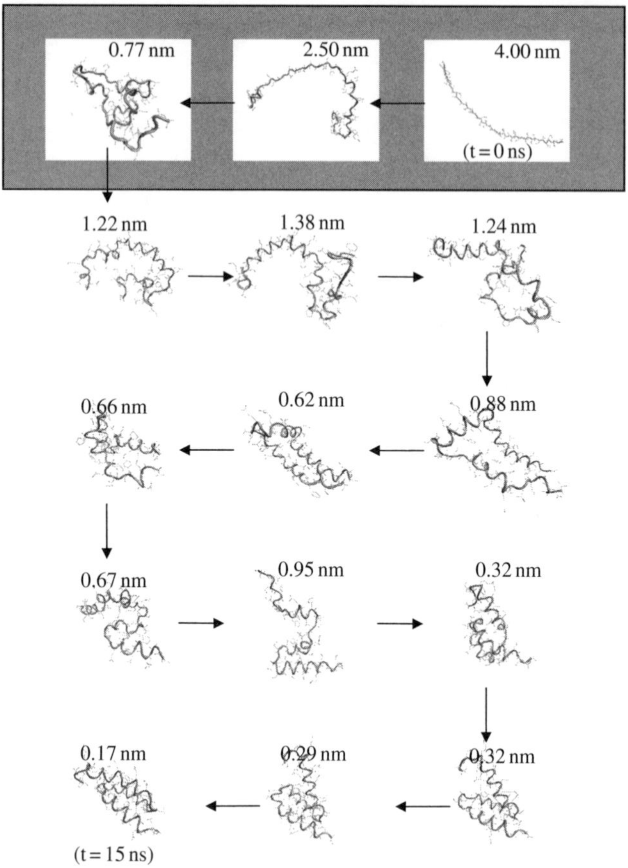

Fig. 3. Illustration of the folding trajectory of one replica. Simulations start from a nearly extended configuration of the protein (RMSD 4.0 nm). The first two nanoseconds of simulation (conformations shaded in gray background) were performed in the absence of water at high temperature (400 K). The other conformations were sampled during the first 15 ns of replica exchange molecular dynamics (REMD) in the solvated system. The backbone conformation closest to the experimental folded state is at an RMSD of 0.17 nm. The simulation has been extended for 26 ns. See text for details about the REMD simulation.

I observed eight folding events (i.e., RMSD <0.4 nm). Extrapolation of this trend indicates that REMD simulation exceeding 400 ns/replica will be needed to see 95% of the replicas fold at any one time and to have sufficient statistics to perform thermodynamic analysis of the folding/unfolding transition.

Fig. 4. Superposition of the best fit between the simulated structure (gray) and the nuclear magnetic resonance (NMR) structure (black) for protein A. RMSDs are calculated for backbone atoms (N, CA, C, and O) for amino acids 11–55, which are ordered in the NMR structure. The main differences in the backbone are in amino acids 1–11, which are formed in the replica exchange molecular dynamics (REMD) calculation and disordered in the NMR structure, and in the turn I between helices I and II.

3. Conclusions

I have shown that all-atom MD simulations of protein folding are possible, although these calculations are extremely demanding, computationally. The force fields that have been developed over many years give a description of protein structure and dynamics which is reasonably accurate. REMD simulations produce structures that fluctuate around 0.17 nm from the NMR-determined folded structure near room temperature. The superposition of the NMR and the simulated structures is shown in **Figure 4**. Modifications of the force fields that reproduce experimental data accurately are being developed by many groups. As faster computers and computational algorithms are being developed, protein folding from first principles is becoming possible. Currently, all-atom simulations are not at the stage of knowledge-based potentials in predicting protein structures, but progress is being made toward this goal. In addition to describing protein folding and dynamics under standard conditions, all-atom simulations are useful to study protein systems under varying solvent, pressure, and temperature conditions.

Acknowledgments

The author gratefully acknowledges support by the National Science Foundation MCB-0543769. The author thanks D. Paschek, H. Herce, and R. Day for valuable discussions.

References

1. Y. Sugita and Y. Okamoto. Replica-exchange molecular dynamics methods for protein folding. *Chem. Phys. Lett.*, 314:141–151, 1999.
2. U.H.E. Hansmann. Parallel tempering algorithm for conformational studies of biological molecules. *Chem. Phys. Lett.*, 281:140–150, 1997.
3. K. Hukushima and K. Nemoto. Exchange Monte Carlo method and application to spin glass simulation. *J. Phys. Soc. Japan*, 65:1604–1608, 1996.
4. A.F. Voter. Parallel replica method for dynamics of infrequent events. *Phys. Rev. B*, 57:R13985–R13988, 1998.
5. M.R. Shirts and V.S. Pande. Mathematical analysis of coupled parallel simulation. *Phys. Rev. Lett.*, 86:4983–4987, 2001.
6. C.D. Snow, E.J. Sorin, Y.M. Rhee, and V.S. Pande. How well can simulation predict protein folding kinetics and thermodynamics? *Annu. Rev. Biophys. Biomol. Struct.*, 34:43–69, 2005.
7. H. Nymeyer, S. Gnanakaran, and A.E. Garcia. Atomic simulations of protein folding, using the replica exchange algorithm. *Methods in Enzymol.*, 383:119–149, 2004.

8. D. Paschek and A.E. Garcia. Reversible temperature and pressure denaturation of a protein fragment: a replica exchange molecular dynamics simulation study. *Phys. Rev. Lett.*, 93:238105, 2004.
9. D. Paschek, S. Gnanakaran, and A.E. Garcia. Simulations of the pressure and temperature unfolding of an alpha-helical peptide. *Proc. Natl. Acad. Sci. USA*, 102:6765–6770, 2005.
10. H. Nymeyer and A.E. Garcia. Simulation of the folding equilibrium of alpha-helical peptides: a comparison of the generalized Born approximation with explicit solvent. *Proc. Natl. Acad. Sci. USA*, 100:13934–13939, 2003.
11. J. Skolnick, A. Kolinski, D. Kihara, M. Betancourt, P. Rotkiewicz, and M. Boniecki. Ab initio protein structure prediction via a combination of threading, lattice folding, clustering, and structure refinement. *Proteins*, Supp. 5:149–156, 2001.
12. W.D. Cornell, P. Cieplak, C.I. Bayly, I.R. Gouls, K.M. Merz Jr., D.M. Fergueson, D.C. Spellmeyer, T. Fox, J.W. Caldwell, and P.A. Kollman. A second generation force field for the simulation of proteins, nucleic acids and organic molecules. *J. Am. Chem. Soc.*, 117:5179–5197, 1995.
13. M.R. Shirts, J.W. Pitera, W.C. Swope, and V.S. Pande. Extremely precise free energy calculations of amino acid chain analogs: comparison of common molecular mechanics force fields for proteins. *J. Chem. Phys.*, 119(11):5740–5761, 2003.
14. E.J. Sorin, Y.M. Rhee, M.R. Shirts, and V.S. Pande. The salvation interface is a determining factor in peptide conformational preferences. *J. Mol. Biol.*, 356(1):248–256, 2006.
15. W.L. Jorgensen, D.S. Maxwell, and J. Tirado-Rives. Development and testing of the OPLS all-atom force field on conformational energetics and properties of organic liquids. *J. Am. Chem. Soc.*, 118:11225–11236, 1996.
16. S. Nosé. A unified formulation of the constant temperature molecular dynamics methods. *J. Chem. Phys.*, 81:511–519, 1984.
17. W.G. Hoover. Canonical dynamics: equilibrium phase-space distributions. *Phys. Rev. A*, 31:1695–1697, 1984.
18. U. Essman, L. Perera, M.L. Berkowitz, T.A. Darden, H. Lee, and L.G. Pedersen. A smooth particle mesh Ewald method. *J. Chem. Phys.*, 103:8577–8593, 1995.
19. H. Gouda, H. Torrigoe, A. Saito, M. Arata, and I. Shimada. Three dimensional solution structure of the b-domain of staphylococcal protein-a-comparisons of the solution and crystal-structures. *Biochemistry*, 31:9665–9672, 1992.
20. K. Witte, J. Skolnick, and C. Wong. A synthetic retrotransition (backward reading) sequence of the right-handed three-helix bundle domain (10–53) of protein a show similarity in conformation as predicted by computation. *J. Am. Chem. Soc.*, 120:13042, 1998.
21. S. Sato, T.L. Religa, V. Daggett, and A.R. Fersht. Testing protein folding simulations by experiment: B domain of protein a. *Proc. Natl. Acad. Sci. USA*, 101:6952–6956, 2004.

22. E.M. Boczko and C.L. Brooks, III. First-principles calculation of the folding free energy of a three-helix bundle protein. *Science*, 269:393–396, 1995.
23. A.E. Garcia and J.N. Onuchic. Folding a protein in a computer: an atomic description of the folding/unfolding of protein A. *Proc. Natl. Acad. Sci. USA*, 100:13898–13903, 2003.
24. D.O. Alonso and V. Daggett. Staphylococcal protein a: unfolding pathways, unfolded states, and differences between the B and E domains. *Proc. Natl. Acad. Sci. USA*, 97:133–138, 2000.
25. W.L. Jorgensen, J. Chandrasekhar, J.D. Madura, R.W. Impey, and M.L. Klein. Comparison of simple potential functions for simulating liquid water. *J. Chem. Phys.*, 79:926–935, 1983.
26. J.P. Ryckaert, G. Ciccotti, and H.J.C. Berendsen. Numerical integration of the Cartesian equations of motions of a system with constraints: molecular dynamics of n-alkanes. *J. Comput. Phys.*, 23:327–341, 1977.
27. S. Miyamoto and P.A. Kollman. SETTLE: an analytical version of the SHAKE and RATTLE algorithms for rigid water models. *J. Comput. Chem.*, 13:952–962, 1992.

Index

Abelsone tyrosine kinase, 142
Ab initio method, 23, 28, 68–69
ABLE, 26
Accessible surface area, 290
ACE, *see* Atomic contact energy
Adenyl kinase, 30
All-α and all-β proteins, 213
AMBER force field, 246–247, 323–324
  physical models for, 248
Aminoacyl-tRNA synthetase (aaRS), 136
Amyloid encephalopathy, 192
Anfinsen's thermodynamic hypothesis, 244
Antibody–antigen interactions, 283
Antibody/Antigen test, ZDOCK and, 296
Artificial evolution method, 22
ASA, *see* Accessible surface area
ASTRAL SCOP genetic domain sequence subset list, 254
Atomic contact energy, 290, 293, 308

Backbone-independent rotamer library, 24
Basic Local Alignment Search Tool, 20, 64, 70, 79, 185
Baum–Welch estimation procedure, 182, 185, 191
Bayesian probability formalism
  knowledge-based scoring functions from, 252–254
  neural network algorithms with, 258–259
Bayesian rules, 96
Bayes' theorem, 252
Bin-based construction method, 228
BLAST, *see* Basic Local Alignment Search Tool
BLOSUM62, 64, 139
Blue Gene project team, 68
BOINC software, 69
Boltzmann's principal, 10
Bond angles/length, 151
Boolean expressions, CAPRI scoring and, 287
Buckingham–Fowler potential, 250

CABS, de novo method, 252
CAFASP, *see* Critical Assessment of Fully Automated Structure Prediction
CAFASP-CONSENSUS, semiautomatic meta-server, 29–30
cAMP-dependent protein kinase, 142
CAPRI, *see* Critical Assessment of Predicted Interactions
C_α RMSD, 287, 299
CASP, *see* Critical Assessment of Structure Prediction
CATH Database, see Class, Architecture, Topology, Homology
Center-star approach, multiple sequence alignment and, 129
CHARMM program, 246–247, 293
Class, Architecture, Topology, Homology, 19, 71
Classification test, protein structure indexing and, 165–166
ClustalX program, 135
Clustering coefficients, 211
CM, *see* Comparative modeling
ColE1 repressor, 230
Collision free path, 221
Comparative modeling, 4, 20–25, 27–28, 45, 49, 52, 64–65
  usefulness of, 54–55
Complex sequence searches, 80–81
COMPOSER, 22
Conformer clustering and high-resolution refinement, decoy selection by, 268–270
Contact capacity score, RAPTOR and, 103–104
CORNET method, 206
Critical Assessment of Fully Automated Structure Prediction, 6, 26–31, 46, 74, 206
Critical Assessment of Predicted Interactions, 287
  ZDOCK/RDOCK performance and, 288

Critical Assessment of Structure Prediction, 5–6, 12–13, 16, 20, 26–31, 44, 50, 53, 62, 69, 74, 188, 206, 270
  docking algorithms performance and, 287
  experiments, 46–47, 50–52, 69
  measures, 47–48
CTSS program, 149
Cyclical search method, 24

Database-based approach, loop prediction and, 23
Dead-end elimination (DEE) algorithm, 24
Decoy filters, design of, 267–268
Decoy selection, by conformer clustering and high-resolution refinement, 268–270
Decoy sets, knowledge-based scoring functions evaluation and, 261–265
De novo protein structure prediction, scoring functions for, 243
  decoy filters, design of, 267–268
  decoy selection process, 268–270
  knowledge-based scoring functions, 251–257, 265–267
  neural network knowledge-based, 257–265
  physics-based energy functions, 246–251
Desolvation free energy, ZDOCK and, 293
DFT, *see* Discrete Fourier Transform
Dihedral angle, 221, 225
Dirichlet distribution, 234
Discrete Fourier Transform, 291
DISOPRED server, 79
Distance-dependent energy, protein threading and, 11–12
Distance matrix (DM), 199–200, 202
Distance-scaled finite ideal gas reference state (DFIRE), 12
Distributed Annotation System, 77
Docking tools, 286
Double-dynamic programing, 29, 67
DSSP, 187

EBI, *see* European Bioinformatics Institute
EM, *see* Expectation-maximization
ENCAD, 246–247
Energy functions
  class I, 247–250
  protein structures, in aqueous environment, 250–251
  protein threading and, 9–14, 95–98
e-Protein project, 77
ESTs, *see* Expansive spaces trees
Euclidean distance, 152, 199–200
Euclidean 3D transformation, 132
European Bioinformatics Institute, 79
EValuation of Automatic protein structure prediction (EVA), 44, 53
E value, of RAPTOR, 116–117
Ewald convergence factor, 324
Expansive spaces trees, 223
Expectation-maximization, 182, 189, 191
Extreme value distribution (EVD), 117

Fast and Accurate Sidechain Topology and Energy Refinement (FASTER), 24
FASTA, sequence-sequence alignment method, 20, 70, 79–80, 159
Fast Fourier Transform, 285, 290–292, 299
FFT, *see* Fast Fourier Transform
Fold and Function Assignment System (FFAS), 70–71, 80
Folding/unfolding transition, REMD simulations and, 316
Fold recognition, 5–6, 9, 12, 20, 27–28, 45, 52, 55, 65–68, 100–101
  RAPTOR and, 107–116
  SVM approach to, 111–112
Fourier component, 317
FR, *see* Fold recognition
FRAGFOLD method, 25, 69, 252
Fragment assembly based methods, 25–26, 28, 55, 83
Frozen approximation, 15, 29
FR potential, 13
FR/statistical significance, of threading alignments, 17–18
FSSP, 19, 108
FTDock, 285
FUGUE, 15, 72, 81, 94

Gap penalty, RAPTOR and, 103
Gaussian distribution, 321, 324
Gene3D, 74
Generalized-Born (GB) approximations, 251
Generalized-Born/surface-area (GBSA) models, 251
Generalized suffix tree (GST) construction, 154–156
3D-GENOMICS, 74
GenTHREADER method, 15, 71–72, 74, 81, 100
Geometric hashing, 130, 148, 161, 163
Gibbs sampling technique, 99
Global Distance Test (GDT-TS), 48, 55, 77

Global minimum energy conformation, 24
Glutathione S-Transferase family, 140
GMEC, *see* Global minimum energy conformation
gp39 mutations, 4–5
Gradient boosting algorithm, 107, 112
FR and, 113–114
GRAMM, 285
GRASP/Delphi, 250
Grid-based shape complementarity, 292–293
Grid technology, 77, 84
GROMACS, 324
GSC, *see* Grid-based shape complementarity

Hash-based index, 149
Heuristic threading programs, 14–16
HGP, *see* Human Genome Project
Hidden Markov Models, 20, 29, 80, 95, 173, 206, 252
  bioinformatics and, 174
  local structure, 189–193
  profile, 193–195
  secondary structure, 187–189
  for transmembrane helices, 175–187
Hierarchical optimization strategy, 23
High-resolution refinement and conformer clustering, decoy selection by, 268–270
HitFinder, 299
HMMs, *see* Hidden Markov Models
HMMSTR, model for local structure, 190
  application of, 192–193
  training and topological modification of, 191–192
HMMTOP model, 184–186
Homology modeling, *see* Comparative modeling (CM)
HOMSTRAD database, multiple sequence alignments and, 138–140
Hooke's law, 247–248
HSP-60 protein fragment, contact maps and, 201
Human Genome Project, 4
Hybrid fold recognition searches, 81
Hybrid methods, 71
Hydration potential, 13; *see also* Energy function, protein threading and
Hydrogen-bonding potential, 13; *see also* Energy function, protein threading and
Hydrophobic–polar (H–P) energy function, 229

IFT, *see* Inverse Fourier Transform
INBGU, 72, 81
Insulin-like growth factors, 6
Interface RMSD (iRMSD), 285, 295
Intermediate Sequence Searches (ISS), 70
Interpolated RAPDF (IRAPDF), 256
Inverse Fourier Transform, 291
Isomorphic mapping, 16

JACKAL Modeling Package, 22
3D-JIGSAW, 22, 65, 82
3DJury, 81

Knowledge-based potential energies, in protein threading, 9–11, 97

α-lactalbumin, wire skeletal model of, 6, 21
LFF profile vectors, 149
Ligand RMSD (lRMSD), 285
LikelihoodRatio, of amino acid, 134
Lindahl's benchmark, 108, 115
Linear and integer programming, RAPTOR and, 104–107
Linear integer programing (LIP) problem, 16
Livebench, 53–54
Local conformation (ψ, ϕ) potential, 13; *see also* Energy function, protein threading and
Local Feature Frequency (LFF) profile algorithm, 149
Local *versus* global contacts, 212–213
Lock-and-key mechanism, 284
Loop modeling method, 23, 92
Lysozyme, X-ray structure of, 6

MALECON method, 130
Markov-chain theory, 230, 233
Markovian State Model, 232–235
MASS method, 130
MAX SNP-hard, 98
MD/MC simulations, 220–221, 225, 232, 234
Mean first passage time, 233–234
Meta-DP server, 79
Metropolis Monte Carlo (MC) simulation, 318
MFPT, *see* Mean first passage time
MODCHECK, 83
MODELLER, homology modeling program, 22, 25, 82

Molecular dynamics (MD) simulations, of protein folding, 220
REMD application and, 316–317, 323–327
replica exchange algorithm, 318–323
Monte Carlo (MC) methods, 24, 26, 209, 220
Monte Carlo simulated annealing (MCSA), 244, 262
MQAPs, 83
MSM, *see* Markovian State Model
Multiple linear regression approach, ZDOCK performance and
high-quality predictions, 309–311
medium quality predictions, 306–309
Multiple sequence alignment, 125
flexible alignment, 128
HOMSTRAD database and, 138–140
MultiProt, 131–133
pairwise alignment cases, 135
partial alignment, 126, 136–138
sequence alignment, 128
sequence order independent structure alignment, 135–136
structural similarity and low sequence identity, 140–141
structure-derived, 133–135
subset alignment, 127–128, 136–138
time efficiency, 128–131
tyrosine kinase, loop movement in, 141–142
MultiProt, 131–133, 135, 139
MUMmer genome, 156
MUSCLE, multiple sequence alignment method, 129
MUSTA algorithm, 130
MUSTANG method, 130
M-ZDOCK, symmetric multimer docking with ZDOCK, 294

National Center for Biotechnology Information (NCBI), 79–80
NEST, 22, 25, 82
Neural network algorithms
with Bayesian probability formalism, 258–259
decoy sets, scoring function evaluation and, 261–265
local structures prediction and, 257–258
training and post-processing of, 260–261
Neural networks, 183, 187–188
New fold (NF) methods, 28, 45, 68–69
NMR, *see* Nuclear magnetic resonance
NNs, *see* Neural networks
Non-complementarity-determining region (non-CDR) loops, 295
Non-HMM methods, 183
Nose–Hoover temperature-coupling, 318, 324
Nuclear magnetic resonance, 5, 22, 192, 244, 284, 323, 325, 327

Optimal sequence alignment methods, 69
Optimized Potential for Liquid Simulations (OPLS) all-atom force field, 23

Pair-wise energy, 11
Pairwise sequence alignment methods, 70, 135
Parinello–Rahman pressure-coupling algorithms, 318
Path-directed subdivision trees, 223–224
PDB, *see* Protein Data Bank
PDSTs, *see* Path-directed subdivision trees
PISCES, 19
Poisson–Boltzmann models, 250
POSA method, 129
Position-Specific Iterative BLAST, 20, 29, 64–65, 70–72, 74, 80, 93–94, 101
3D-position-specific scoring matrix, 15, 72, 81, 94, 101
Posterior probabilities, HMMs and, 178–180
PriSM, 22
Probabilistic Roadmap Methods (PRMs), 222–223, 225
for protein-folding pathways, 225–228
Probability density function (PDF), 116
PROFcon, 206, 214
PROFESY, 26
Profile-based methods, 20–21
Profile HMMs, 193–195
Profile-profile methods, 13, 77
ProGreSS, 149, 161, 163, 164, 166–167
PROQ, 83
PROSAII, 82
PROSPECTOR, 95
PROSPECT, protein threading program, 16, 29, 81–82, 94–95, 99
ProtDex, SSE-based method, 149
Protein contact maps (CMs)
HSP-60 protein fragment and, 201
prediction, 205–207
properties of, 203–204
pros and cons of, 202
protein structures and, 199–203

reconstruction of protein structures and, 204–205
small world and, 207–215
Protein Data Bank, 5, 7, 19, 25, 30, 53, 65, 79, 91–92, 230, 262, 290
in regression analysis data set, 303
Protein domain swapping, 126
Protein-folding mechanisms, 225
Protein folding, molecular dynamics (MD) simulations of, 315–318
REMD application and, 316–317, 323–327
replica exchange algorithm, 318–323
Protein folding, roadmaps for, 219, 224
Markovian State models, 232–235
pathways, PRMs for, 225–228
Stochastic RoadMap Simulation (SRS), 229–232
Protein–protein complexes, attributes of, 304
Protein–protein docking, 283–285
benchmark for, 289–290
CAPRI experiment and, 287–289
predicted complexes, accuracy measurement of, 285–287
protein complex characters, ZDOCK performance and, 298–311
ZDOCK/RDOCK/M-ZDOCK approach, 290–298
Protein representation, roadmap methods and, 220–221
Protein secondary structure prediction, methods for, 187–189
Protein structure indexing, suffix tree and, 147, 149
classification test, 165–166
database indexing, 148
GST construction, 154–156
local feature extraction, 150–153
multiple structure alignment approaches, 148
normalization, 153–154
pair-wise structure alignment methods, 148
performance test, 166–167
PSIST, 150
querying, 156–160
retrieval test, 161–164
Protein Structure Indexing using Suffix Trees, 150, 156, 161–164, 166–167
Protein Structure Initiative, 92
Protein structure prediction, template-based
Ab initio methods, 5
assessment of, 43–55
CASP/CAFASP, 26–31
comparative/homology model building, 21–25
de novo methods, 4
fold recognition (FR) (*see* Fold recognition)
fragment assembly, 25–26
future of, 83–84
key steps for, 7
major milestones in, 8
physics-based methods, 4
sequence-based alignment methods, 19–21
threading (*see* Threading, protein structure prediction by)
Protein structures
in aqueous environment, 250–251
reconstruction, from contact maps, 204–205
small world and, 209–211
Proteomes, databases serving structural annotations for, 74–77
PROTINFO, de novo method, 252
PSI-BLAST, *see* Position-Specific Iterative BLAST
PSIPRED server, 72, 75, 94
PSIST, *see* Protein Structure Indexing using Suffix Trees
PSSM, *see* 3D-position-specific scoring matrix

RAPDF scoring function, 255–256, 262, 265
Rapidly exploring random trees, 223
RAPTOR, optimal protein threading, 16, 18, 20, 29–30, 94–95
E value of, 116–117
fold recognition, 107–116
integer programming formulation, 104–107
scoring function, 101–104
threading assumptions and model, 104
RDOCK, refining ZDOCK predictions, 294
Relaxins, 6
REMD alogrithm, *see* Replica exchange molecular dynamics algorithm
Replica dynamics (RD) methods, 316–317
Replica exchange molecular dynamics algorithm, 315–316, 320
three-helix bundle protein and, 323–327
Replica exchange (RE) algorithm, 318–323
Retrieval test, protein structure indexing and, 161–164
Rigid-body method, 21–22, 65, 284, 299
Rigidity-based sampling, 228
Rigorous threading algorithms, 16
RMSD, *see* Root mean square distance
Roadmaps
construction method, 227
for protein folding, 224–235
protein representation, 220–221
robot motion planning, algorithms for, 221–224

Robot motion planning, roadmap algorithms for, 221–224
Root mean square distance, 24, 204–205, 262, 268–269, 285, 295, 299, 315, 323–325
Rop protein, alignment of, 135
ROSETTA method, 25–26, 69, 252
RRTs, *see* Rapidly exploring random trees

SAG1–antibody complex structure, 289
SAM method, 80
SAM-T98, 70–71
Scale-free networks and contact maps, 213–214
SCATD, 24
SCOP, *see* Structural classification of proteins
Scoring functions, knowledge-based, 251, 265–267
  from Bayesian probability formalism, 252–254
  compilation of probabilities, 254–255
  pairwise distance, 255–257
SCWRL method, 24, 82
Secondary structure elements, 148–149
Segment-matching method, 22
seg method, 79
SegMod/ENCAD, 22, 25, 82
Sequence alignment to known structure, steps for
  models building, 82–83
  sequence preparation, 77–79
  structural annotation databases for model, 77
  template structure, 79–81
Sequence-based alignment methods, protein structure prediction and, 19–21, 128
Sequence-based search methods, 70
Sequence Derived Properties (SDP), 72
Sequence searching improvements, 70–71
Sequence-structure alignment algorithms, 14–17
SETTLE, 324
SF sequences, 154, 156, 158
SHAKE, 324
3D-SHOTGUN, 74
Side-chain prediction methods, 23–24
Side-chain–side-chain interaction potential, 13; *see also* Energy functions, protein threading and
SignalP, 186
Signal peptides prediction, HMMs and, 186–187
SIMFOLD, 26
Simple sequence search, 79–80
Singleton energy, 10–11, 18
Small world, contact maps (CMs) and, 207–208
  all-α *versus* all-β contacts, 213
  local *versus* global contacts, 212–213
  properties, 214–215
  and protein structures, 209–211
  scale-free networks and, 213–214
Smith-Waterman algorithm, 20, 64, 69, 160
SNAPP potential, 265
SPARKS, 20, 81
SPratt2, 130
SRS, *see* Stochastic RoadMap Simulation
SSEARCH, 70, 80
SSEs, *see* Secondary structure elements
STACCATO, 139–140
Stochastic RoadMap Simulation, 229–232
STRIDE, 187
Structural classification of proteins, 19, 29, 70–71, 93, 150, 160, 166, 213, 262–263, 290
Structural superposition methods, 15
Structure prediction from sequence, techniques for, 62
  ab initio and new fold methods, 68–69
  comparative modeling, 64–65
  fold recognition, 65–68
Support vector machine, 18, 107, 109, 252
  linear SVM regression, 110–111
  nonlinear SVM regression, 111
Surface Generalized Born implicit solvent model, 23
SVM, *see* Support vector machine
SVM regression, 115–116
  FR and, 111–112
  linear, 110–111
  nonlinear, 111
SWISS-MODEL, 22, 65, 82

TASSER method, 30, 252
T-COFFEE method, 82, 130
Template structure library, 19
THREADER method, 67, 69, 71, 81
Threading, protein structure prediction by, 8, 91
  branch-and-bound algorithm, 99
  divide-and-conquer algorithm, 99
  dynamic programming algorithm, 98
  energy function, 9–14, 95–98
  fold recognition (FR) (*see* fold recognition (FR))
  FR/statistical significance of threading alignments, 17–18
  RAPTOR (*see* RAPTOR, optimal protein threading)
  sequence-structure alignment algorithms, 14–17
  sequence-template alignment problem, computational complexity of, 98
  sequence to structure alignments and, 69–70

targets and templates, representation of, 93–95
template structure library, 19
tree decomposition-based algorithms, 99
Threading scoring function, RAPTOR and, 101
contact capacity score, 103–104
environmental fitness score, 102
gap penalty, 103
mutation score, 102
pairwise contact score, 103
secondary structure score, 103
TMHMM model, 183–186
Torsion angles, 151–152
Transition state ensemble (TSE), 231
Transmembrane helices, HMMs for, 175–177
more complex models, 180–181
parameter estimation, 181–183
posterior probabilities, 178–180
signal peptides prediction, 186–187
topology of, 183–186
Viterbi algorithm, 177–178
Tree decomposition-based algorithm, 16, 24, 99
for threading problem, 16–17
Tree-progressive approach, multiple sequence alignment and, 129–130
Triose-phosphate isomerase (TIM) barrels, 7
Two-dimensional robotic configuration space, 221–222
Type-1 beta hairpin, 190
Tyrosine kinase, loop movement in, 141–142

UNDERTAKER, 26
Uniform distribution model, 12

ϕ-values, 231–232
van der Waals (vdW) potential, 292
VERIFY3D, 82
Viterbi algorithm, 174, 177–178, 181

Wire skeletal model, of α-lactalbumin, 6, 21
WU-BLAST, 70

X-ray crystallography, 5, 48, 192, 284

ZDOCK, 285
FFT-based initial stage docking algorithm, 290–294
M-ZDOCK and, 294
performance and protein complex characteristics, 298–312
performance on benchmark 2.0, 295–298
RDOCK and, 294
ZDOCK performance
on benchmark 2.0, 295–298
versus interface curvature, 305
multiple linear regression approach, 306–311
and protein complex characteristics, 298–312
regression analysis of, 302–303
simple linear regression approach, 303–306
ZDOCK/RDOCK performance, CAPRI experiment and, 288
Z-score scheme, of threading, 18, 100–101, 115

PRINTED IN THE UNITED STATES OF AMERICA